Praise for *The Ecological Far*

"For 40 years, Helen Atthowe has followed a relentless calling to combine her deep understanding of ecological systems with her love of farming. Now, she shares the best of her insights and methods in *The Ecological Farm*. This classic volume will guide all of us as we learn to farm in harmony with an ecosystem and to become obedient to the whole rather than being distracted by the urge to tinker with the parts."

—**Wes Jackson**, cofounder and president emeritus, The Land Institute

"Helen Atthowe's book takes ecological farming to the next level. It is packed with useful, field-tested, innovative techniques for farming more gently without sacrificing productivity. Atthowe effectively makes the case that, with a nature-based and minimalist approach, farmers can achieve more by doing—and spending—less. This is the future of farming. I highly recommend this book."

—**Ben Hartman**, author of *The Lean Farm* and *The Lean Farm Guide to Growing Vegetables*

"Helen Atthowe has a rare gift: She knows how to listen to scientists, but she speaks 'farmer.' Her knowledge comes from observation and practice as if decades matter, not just seasons. Helen applies her intuitive ideas to complex whole-systems organic agriculture, with a special focus on growing the fertility of the soil. Most importantly Helen engages in her work with a sense of joy and celebration. She's a born teacher who retains her sense of wonder that there is so much more to learn. Oh, and I think Helen must never sleep. The depth of the material she presents in *The Ecological Farm* and the citations she offers in support of her work is beyond my own comprehension!"

—**Bob Scowcroft**, cofounder, Organic Farming Research Foundation; board member, Nell Newman Foundation

"In *The Ecological Farm,* Helen Atthowe shares the practical knowledge she acquired over many years through experimentation on her own low-input, high-output ecological farm. Her book makes an exceptional and timely contribution to addressing interconnected global crises for which hands-on solutions are badly needed. Helen's work will also be invaluable to smallholder producers who wish to transition to ecologically based, sustainable, and profitable organic production systems, also known as organic Conservation Agriculture."

—**Dr. Amir Kassam**, visiting professor, School of Agriculture Policy and Development, University of Reading, UK; former senior technical officer, United Nations Food and Agriculture Organization; editor, *Advances in Conservation Agriculture,* volumes 1–3

"The story of Helen Atthowe's farming journey has stuck with me for many years, and now being able to hold it, savor it, and dive into the many nerdy details is an absolute gift. The combination of decades of experience and loads of data make her book incomparable: It is well researched, well written, and endlessly idea provoking. There are a thousand 'aha!' moments, with each section offering a new applicable insight or concept for improving or understanding soil health. *The Ecological Farm* will teach you how to rebuild soil, minimize tillage, grow your own garden fertility, improve ecology, and much more. Quite frankly, Helen's book belongs on the shelf of every serious grower."

—**Jesse Frost**, author of *The Living Soil Handbook*

"Helen Atthowe employs her high skill set and shares the full depth and length of her experience in *The Ecological Farm*. The ecology she describes is beautiful to look at and a powerful tool for maintaining balance on the farm or in a garden. Helen guides readers through many methods, backed up by decades of results."

—**Charles Dowding**, author of *Charles Dowding's No Dig Gardening*

"Helen Atthowe is a rare, knowledgeable grower of both vegetable and fruit crops, one who also knows insects, plant diseases, and soil science. She is one of those people who never stop searching for better, Earth-friendly ways of growing food. In *The Ecological Farm,* Helen takes us on her journey into deep organics. Read this book, then keep it on your essential reference shelf next to Eliot Coleman, Michael Phillips, JM Fortier, Ruth Stout, and Louis Bromfield. Refer to it often. Use it to launch your own deep ecology journey."

—**Brian Caldwell**, organic farmer; National Organic Standards Board member; former field manager and researcher, Cornell Organic Cropping Systems Project

"*The Ecological Farm* beautifully articulates the principles of holistic growing. Drawing on her wealth of practical experience, research, and years of observation, Helen Atthowe has distilled the infinite biological complexity of a farming system into some relatively simple principles. There's plenty of soil and plant science in there for us geeks but also a wealth of technical detail and cultivation tips for soil health and individual crops. This book is essential reading for anyone starting a new growing enterprise, but even the most experienced farmers and gardeners will find something new."

—**Ben Raskin**, head of horticulture and agroforestry, The Soil Association; author of *The Woodchip Handbook* and *Zero-Waste Gardening*

"I think about my farm's fertility program nearly every day, and the tenets that Helen Atthowe lays out in *The Ecological Farm* are spot on. We can grow our own nitrogen and build the carbon resources of our soil without resorting to energy-intensive and expensive off-farm inputs. Farm ecosystems should be *net producers* of energy and nutrients, and Helen spells out pragmatic ways that a farm can simultaneously sponsor its own fertility, be productive, *and* build ecosystem health."

—**Steve Ela**, owner, Ela Family Farms; former chair, National Organic Standards Board

"For farmers and gardeners wanting to go beyond the basic standards of certified organic farming and the superficial platitudes of 'sustainable methods,' Helen Atthowe's book is a breath of fresh air. Over decades of experimenting with crop systems, she and her late husband Carl Rosato have pioneered new ways of thinking about crops, weeds, pests, and soils that challenge many of our assumptions regarding carbon/nitrogen ratios and the use of compost, animal manures, and imported soil amendments in general.

"I have grown skeptical of academic experts with their analytical, reductionist approach to food and soil issues. Atthowe has an advanced degree and has worked as an extension agent, but despite that, she has an impressively holistic view of things organic. While not strictly no-till, she has devised strategies to greatly minimize disturbance of the soil community and encourage helpful pollinators and predators. She has lots of answers, yet I'm reassured that she frequently comments: 'I'm still working on that.'"

—**Will Bonsall**, author, *Will Bonsall's Essential Guide to Radical, Self-Reliant Gardening*

"*The Ecological Farm* offers much for beginning and experienced farmers as well as gardeners. Helen eloquently summarizes her deep ecological farming experience and knowledge as well as insights gained from her partnership with Carl Rosato, providing practical examples of how farmers can work with nature to support diverse ecologies both below and above ground. She explains the evolution of her understanding of the relationships among plants, microbes, the soil, and insects, and how to manage them ecologically in an agricultural system. It's both a great read and excellent reference book!"

—**Rex Dufour**, senior fellow, National Center for Appropriate Technology

"Drawing on decades of experience as a farmer and researcher, Helen Atthowe has developed an agroecological approach to growing organic food that is science-based, practical, and adaptable to site-specific conditions. Moving beyond input substitution, she offers a roadmap to minimum-input, soil- and Earth-friendly organic production that gardeners, homesteaders, and USDA-certified organic farmers can easily implement and adapt to their locale. This volume belongs on the bookshelf of all who seek to make a living in mutualistic partnership with the land."

—**Mark Schonbeck**, research associate, Organic Farming Research Foundation

"With the concepts she lays out in *The Ecological Farm*, Helen Atthowe gifts us with a practical and scalable approach to cultivation that produces nutrient-dense food with an absolute minimum of external inputs. Full of stories and wisdom gathered from 40 years of hands-on growing, this book is a rare synthesis of careful scientific research, long-term observation, and deep intuition developed through decades of listening to the land."

—**Alan Booker**, executive director, Institute of Integrated Regenerative Design

"In *The Ecological Farm*, Helen Atthowe shares decades of hard-learned lessons and keen observations. She is an inveterate tinkerer, experimenter, and researcher and has refined her organic production through the years. Atthowe has an immense respect for the role of biodiversity in the soil system. A fungi advocate, she guides the reader to reduce soil disturbance and feed the soil carbon. She writes as both a teacher and learner—as she tells readers, 'I am still learning.' Luckily she has taken a break from learning to share her accumulated knowledge and tips. While 'more is better' is often the strategy in our agriculture, Atthowe provides evidence and inspiration for selective and judicious management strategies to enhance your ecological farm."

—**Dr. Douglas Collins**, extension specialist and soil scientist, Washington State University

"This inspiring book from Helen Atthowe demystifies some of nature's critical interactions, helping farmers and others work with and support our natural world."

—**Jo Ann Baumgartner**, executive director, Wild Farm Alliance

The Ecological Farm

The Ecological Farm

A Minimalist No-Till, No-Spray, Selective-Weeding, Grow-Your-Own-Fertilizer System for Organic Agriculture

Helen Atthowe

Foreword by **Paul Muller**, Full Belly Farm

Chelsea Green Publishing
White River Junction, Vermont
London, UK

Project Manager: Rebecca Springer
Editor: Fern Marshall Bradley
Copy Editor: Laura Jorstad
Proofreader: Nancy A. Crompton
Indexer: Shana Milkie
Designer: Melissa Jacobson

Printed in the United States of America.
First printing June 2023.
10 9 8 7 6 5 4 3 2 1 23 24 25 26 27

Our Commitment to Green Publishing

Chelsea Green sees publishing as a tool for cultural change and ecological stewardship. We strive to align our book manufacturing practices with our editorial mission and to reduce the impact of our business enterprise in the environment. We print our books using vegetable-based inks whenever possible. This book may cost slightly more because it was printed on paper from responsibly managed forests, and we hope you'll agree that it's worth it. *The Ecological Farm* was printed on paper supplied by Versa that is certified by the Forest Stewardship Council.

Library of Congress Cataloging-in-Publication Data

Names: Atthowe, Helen, author. | Muller, Paul (Organic farmer), author of foreword.
Title: The ecological farm : a minimalist no-till, no-spray, selective-weeding, grow-your-own-fertilizer system for organic agriculture / Helen Atthowe ; foreword by Paul Muller
Other titles: Minimalist no-till, no-spray, selective-weeding, grow-your-own-fertilizer system for organic agriculture
Description: First printing | White River Junction, Vermont ; London, UK : Chelsea Green Publishing, 2023 | Includes index.
Identifiers: LCCN 2023003879 | ISBN 9781645021810 (paperback) | ISBN 9781645021827 (ebook)
Subjects: LCSH: Organic gardening. | Agricultural ecology. | Permaculture. | Handbooks and manuals. | BISAC: TECHNOLOGY & ENGINEERING / Agriculture / Organic | GARDENING / Vegetables
Classification: LCC SB453.5 .A88 2023 | DDC 631.5/84—dc23/eng/20230215
LC record available at https://lccn.loc.gov/2023003879

Chelsea Green Publishing
White River Junction, Vermont, USA
London, UK

www.chelseagreen.com

For Carl Rosato, my farming partner, save-the-world collaborator, soil revolution accomplice, greatest love, and inspiration. Carl did not live to see the completion of this book, but without him it could not have come to fruition.

Thanks to my parents, Jean and John Atthowe, who started me gardening before I was old enough to object and were proud of all the strange farming adventures I took. Thanks to Joni Peterson who helped me on my farm in Montana for 17 years, supported my crazy ideas and experiments, and continues to be my best friend.

Contents

Foreword

Nearly 40 years ago, a new agrarian movement arose in the United States out of concerns about impacts from the effective but harmful chemicals that were the tools of modern mainstream agriculture. Sixty years of conventional food production had created profound problems: contaminated groundwater, persistent pollution, pesticide-resistant insects and plant diseases, soil erosion, and declining soil fertility. There was a hollowing out of rural America as well. Farm failures and the deconstruction of rural communities were driven by a pattern of extraction from rural places.

Organic farming found its beginnings as a collaboration between new farmers and consumers. They shared a vision for a safe food system built upon whole-system thinking: a new agricultural paradigm that was environmentally sound, economically fair, and socially just.

In the beginning, the land-grant colleges and agricultural universities scoffed at the idea of organic or ecologically sound farming. The body of information supporting this new agrarian movement was unformed. Farmers had to struggle to develop systems to grow healthy crops.

Some organic farms simply adapted the paradigm of conventional farming that calls for relying on a specific product to solve a particular problem. These farmers strove to replace toxic pesticides with naturally occurring and safer alternatives. This approach is more mindful of the whole than conventional farming, but the evolution in thinking in organic agriculture continues as science now offers us insights into the complexities of the interdependent ecological functions within a landscape.

Even 40 years ago, it was clear to many new organic farmers that in order to create a new paradigm, they needed to forge a new pattern of farming. Rather than analyzing crop production as components—weed science, soil science, plant science, insect control, or even rural sociology—the new organic model would focus on the health of all parts and valuing the relationships among living organisms above, below, on, and around a farm. This new thinking centered on building soil health, growing fertility on-farm with cover crops, and working with livestock and utilizing manures for soil fertility. Healthy, vibrant soil would create healthy plants that are resistant to disease and insect attack and would in turn create more nutritious food. That food would drive consumer confidence and help to grow what is now a thriving organic marketplace.

In the late 1970s and early '80s, the empowerment of a new rural people sought to stem the tide of farm failures and reverse the trend of farming as an occupation with no future. Those years marked a time of enthusiastic hopefulness, empowerment, and self-reliance. Full Belly Farm began in Northern California during that period. It was also the time when Helen Atthowe and her late husband, Carl Rosato, began their respective farms.

I knew Carl as a friend and a mentor. He was committed to enhancing the farm ecosystem as a productive whole. He simply intuited the steps and processes for growing peaches by balancing complex ecology around his fruit trees, and great peaches were the reward for tending a farm that grew more vibrant and productive over time. His key was soil health driven by a diverse plant community above ground feeding a diverse soil microbial community below ground. His system of balancing minerals and enhancing soil ecology was validated by a loyal farmers market following of patrons who could taste the complex, sweet quality of his fruit. He enthusiastically shared his techniques and insights as a teacher and tireless promoter of organic farming. His passing was a tragic loss.

Helen and Carl were cut from the same cloth. Helen's decades of farming and farm advising had started from much the same point as Carl's, and they both developed insights into farm management for capturing optimal total energy. The new Woodleaf Farm that Carl and Helen built together in Oregon manifests their application of learned principles and design for a larger pattern of production and loving stewardship.

Helen also has a deep affection for the remarkable expressions of life that can be nurtured on a farm. For more than 40 years, she has been gathering evidence and planting the ideas you will read about in this fine book that represents cutting-edge farming knowledge. *The Ecological Farm* is a compendium of observation, experimentation, design, acquired knowledge, and experiential realizations. Helen's journey through her farm work, action on the ground, trial and error, and serendipity are a remarkable example of a farmer's integration within a whole farm system. Helen offers us her collected insights and general principles for mindful stewardship and care, for doing the least harm, and for planning one's work on a farm to enhance the whole.

Helen was an early adopter of organic principles, but her level of care in her approach to farming was not typical. Early on she had the insight and courage to think and act differently from what was commonly expected. She saw the subtle relationships between what she did on the land and how her crops changed in appearance, how they grew and fruited more fully and consistently, and how the soil absorbed water or breathed. She applied scientific processes to measure these changes. She took samples and calculated what her impact was on the land. She had a commitment to choosing a different pathway for achieving more with less, and then chronicled how that principle can be realized through her experiments.

Helen's approach to farming has always been grounded in the goal of producing great-tasting, safe, and abundant food while reducing or eliminating off-farm inputs. She shows us that a deep, committed relationship to land arises not only from observation, measurement, and trials but also by breathing in morning dews or the very dust of the place, witnessing life that can be engendered through design, listening to the hum of bee activity or bird song, feeling the sponginess of good soil. Each year is an opportunity to get closer to feeling the rightness of one's labor.

Caring, affection, and gratitude—sentiments at the heart of the organic movement—were manifest in Carl and Helen's farm design. The love of farming goes beyond yield of marketable crops; it engages the heart and senses of the practitioner rooted into a place. Over time, the spiritual and physical connection deepens. Agricultural knowledge evolves as each crop year allows insight into the working relationships established on a farm. For a farmer whose heart and place grow intimate, each springtime is an opportunity to take learned principles, advice, and ruminations and ideas borne from experience into the field and orchestrate an effort that will in turn set the parameters for future crops. This is how our journey with Helen begins in *The Ecological Farm* as she defines the new farming revolution of our time: A focus on all parts of the farm as relationships built upon interdependent ecological systems. Readers learn alongside Helen as she leads us through the observations and measurements over her years as a farmer-scientist that resulted in her progressive realizations about the importance of a systems approach. Helen asserts that the whole is the whole—its own complexity is the near-miraculous integration of everything all at once as one.

The Ecological Farm is also a practical guide to growing nutrient-rich fruits and vegetables. Helen describes how to bring greater insect balance into a farm or garden, and which insects or birds thrive when practices are mindful of their well-being. She presents soil ecology as a function of diversity in design principle, careful choosing of tools and micronutrients to optimize health, and careful decision-making regarding tillage and interventions.

As Helen points out, the dynamic relationships found in healthy soil can provide most, if not all, of the fertility that a crop needs. The ecology of the billions of microorganisms that thrive in the rhizosphere of a plant or in a community of plants is driven by a seamless association as all parts sense, inform, feed, and support all other parts. But one of the profound insights that Helen presents, and that many new organic farms are realizing, is that building fertility is a multiyear process. This year's harvest is in part determined by the inputs provided for crops during the growing season, but the yield is also built on a continuum of investment in ecological health. Feeding the soil is a long-term commitment, and so is increasing farm resilience and enhancing water capture. The results we see in our fields and crop beds now are, to a large extent, determined by the practices of past years.

Helen offers farmers and gardeners ten ecosystem management principles to holistically guide their practices and decisions for developing healthier soil by identifying the direct relationships among the soil, plants, insects, and microorganisms. It is a great leap forward to bring these principles together with background and evidence to support the power of interdependent systems. Growing one's fertility, gaining greater water use efficiency, and utilizing the vitality of living organisms engendered on the farm through diversity are key lessons for greater resilience as all farms face challenges posed by a changing climate.

Helen welcomes us to understand her caring and affection built upon a lifetime of learning and observation as a vegetable and fruit grower and a scientist. Her first measurements are tactile. She puts her hands in soil, feeling its texture, moisture, or crumbly character; smelling that scent of fertility, the redolent funk of fungus and bacteria wildly coexisting with billions of other soil microbes; observing the infinite greens of diverse plants with their tangle of living roots; hearing the buzz, song, and hum of life above ground; and delighting in the art of managing this symphony. Her second tier of measurements is built on practice and the questions she pursues. As she observes the changes in soil once tilling is stopped, she takes us through the strategies she explored for optimizing what can be grown in place, from fertility to mulches to beneficial insects to cycling nutrients. She, like all farmers, participates in an annual calculation of when and how to act on the land. Her work yields a set of guidelines that have been tempered by experience and affirmed with science and careful measurement.

Here is one of the insights that Helen has contributed to an integrated view of farming: That a farm is not the sum of its individual parts but rather is an indivisible whole in a continuum of choices *made over time*. Most science chooses to take the parts and study the whole. There is a next step, wherein the whole is known by being the whole. This is the beauty realized in an ecological farming system. It is a reawakening to an indigenous mind where the spirit of the land resides in the very being of the farmer/dweller on the land—integrated, seamless—one organism. To farm is to be part of a continuum of care, the history of past hands on the land, the choices made as to what will be taken and what will be added back.

Farmers and gardeners are at a wonderful and exciting juncture as we are reintegrating our parts and making a more wildly dynamic whole. *The Ecological Farm* will serve as a fundamental guide and practical reference as we embrace a new agrarian pattern of creating whole mind, whole harvest, whole nutrient cycling—which in turn will support the ecological community regeneration that is at the heart of human community revitalization.

—**Paul Muller**

Full Belly Farm, Capay Valley, California

Figure 0.1. Ecological decision making means taking into consideration the health and habitat of beneficial insects and pollinators, like this swallowtail butterfly pollinating peach blossoms.

INTRODUCTION

My Farming Revolution

When I began farming organically and studying horticulture in the early 1980s, I focused on the specific needs of individual crops. I researched and learned to manage each crop's nutritional needs; light, temperature, water, and soil requirements; insect and disease pests; and special growth preferences. I studied the ecology, etiology, and phenology of individual crop diseases, insect pests, and weeds. I learned the best timing to control and kill insect, disease, and weed pests. It was a classic reductionist scientific approach that quantified a crop's needs in isolation from its environment and all the ecological relationships affecting it.

Over the years, as I farmed and gardened in different states, climates, and soils, as well as at different scales of production, there was a slow, steady revolution in my thinking about how to grow crops. I began to conceptualize plants as parts of whole systems, and I puzzled to figure out ecological functions and interactions within plant communities. Rather than a targeted, direct stream of light focused on a single crop, my approach is now more like a diffuse glow that illuminates all the relationships and connections that make up a cropping ecosystem. My vision is to discover how to more closely mimic natural ecosystems and how to farm well by doing *less* rather than *more*.

In 2010 I met California fruit farmer Carl Rosato, who was evolving a similar farming perspective, methods, and strategies, based on 30 years of lessons from his certified organic orchard. Carl and I married, and through farming together we realized that our ecological approach revolved around *managing relationships* rather than simply managing crops for the highest yields.

In recent years, many researchers in both ecological and agricultural science are also studying ways to utilize plant, soil, microorganism, insect, bird, and other animal connections in order to farm with fewer chemical and off-farm inputs. This book is a review of that science and a culmination of Carl's and my 40 years of farming and gardening lessons, successes, and mistakes. It focuses on soil building and habitat building for below- and aboveground beneficial organisms and on systems thinking for gardeners and farmers. The first goal is to identify and understand ecological functions within plant, soil, microorganism, and insect communities. I want to make complex, sometimes messy, ecological concepts and practices accessible to all of us who grow food. The second goal is to describe in detail how to manage the many connections that make up a healthy cropping ecosystem, one that does not require use of outside inputs, not even organically certified pesticides or fertilizers.

About This Book

The book is set up in two distinct parts. Part 1 focuses on why a systems approach matters and how to develop your own farm or garden system. I start with a look at the end goal—a developing or developed cropping ecosystem—and I describe the "recipe" Carl and I came up with for moving toward and eventually reaching this goal. The recipe consists of 10 principles for managing ecological relationships—principles that can be applied wherever you farm or garden.

In chapters 2 through 8, I build out the case for managing ecological relationships rather than just crops, using practical details from our farm operations and lessons learned from our on-farm research. In chapters 2 through 4, the focus is on the soil. Then in chapters 5 through 8, I move on to examine aboveground relationships, but ever mindful of all that is happening belowground.

I cover multiple methods for building a soil organic matter system to feed crops with minimal outside inputs, explain how to build habitat for beneficial organisms that naturally suppress pest insects and diseases, and discuss strategies to shift the competitive balance toward our crops, including a chapter on "plant competition" (yes, that includes weeds) and how it can benefit a farm ecosystem when properly managed. This systems approach to improve soil and habitat also improves crop health and nutrition through intricate interactions that I am still learning about. For example, tree fruit crops growing within my own farm's no-till soil organic matter system, and receiving *no* pest sprays, are higher in human-healthy antioxidants. It turns out that these same antioxidants may also help the fruit trees suppress diseases, such as peach brown rot. In fact, new research suggests that the kinds of soil fertility systems I present in this book—which depend on the soil microbial community to cycle and recycle nutrients instead of on inputs of quick-release fertilizer—produce crops with higher antioxidant content. For example, both mycorrhizal fungi associated with plant roots and plant-growth-promoting bacteria (which I introduce in chapter 2) increased the antioxidant anthocyanin in strawberry fruits.[1] Blackberry roots associated with another beneficial bacterium showed increased flavonoid antioxidants in fruits; simultaneously, blackberry fungal disease decreased.[2] This same beneficial bacterium also increased antioxidant content in vegetable crops.[3] Hence the beauty of the minimalist systems approach is that plants grown this way, with fewer chemical and fertilizer inputs, have strong immune systems and also contain more of the minerals, vitamins, and antioxidants that humans need to resist stress and disease and stay optimally healthy.

The ecological and human health benefits of this systems approach are significant also because *suppressing* rather than *killing* pest insects and disease organisms allows native pollinators and natural enemies of crop pests to flourish. Simultaneously, a systems approach decreases human and soil microorganism exposure to toxic pesticidal chemicals. And keep in mind that even some pest control materials approved for use on certified organic farms, such as copper and lime sulfur, are caustic and can harm humans and beneficial organisms. Reducing their use is a big part of the goal in an ecological approach to farming. As farmers and gardeners, we also benefit from working amid beautiful surroundings and from the joy of blending our farms and gardens into the inclusive, connected natural world around us.

Part 2 of the book moves from the ecosystem perspective to individual crop details, including "interventions" for specific crops if the cropping ecosystem gets out of balance and troubleshooting is needed to manage problems such as insects, diseases, weeds, or poor soil fertility and crop growth. You'll discover, though, that I don't start by prescribing solutions for individual "problems." And I don't use the phrase *control pests* because, in my framework, it almost always takes an ecosystem approach to truly solve problems! The goal is not simply to kill or control a specific pest. Instead I present options for suppressing organisms that

Figure 0.2. With an ecological systems approach, a farm or garden is a beautiful, diverse, integrated planting that can yield some of the most nutritious food you've ever eaten.

thwart crop growth and for enhancing optimal crop growth and natural soil nutrient cycling, so that our crops' own defensive systems allow them to thrive within a larger ecological system.

Information in part 2 is organized by individual crops, insects, and diseases, but you'll find that the intervention recommendations for specific pests and problems are ranked from minimal ecological impact to heaviest ecological impact. Thus I first list methods such as identifying effective biological suppression and beneficial organisms for a specific pest, creating habitat to encourage these organisms, and cultural options, including when and when not to disturb habitat or irrigate. In the sections describing heaviest-impact interventions, I do make recommendations for use of pest management materials and methods that are deemed acceptable for use on certified organic farms. But they should be used only as the last resort, and I explain the ecological impacts of these methods and materials, using examples both from my practical experience and from research data.

When you are feeling troubled, fearful, or uncertain, there is a certainty in tending plants and growing your own food. Nurturing the healthy goodness in a field of multicolored kale and lettuces or coaxing so much sweetness from a special peach tree: It all makes me feel I belong, safely, wherever I am. The process of creating farms and gardens opens my eyes to awe, tunes my ears to listening, and offers the gifts of curiosity, discovery, and deep connection. Being part of this farming revolution, I reap my own personal revolution for free: Healthy farming, in which nature is invited to participate, grows healthy food that in turn grows healthy humans, resistant to disease and stress. These reconnected healthy farms and gardens are, in turn, resilient and resistant to plant disease and stress.

PART ONE

Farming at the Intersections of Ecological Relationships

I am about to introduce you to a different way of thinking about farm and garden success and sustainability, a way of thinking based on ecological functions and principles. By *ecological functions* I mean the roles or services that species (such as plants or insects or microorganisms) perform in the ecosystems in which they live. My view arises from the understanding that all organisms in an ecosystem are connected and interact within a complex web of overlapping relationships.

Figure 1.1. These bins of onions, shallots, and winter squash show that high yields are possible even with Woodleaf's low-input systems. Clover living mulch has grown into this onion row from the row middles, but the onions we are harvesting are large and pest-free.

CHAPTER 1

Principles for Managing Ecological Relationships

The revolution in the way I have come to think about farming starts in the earth just below our feet, at the intersection where plant roots, the microbes that cling to those roots, and the soil itself mingle and evolve together. It spreads out to minerals, biochemicals, complex sugars, and carbon forms that cycle from one soil particle to the next among a web of soil fungi, bacteria, protozoans, and nematodes. It moves on to earthworms, mites, beetles, and other small animals who bring nutrients from deep in the soil up to the soil surface. From there, the millions of connections spread everywhere and include everything that affects plant growth. My farming-systems thinking encompasses the microorganisms in the soil and on leaf surfaces that cause and inhibit plant disease. The insects that attack plants and the ones who eat plant-attacking insects. The weed seeds that can cause weed outbreaks but also feed beneficial insects and birds. There are myriad ecological relationships that enhance or thwart crops growing in our backyards and fields. Our human bodies are part of this revolution, too, when we ingest the highly nutritious food that results from the reconnecting of farming to ecological relationships.

Figure 1.2. In August the harvest at my farm provides a rainbow of deeply colored, antioxidant-filled, vitamin-rich fruit.

Systems Thinking for Farmers and Gardeners

Definitions are difficult when you are describing a moving target or, better yet, an ever-changing system of interactions. Farming and gardening systems are not static. Some farms appear to be static, continually producing uniform results, but that is only because they are artificially maintained through significant inputs of fertilizer, pesticides, and tillage.

I define a *cropping ecosystem* as an attempt to mimic natural plant communities and the ecological processes and synergies that keep the system functioning. It is a human-created system undergoing human-induced progressive change. Because nature does not stand still, design is only the first stage of a cropping ecosystem. After the initial design, our systems thinking and the managing of relationships really begin. All biological systems change, moving regularly back and forth through *at least* three stages: the developmental stage, dynamic equilibrium, and senescence.

DEVELOPMENTAL STAGE

The developmental stage is a stage of rapid changes. Farmers and gardeners disturb the original plant system present and help to direct *succession*, which is progressive change in an ecological community. The focus during this stage is to build soil and habitat. As the developing system establishes and ecological relationships form, growers will likely need to use specific organic fertilizers and intervene to suppress pests with inputs such as row cover to keep flea beetles off susceptible arugula or radishes, or as a last resort an insecticidal spray such as insecticidal soap or neem if an insect pest is causing economic damage. Initial land preparation and tillage also disturb the soil ecosystem, and soil microbe density and diversity and populations of natural biological control organisms (such as carabid beetles, spiders, and beneficial fungi). Diverse populations of soil microbes and biological control organisms eventually rebuild during this stage. It takes patience to allow ecological relationships to build and form.

Figure 1.3. First harvest from this 5-year-old peach tree at my farm in 2020. The trees and the living mulch ground cover, microorganisms, insects, birds, and wildlife are forming and re-forming relationships as this orchard enters a dynamic equilibrium stage.

DYNAMIC EQUILIBRIUM

The cropping ecosystem reaches a dynamic but relatively steady state. The level of change is less drastic, and most of the changes are seasonal and slow, occurring over a longer period of time. Gardeners and farmers find they can generally intervene less often, using fewer but more targeted fertilizing, weeding, and insect/disease inputs that affect many parts of the system at once. Soil fertility management begins to more closely mimic nutrient cycling and recycling in a natural plant community, where decomposing plant residues are tightly coupled with soil microorganism release of plant nutrients into the soil near crop plant roots. Soil building and habitat building become closely coupled. A multifunction method such as maintaining crimson clover strips between or near crop plants (providing bloom for beneficial insects and soil nutrients/carbon when mowed) may serve to promote both of these processes.

SENESCENCE

During senescence, the cropping ecosystem is aging out of the dynamic steady state and requires more farmer or gardener input. Senescence stages flow naturally back into developmental stages. Senescence happens, just like you-know-what happens. It is not a bad thing! It is working with nature, which is changing and reestablishing new connections all the time.

10 Principles for Managing Ecological Relationships

Before I met and married Carl, he farmed in the Sierra foothills of California on his Woodleaf Farm and I farmed in western Montana on my Biodesign Farm. I sold Biodesign Farm and moved to California to help Carl manage Woodleaf Farm. We made enough money to sell that farm and semi-retire to a new farm in eastern Oregon, which we also named Woodleaf Farm. There we continued to experiment and develop our ecological methods. Heartbreakingly, Carl suffered a farm accident and died in 2019. But our years of farming alone and farming together in different climates had given us ample opportunity to evolve systems management methods—methods I am still evolving.

These principles came from paying attention to the land, watching closely, helping our farm systems change and develop over time, and learning from our mistakes. The first 3 of our 10 principles were inspired by the principles of soil health formulated by the Natural Resources Conservation Service (NRCS). The Food and Agriculture Organization of the United Nations also describes 3 interlinked principles called "Conservation Agriculture" that embody most of our 10 principles. There are many ways to garden and farm, and each garden or farm is the creative blending of the gardener or farmer and their creation!

Here are the 10 principles that Carl and I developed:

1. **Create above- and belowground diversity.** Feed the soil microbial community with as much crop and ground cover, root and whole plant, above- and belowground diversity as is economically possible.
2. **Minimize soil disturbance.** Disturb the soil as little as possible; create year-round undisturbed refuges for above- and belowground natural enemies (biological control organisms).
3. **Maintain growing roots year-round.** Avoid bare soil; instead, maintain plants with living roots to keep the soil covered year-round and to feed the rhizosphere (the interface area around plant roots) a steady, balanced diet of carbon and nutrients.
4. **Grow your own carbon.** Grow or obtain your carbon source where you are, rather than importing carbon from a distance.
5. **Add organic residues all season.** Add organic residues regularly throughout the year, not all at once in the spring.
6. **Focus on carbon fertilizers.** Prioritize slow-release, plant-based carbon fertilizers rather than fast-release nitrogen fertilizers. (Carbon fertilizers are plant residues that contain a relatively high ratio of carbon to nitrogen.)
7. **Recycle rather than import nutrients.** When possible, recycle minerals from on-farm sources (such as cover crops, living mulches, and chipped branch wood) rather than importing mined mineral nutrients from somewhere else and applying as fertilizer.
8. **Fertilize selectively.** Fertilize each crop selectively; avoid fertilizing the whole garden or field.
9. **Weed selectively.** Choose a ground cover mix; then identify and "chop and drop" the plants or "weeds" that are competitive with specific crops in specific growing situations and climates.
10. **Create habitat in the field.** Preserve wild habitat or create and maintain undisturbed habitat for natural enemies interspersed with or as close to crops as possible.

As I developed these principles over time, I made my share of mistakes. My biggest mistakes have been doing too much, but also too little, during the developmental stages of my gardens and farms.

Table 1.1. Strategies for Applying Ecological Principles at Biodesign Farm and Woodleaf Farm (CA)

Ecological Principle	Biodesign Farm	Woodleaf Farm (CA)
1. Create above- and belowground diversity.	Vegetable species and variety diversity. Living mulch between crops consisting of shallow- and deep-rooted annual and perennial legumes and weeds. Landscape-level diversity provided by perennial pastures, native plant hedgerow, and native grassland on crop field perimeter.	Fruit tree diversity: 10 species, 85 varieties. Living mulch between trees consisting of shallow- and deep-rooted annual and perennial grasses, legumes, and weeds. Landscape-level diversity provided by native oak/pine forest on crop field perimeter.
2. Minimize soil disturbance.	Minimal tillage in crop row; no tillage in row middles. Perennial living mulch to provide winter shelter and season-long undisturbed bloom for beneficials.	No tillage; minimal tillage when renovating orchard. Perennial living mulch to provide winter shelter and undisturbed season-long bloom for beneficials.
3. Maintain growing roots year-round.	Year-round annual and perennial legume and weed root diversity providing diverse root exudates.	Year-round perennial and annual grass, weed, and legume root diversity providing diverse root exudates.
4. Grow your own carbon.	Living mulch mowed 4 or 5 times per year.	Perennial living mulch mowed 4 times per year. Chipped branches from 2 tree prunings per year.
5. Add organic residues all season.	Regular surface application of mowed legume living mulch throughout the growing season. Spring application of on-farm-made straw, clover, and manure compost (early years) or alfalfa meal (later years) applied in crop rows only.	Regular surface application of plant residues throughout the growing season: mowed living mulch, raw and composted chipped branch wood.

Intervening too much or too little can delay or prevent a system from reaching dynamic equilibrium.

The rest of part 1 of this book is a deep dive into these principles. Throughout I use examples from Carl's and my farming experience and the on-farm research we conducted over the years. To help you get oriented, here is a brief overview of the three farms I refer to in the book and some of the key methods used on each one.

BIODESIGN FARM, MONTANA

Biodesign Farm is about 30 acres (12 ha) of land that I owned and managed as a certified organic mixed vegetable crop farm from 1993 to 2010. It is located in western Montana, USDA zone 5b, with rocky, sandy loam soils, rated by the USDA Natural Resources Conservation Service as "poor" for agricultural use. This part of Montana is semiarid, receiving approximately 13 to 16 inches (33–41 cm) of annual precipitation. Spring (May and June) is the wettest time of the year. It has a short growing season, with a frost-free period of only 100 to 115 days. Summer temperatures can reach the high 90s F (35°C or hotter), winter lows are regularly below zero (–18°C or colder), and the ground reliably freezes each winter.

On that farm, I developed an annual (and later a perennial) living mulch system between crop rows to manage weeds, build soil organic matter, cycle

Table 1.1 *continued*

Ecological Principle	Biodesign Farm	Woodleaf Farm (CA)
6. Focus on carbon fertilizers.	High-nitrogen compost (early years) plus mowed and annually incorporated lower-carbon living mulch. In later years: alfalfa meal plus mowed perennial living mulch.	Diverse high-carbon/low-nitrogen organic soil amendments: mowed perennial grass living mulch; chipped branches; off-farm high-carbon chipped wood yard-waste compost.
7. Recycle rather than import nutrients.	Mowed perennial living mulch. Annual compost or alfalfa meal addition. No other minerals applied to soil or foliage.	Mowed perennial living mulch. Boron, manganese, sulfur, and gypsum applied to soil and as a Mineral Mix Bloom Spray at tree fruit bloom.
8. Fertilize selectively.	Compost or alfalfa meal applied to crop rows only, plus mowed perennial living mulch.	Depending on soil test results, individual minerals applied around each tree (soil mineral balancing). Annual application of Mineral Mix Bloom Spray.
9. Weed selectively.	Mowed living mulch selectively based on maximizing airflow and reducing disease infection during rainy, wet periods; reducing vole habitat; and cycling nutrients. Selective wheel-hoeing and hand-weeding of weed-sensitive vegetable crops.	Selective mowing based on avoiding beneficial insect disruption in spring; maximizing airflow and reducing disease infection during rainy, wet periods; cycling nutrients; and facilitating harvest in summer/fall.
10. Create habitat in the field.	Perennial living mulch to provide diversity, season-long food source of pollen/nectar/seed, and winter cover. Selective mowing of perennial living mulch to avoid disturbance of beneficials at key pest pressure times.	Small crop fields bordered on 3 sides by native forest. Perennial living mulch for diversity, season-long food source of pollen/nectar/seed, and winter cover. Selective mowing of perennial living mulch to avoid disturbance of beneficials.

nutrients, and create habitat for natural enemies. I planted legumes (mostly clovers) and mowed them regularly, maintaining an undisturbed yearlong cover crop during the growing season and winter. In the spring I tilled the whole field, including the row middles, which became planting beds for the next growing season. Each spring, I seeded a new legume living mulch or cover crop between the new crop rows. I made compost on the farm, which I added to the entire field and incorporated every spring. Over time, I observed huge increases in soil nutrient levels, soil organic matter, and abundance and diversity of natural enemies (especially predator and parasite insects), as well as improved insect pest suppression. Over time, my practices evolved toward less tillage and leaving more and more habitat undisturbed, using less compost, and selective mowing techniques. By the time I sold Biodesign Farm and moved to Carl's farm in California, I felt I had reached an in-depth understanding of the system components of my farm, but my understanding of managing relationships was still evolving.

In 2005, I began to experiment with further minimizing soil disturbance, by implementing strip tillage only in crop rows and planting a perennial legume living mulch in row middles. The perennial legume living mulch provided permanent growing roots in the soil, nutrient cycling, and winter soil

Figure 1.4. In this scene of transplanting eggplant at Biodesign Farm in 2006, note the perennial red clover in between crop rows. This living mulch minimized soil disturbance and provided soil nutrients and winter cover as well as habitat for natural enemies.

protection. I tilled the 4-foot-wide (1.2 m) crop rows only, leaving the row middles untilled and unfertilized. I discovered that I could reduce fertilizer use and stop applying all insecticidal sprays, even those allowable for use on organic farms. Overall, I intervened less and less and began to see what a farm ecosystem in dynamic equilibrium looks like. After I stopped spraying, I wanted to test the biological control within my system, so I set up natural pest suppression trials for cabbage worm predation and parasitism by natural enemies, which I describe in detail in chapter 6.

I also investigated how living mulch or weed competition and various in-crop-row tillage and weed management strategies might affect both crop yield and long-term soil health. I cover the results of these trials in chapter 3. These experiment results helped me to discover compromises between good yields and keeping soil microorganisms and natural enemies

Figure 1.5. Harvesting peppers at Woodleaf Farm (OR) in 2020. Perennial grass, weed, and clover living mulch covers the soil between crop rows, and its roots feed the rhizosphere a steady, balanced diet of carbon and nutrients. The mowed mulch blown into the crop row serves as a slow-release carbon fertilizer, habitat for natural enemies, and weed suppression.

happy. After I moved to Woodleaf Farm in California, this vegetable cropping system slowly evolved, and I further refined it to reduce tillage and inputs when we moved to our new farm in eastern Oregon.

WOODLEAF FARM, CALIFORNIA

Woodleaf Farm is 26 acres (10.5 ha) in the Sierra foothills of California, just above the vast agricultural area of the Sacramento Valley. There are 8 acres (3.2 ha) in crops, split into seven fields nestled among 18 acres (7.3 ha) of native oak and pine forest. The farm lies in USDA zone 8b–9a, with heavy, clay loam, sloping soils, rated by the USDA Natural Resources Conservation Service a as "poor" for agricultural use. This part of California is Mediterranean (wet winters, dry summers). It is possible to grow crops almost year-round, with cool-season crops grown during the winter. Summer temperatures can reach above 100°F (38°C), winter lows rarely go below 20°F (–7°C), and the ground almost never freezes in the winter. Annual precipitation is 35 inches (89 cm). The farm has been certified organic since 1982, and Carl farmed it continuously until 2015 (I joined him there in 2010). For more than 30 years, Woodleaf used no-till, diverse living mulches to cycle nutrients and create habitat for natural enemies. We sold the farm to new owners in 2016, and they continue to use our ecological methods.

The soil fertility system involved a diversity of high-carbon, low-nitrogen plant-based fertilizers, which I describe in chapter 12. This system simultaneously provides both increased soil organic matter and habitat for soil microorganisms and year-round cover and season-long flower pollen and nectar for beneficial insects (details are in chapter 5).

The farm's long-term records document the ecosystems approach success: Loss of fruit yield and quality due to insect damage decreased over

Figure 1.6. Blooming peaches and cherries and unmowed perennial living mulch ground cover at Woodleaf Farm (CA) provide undisturbed refuges for above- and belowground natural enemies. Selective mowing of the living mulch during the growing season cycles and recycles nutrients within the agroecosystem.

Figure 1.7. Woodleaf peaches for sale at a farmers market in the Bay Area of California. The farm had a reputation of consistently providing very tasty fruit.

30 years along with certified organic insecticide use, which was eliminated altogether in 2013. Weekly field monitoring recorded high, season-long diversity and abundance among ground-dwelling predators and nectar-requiring predators/parasites. Even though we applied no pest management materials, in some years we saw less than 10 percent total average annual fruit damage to most crops. Over time, average soil organic matter increased from 2.2 to 5.1 percent. Our soil fertility system resulted in 10 to 20 inches (25–51 cm) of annual tree growth and good yields of high-quality, mineral-filled, flavorful fruit.

THE NEW WOODLEAF FARM, OREGON

In 2016 Carl and I moved to a hay-producing ranch in eastern Oregon. Our plan was to put in an orchard and begin producing vegetables, dry beans,

Figure 1.8. A big crop of apricots at the new Oregon orchard growing using no pest sprays and no off-farm fertilizer.

and grains while further reducing tillage and minimizing off-farm fertilizer/pest management inputs and weeding labor. This farm is located at the base of the Wallowa Mountains in a narrow canyon along a creek. The area is USDA zone 6a–6b, with deep river bottom, sandy loam soils. The soils are rated by the USDA Natural Resources Conservation Service as having "some limitations for agricultural use," but they are better than the soils we began with on our farms in Montana and California. This part of Oregon is semiarid, receiving approximately 15 to 18 inches (38–46 cm) of annual precipitation, and spring is the wettest time of the year. It has a relatively short growing season, with an average of 126 frost-free days. Summer temperatures can reach close to 110°F (43°C). Winter lows are usually, but not always, below zero (–18°C and colder), and the ground reliably freezes each winter. Carl and I were starting with deeper, better soil than ever before, but soil tests showed low fertility, which was due to continuous haying.

Right from the start we began to experiment with no-till, plant-based soil fertility systems using materials grown on-farm, and also with reduction of off-farm material inputs and labor. How far could we push our minimalist natural farming theories? we wondered.

I continue to manage this farm on my own, using perennial, no-till living mulches in between crop rows and no-till methods in the orchard. In the vegetable crops, it's a minimal-tillage approach, strip-tilling the soil in the spring in the crop rows only. I use "grow-my-own-fertilizer" hay mulch consisting of mowed grass, legume, weed living mulch blown into rows from row middles and sometimes a mulch of hay cut from other fields and applied to crop rows. I explain the details of this system in chapter 4.

MANAGING RELATIONSHIPS IN PRACTICE

Ecological relationship management, including pest suppression rather than pest control, takes far fewer material inputs, saving both labor and money, but it requires more brain power and figuring things out. It requires lots of looking and listening. We first prioritize the importance of various ecological functions within the system *at the moment* and then determine how and when to intervene. Those functions relate to soil fertility, plant competition, and management of pest insects, diseases, and animals. For example, our "selective mowing" farming strategy sounds simple, but it shows the complexity involved in managing relationships, including relationships among mowing for nutrient cycling, ease of operations, raptor hunting behavior, rodent populations, and crop-damaging frosts. There are other relationships between beneficial insect habitat and higher-carbon versus higher-nitrogen nutrient cycling/recycling. Our management approach is to prioritize among all of these interconnected and sometimes-competing relationships based on weather conditions and biological and economic goals. I am still learning the best timing for specific management strategies and how to prioritize which relationship to manage when.

Figure 1.9. Cranberry dry beans in July at the new Woodleaf Farm (OR), growing in a strip-tilled row between no-till living mulch row middles. Mowed living mulch blown into the crop row provides fertilizer and habitat and suppresses weeds.

Mowing the living mulch in the summer cycles organic residues that are relatively high in carbon content. These surface-applied residues slowly cycle/recycle nutrients back into the soil to feed soil microorganisms. But in the spring and early summer, I leave at least 30 percent of the living mulch unmowed to maintain undisturbed cover for ground-dwelling predators (spiders and beetles) as well as pollen- and nectar-providing flowers for pest-attacking parasitic wasps and syrphid flies. I want as much habitat as possible interspersed within the crops so that beneficial insects and birds can live where they work, rather than having to commute into the crop fields from a distance. Whenever possible, for crops with challenging pests (like apples), I try to leave 50 percent of the living mulch unmowed to enhance beneficial insect habitat during times when specific pests are hatching into predator- and parasite-vulnerable larvae or are in an underground stage of their development and vulnerable to spider and carabid beetle predation. During cold springs, however, I sometimes mow some of the succulent higher-nitrogen living mulch in order to promote better airflow and frost protection as well as for more rapid nutrient cycling, even though it disturbs beneficial insect habitat at a time when orchard pest insects are most damaging. And when fewer insect pests are present in late summer, I mow the orchard ground cover shorter for ease of harvest and to enhance nutrient cycling. Then raptors and owls can also better hunt for voles. In each chapter in part 1, I focus on specific ecological interactions like these within gardens and farms and how to manage them with ecological and systems thinking.

Figure 1.10. Orchard predators include this pygmy owl, which landed in a peach tree to consume a just-caught vole.

Figure 1.11. This bee-like insect is a syrphid fly. It's a welcome sight in an orchard because it provides pollination and its larvae feed on aphids, scale insects, leafhoppers, and small caterpillars like newly hatched codling moth larvae.

CHAPTER 2

The Soil Revolution

When I first started farming, I thought and acted like a nitrogen hoarder. To evaluate soil health, I conducted soil tests to see how much nitrogen my soil provided. Then I calculated how much nitrogen each specific crop needed for optimal growth, and I tried to add that amount to my soil by applying my farm-made compost (I measured the nitrogen content of the compost, too) and by growing cover crops. The form of nitrogen I measured at that time was nitrate-nitrogen, which is the common form that most researchers and farmers were (and still are) measuring. I didn't think much about carbon back then, but during the dramatic, although prolonged, revolution in my farming methods, I rethought my whole approach to nitrogen—and to carbon, too. (Carbon is what makes up the bulk of plant tissues and gives soil microbes the energy to do their ecological jobs. Nitrogen is what makes up proteins in plants, animals, humans, and soil microbes. Both are vital, but traditionally farmers and gardeners have focused on nitrogen.)

My soil management revolution is a three-part evolution:

- First, I learned to view nitrogen less as a crop requirement and more as a sign of effective, balanced, and active soil microbial activity that leads to optimal organic residue decomposition and nutrient cycling. The amount of nitrogen (or other nutrients) needed for any particular crop is not as important as the abundance and diversity of soil microbes within my soil.
- Second, soil organic matter and nutrient cycling potential are more than just discrete soil health measurements; they are a part of a complex connection of ecological relationships, the *soil organic matter system*.
- Third, soil carbon is more complex than I thought. Soil microbes are a vital component of the soil organic matter system, and soil microbes need lots of carbon for their fuel. Nowadays when I test my soil, I measure microbial biomass and microbially active carbon, rather than just the percent organic matter that I (and most organic farmers and gardeners) have been measuring for the past 50 years.

In this chapter I tell the story of how I came to understand that a soil organic matter system is the foundation of low-input ecological farming and gardening. It's a revolution that includes changing the ways we measure soil health, build the levels of carbon in our soils, optimize organic matter decomposition, and encourage nutrient cycling.

The Reformation of a Nitrogen Hoarder

I like to begin any discussion of soil solutions with a practical story of what happens over time when a farmer (me) acts like a nitrogen hoarder who

Figure 2.1. White clover living mulch growing between rows of trellised cucumbers. The nitrogen-fixing clover was part of my strategy to build soil nitrogen, and it also provided season-long ground cover and bloom at Biodesign and Woodleaf (CA) farms.

monitors soil health solely by measuring soil nutrient levels, rather than managing the soil organic matter system. This story reveals, among other things, the problems that can result from adding too much compost. My practical story includes a lot of nerdy details and revelations from on-farm research. Experienced farmers and hands-on learners, these on-farm research details are for you! If you are a beginning gardener or farmer, you may want to skip ahead to the next section, "Exploring New Methods," on page 21. Or if you do read this section, don't worry too much about trying to understand every detail or studying every figure.

When I started farming in Montana, where springtime is notoriously cold, my main concern was how to rapidly get enough nitrogen and phosphorus into my crops to ensure good yields and early maturity. Cold soils slow the activity of soil microbes that decompose organic residues leading to release of nitrogen and phosphorus. Slow nutrient release usually means slow early crop growth. I wanted to have the first tomatoes and sweet red peppers for sale at my local farmers markets. And, you see, when I was in graduate school studying horticulture and plant physiology, my professors told me that fertility systems that relied only on organic residues could never provide nitrogen rapidly enough to produce good commercial yields. Because of their warnings, I designed a fertility system with lots of high-nitrogen organic residue inputs, including compost made using animal manures.

Over time, my zeal to prove my professors wrong did indeed raise the nitrogen levels in my soil. In fact, my primary challenge became how to manage having *too much* nitrogen, phosphorus, and potassium in my farm system! This led me to explore how to move toward using soil amendments that were higher in carbon and lower in nitrogen. I

Understanding Nitrogen

Let's take a minute to review the forms of nitrogen that can be present in soil. Nitrogen is available in the environment in many chemical forms. Atmospheric nitrogen (N_2) is a gas, of course, and as mentioned below, nitrogen can also take the form of gaseous ammonia (NH_3). In the soil, however, nitrogen generally exists mainly in three forms: ammonium (NH_4^+) ions, nitrate (NO_3^-) ions, and organic nitrogen compounds.

The bulk of the nitrogen that crops use is inorganic ammonium and nitrate. Most soil tests measure nitrate-nitrogen because it is the form most readily and easily taken up by crops (at least in the way farmers have traditionally grown and fertilized them).

By contrast, the majority of potentially available nitrogen in the soil is in organic forms: plant or animal residues, soil organic matter, growing roots and root exudates, and especially the bodies of soil microbes. This includes all the forms of nitrogen available in the soil organic matter system, from simple amino acids and proteins to complex macromolecules. It is generally not directly available to plants, but a small amount of organic nitrogen may exist in the soil as soluble organic compounds (such as amino acids, proteins, and urea) that can be taken up by plants directly, especially when the amount of easily available inorganic nitrogen forms are low. Also, soil microorganisms convert organic nitrogen to the inorganic forms.

gradually became the "anti-compost" and a systems thinker who grows her own fertilizer right on her farm and creates a fertility system that multi-tasks as a habitat-building system. Other researchers, and a few farmers, were also talking about feeding and relying on the soil microbial community to cycle nutrients and about using high-carbon cover crops rather than applying a lot of high-nutrient compost. However, even today it is common for organic farmers—especially those working to develop no-till methods—to apply a lot of compost rather than to incorporate high-carbon fertilizers like cover crops, living mulches, or hay mulches.

In the early years at Biodesign, I planted a clover living mulch that I incorporated into the soil each spring, along with compost made on the farm from sheep manure and clover clippings. I wasn't shy about the compost: I applied 7 to 10 tons per acre (18–25 tons/ha). I also mowed the clover living mulch in the row middles between crop rows throughout the growing season. I left the mowed material in place, and thus it was subject to *nitrogen loss*—some of the nitrogen in the clover could volatilize and escape into the atmosphere in the form of ammonia gas. Because of this, I decided not to include the nitrogen content of the mowed clover in my fertilizer and soil nutrient level calculations. All my soil science colleagues agreed, and my coursework confirmed, that surface-applied organic residues lose a lot of nitrogen to the atmosphere and hence do not add much nitrogen to the soil. However, after several years of this management regimen, actual soil tests indicated that despite my careful calculations, the nitrogen, phosphorus, and potassium levels in my soil were becoming too high.

The rise in nitrogen levels was rapid and excessive when I added high levels of compost in the early years. Soil nitrogen decreased as compost additions also decreased. I was able to see this even more dramatically because I compared soil nitrogen levels in two fields that I managed in different ways, one with high levels of manure-based compost, and the other primarily by mowing and surface-applying clover living mulch between strip-tilled crop rows. I called the two fields the Old Field and the New Field, and figure 2.2 shows the differences in nitrogen levels between the two fields over time.

I applied the high levels of manure-based compost from 1993 to 1996 in the Old Field. Notice that these higher applications resulted in a nitrate-nitrogen peak in 1996 (the orange line in figure 2.2). The soil nitrate-nitrogen levels in the field were ridiculously high (over 100 ppm). From 1997 to 2002 I reduced and then eliminated (2002) the compost applications. As I reduced the quantity of compost applied, soil nutrients decreased to acceptable levels.

The purple line in the figure represents the New Field, where the main source of fertilizer was mowed perennial clover living mulch in the row middles and blown into crop rows and strip-tilled clover and weeds in the crop row. Nitrate-nitrogen in the Old Field never reached the high levels of the New Field, though I still applied very low rates of manure compost there in 2006 and 2007. Like the nitrogen addict that I was, I applied compost even though soil tests indicated that nitrogen levels were perfectly sufficient and I did not need to add any more nitrogen at all (my nitrogen-hoarder mentality was still driving me then). After 2007 I found the courage to finally stop depending on my compost. Between 2008 and 2010 in the New Field, I applied no compost. I did apply alfalfa meal, but only in the crop rows. I also mowed perennial red clover living mulch and blew it into the crop rows as a surface-applied mulch that would supply some nitrogen to the soil system as well.

I had not believed that I could get great yields without relying on lots of nitrogen and applying compost, but this long-term data showed me I was wrong. I call the sharply rising, pyramid-shaped orange line representing nitrogen level excess and decline in my Old Field my learning curve. I learned that I did not need to apply as much nitrogen as I had assumed to achieve good crop yields and

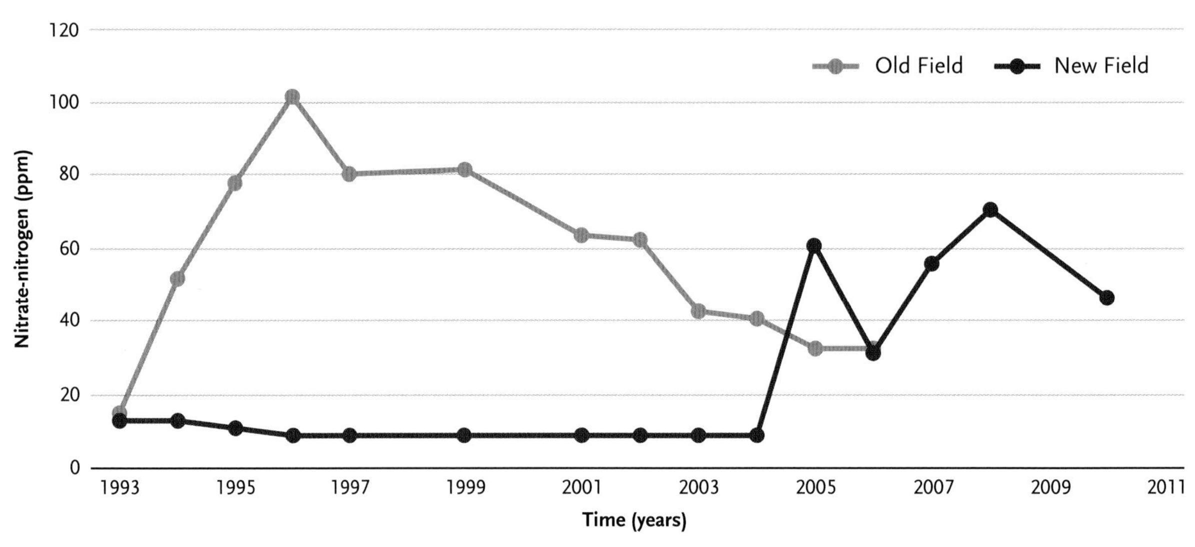

Figure 2.2. Nitrate-nitrogen levels (PPM) in Old Field and New Field at Biodesign Farm from 1993 to 2010. Notice the sharp increase in nitrogen in the Old Field, followed by a decline (when compost application decreased and stopped), as well as the slow increase in the New Field (where mowed living mulch was the main fertilizer).

quality. I began to appreciate how much my surface-applied, mowed living mulch was contributing to nutrient cycling in my soil fertility system.

Even though I decreased my nitrogen fertilizer applications beginning in the 2000s, I was still a recovering nitrogen hoarder (and secretly proud of my high soil nutrient levels because my professors had told me that it was impossible to get high levels using only organic fertilizers). It was hard to let go of my "nitrogen fix." I did not realize or yet understand the changes that occurred among the soil particles, roots, and soil microorganisms in the rhizosphere when I reduced tillage and maintained plants with growing roots in the soil year-round. Setting a goal of maintaining growing roots to support microbial communities changed all the rules I had been following about soil fertility! But despite my new intellectual understanding, it took years to get over my nitrogen addiction.

Figure 2.3. As I started using new methods at Biodesign Farm, it was exciting to see the quality of my vegetables improve. I started to see less blossom-end rot on tomatoes and less sunscald on sweet peppers.

Exploring New Methods

All organic fertilizers feed soil microbes—we can think of these fertilizers as "microbial food." Farmers and gardeners can choose between unprocessed microbial foods or processed microbial food. Unprocessed microbial foods include a tilled-in cover crop, full of a diversity of organic compounds, some of which are easy to break down, and others that aren't. Processed fertilizers are organic materials that have already been partially broken down, such as soybean meal. Processed microbial food (like processed human food that removes fiber and/or other parts of a whole food) may not supply as complete a diet for as diverse a microbial community. This is important because the way we fertilize our plants affects the whole farm/garden agroecosystem, including the taste and nutrient balance of the food we grow. In fact, in one field study the quality of tomato fruits with respect to vitamin C and total antioxidants improved after both inoculation with root-associating beneficial bacteria and fungi and reduction of high-nitrogen fertilizers.[1]

Once I fully realized that my reduced-tillage and living mulch system (with year-round growing roots) was effectively feeding the soil microbial community and also *recycling* nutrients, I began trying new methods to match the supply of nutrients in my soils to the levels that the crops needed for optimal growth.

LINKING NUTRIENT SUPPLY TO NUTRIENT NEEDS

I began to investigate how to manage organic residues in a way that would provide enough nutrients for good and early yields and simultaneously link *mineralization* to soil organic matter decomposition. Mineralization is the microbial conversion and release into the soil of nutrients such as nitrogen and phosphorus from organic forms tied up in plant and animal residues into forms that plant roots can absorb. The challenge was: How to link the slow process of organic residue decomposition with nutrient release rapid enough for good yields? The answer I came up with evolved into what would become one of my most important ecological farming principles:

Maintain growing roots year-round. Avoid bare soil; instead, maintain plants with living roots to keep the soil covered year-round and to feed the rhizosphere a steady, balanced diet of carbon and nutrients.

With this revelation, my soil fertility management system evolved. I moved away from relying on incorporating compost and cover crops into the soil every spring. I tilled less and less and instead left regular additions of organic plant residues, such as mowed legume living mulch, on the ground surface to decompose. And the perennial living mulch supplied living roots in the field at all times. Another revelation was when I began to think about applying living mulch residues at a range of maturity stages (succulent, green, and young versus dried, browning, and old plant residues). The stage of maturity affects the amount of nitrogen in a material compared with the amount of carbon, which is expressed as the *carbon-to-nitrogen (C:N) ratio*. The idea of changing my management approach to use organic residues with higher or lower C:N ratios at different times of year to produce different results was another transformation in my soil revolution. I became less obsessed with high-nitrogen legumes as living mulches. I was no longer afraid that my crops would go hungry if I fed them dried organic residues that were lower in nitrogen and higher in carbon. This revelation led me to adopt two more ecological farming principles:

Add organic residues all season. Add organic residues regularly throughout the year, not all at once in the spring.

Fertilize selectively. Fertilize the crop selectively; avoid fertilizing the whole field or garden.

Organic residues such as manure, compost, grass and/or legume cover crops, mulches, and woody residues (like chipped branch wood) vary in carbon and nitrogen content and in *lignin*, which is a complex, very slow-to-decompose organic material found especially in wood and bark. Organic residues also vary in other nutrients, such as phosphorus and micronutrients including manganese or iron.

I dove into the challenge of matching crop nutrient demand with nutrient release from organic residues rather than just adding more inputs. I also wanted to try to quantify nutrient release both from the organic residues I applied and from the store of soil organic matter. In other words, I try to match the rate of nutrients becoming available to crop roots from slowly decomposing plant residues with the rate at which my crops need to take up and use nutrients for optimal yield. And I want to do this while still slowly building soil organic matter to match any I use during my annual minimal tillage (even minimal tillage decreases soil organic matter). This matching achieves a more idealized steady-state condition; it equalizes organic residue decomposition with nutrient cycling into crop roots.

When the soil reaches a steady state, organic nitrogen and other organic nutrient levels remain relatively constant in the soil. There are no dramatic flushes or decline of nutrients. In other words, nutrients become available at about the same time that plant roots need them, so excessive levels don't build up. Some soil scientists suggest that this approach is likely to be superior to spreading and incorporating compost in the spring, which leads to a sudden flush of nutrients in the soil, with the possible loss of nutrients into groundwater.

However, the same soil scientists also suggest that the ideal steady-state condition may be an unachievable goal. This is because nutrient release rates vary considerably with moisture and

Why C:N Ratios and Organic Nitrogen Matter

When I stopped relying on materials that provide easily available forms of soil nitrogen—including composts, processed fertilizers (like alfalfa meal), and tilled-in legume cover crops—I had to learn how to manage the carbon and organic forms of nitrogen present in the grass, weed, and legume living mulches I was experimenting with as fertilizer. The surface-applied foliage and growing roots of these living mulches feed the soil microbial community in a more continuous, slow, and stable way than an annual application of manure-based compost does. (Many people argue, incorrectly, that manure-based composts are high-carbon fertilizers.)

Working solely with plant-based fertilizers is truly "feeding the soil" rather than just feeding crops, and feeding the soil is what I aspire to. It takes courage to make the switch, though, because it requires learning how to account for sharing the nutrients intended for crops with the soil microbial community. It takes time for soil microbes to break down organic residues and unprocessed fertilizers. And during the decomposition process, the microbes "tie up" plant nutrients, causing *immobilization*.

Immobilization is the opposite of the microbial conversion process of mineralization, in which inorganic nutrients are released from their organic forms. During immobilization, nutrients in inorganic forms are converted back to organic forms and are absorbed into the bodies of soil microbes as they use them for their own growth. This temporary tying up of nutrients is problematic because it means the nutrients are unavailable to plants. Immobilization occurs due to organic residues with high carbon-to-nitrogen and/or carbon-to-phosphorus ratios. Both immobilization and mineralization are dependent on the abundance and diversity of the microbial populations in soils and all the environmental conditions that favor or thwart microbial activity and chemical reactions: soil moisture, temperature, oxygen, and pH, and the amount of organic residues to decompose.

During immobilization, microorganisms actually remove nitrate-nitrogen from the soil's available nitrogen store. This is not good for crop plants that want that available soil nitrogen. Fortunately, nitrogen is made available to crops when microorganisms break down an organic residue source *that contains more nitrogen than the microbes can use at the time*, such as a succulent, green legume residue. And as some microorganisms within the community die, more nitrogen is added to the soil's store of available nitrogen. Nitrogen derived from dead microbial organisms within the microbial community may account for as much as 60 percent of soil organic nitrogen. This is the process minimalist farmers and gardeners need to understand and encourage. If we can encourage steady decomposition of organic residues, provide a continuous supply of slowly decomposing residues, and increase microbe populations and diversity, the breakdown process does not result in immobilization and may actually release plant-available organic nitrogen as well as more available forms of inorganic nitrogen, including nitrate-nitrogen. Organic residue

C:N ratio matters because microbial communities respond differently when a fertility system relies on organic residues that are predominantly high in nitrogen rather than high in carbon. When we add only a lot of high-nitrogen, low-carbon organic fertilizers, plants stop taking up organic nitrogen. Researchers have found that plants may rely more heavily on organic nitrogen sources under low-input farming systems with lower soil nitrate-nitrogen levels than under highly fertilized systems. They suggest that in production systems that receive large amounts of higher-nitrogen, lower-carbon fertilizers, there can be a potential mismatch between the supply of soil nitrogen and the capacity of plant roots to take it up. Learning to use less nitrogen fertilizer inputs means learning to use residues with higher C:N ratios in combination with strategies to enhance microbial activity while at the same time enhancing decomposition.

temperature and also with plant residue maturity. This brings us back to why C:N ratios matter so much in my low-input system. We can quantify the maturity of plant tissues in terms of C:N ratios and also lignin content. Both lignin content and C:N ratios play a big ecological role in a soil organic matter system designed to match crop nutrient demand with nutrient release. Generally, materials with a high C:N ratio decompose and release nutrients into the soil more slowly than do organic residues with lower C:N ratios. Materials with high lignin content break down even more slowly than high-carbon materials. It's not surprising that lignin is very slow to decompose, because it is the component of plants that make them structurally sound.

I began to realize that the main obstacle to achieving a steady-state condition of available soil nutrients was my earlier lack of understanding of:

- The carbon-to-nitrogen (C:N) ratios of organic residues I was applying.
- The fate of specific plant nutrients in *surface-applied* organic residues.
- The predictability and speed of nutrient availability to crop plants as various organic residues decompose.

This led me to some new experiments, because I wanted to know how my management methods affected nutrient release from many kinds of decomposing organic residues. For example, clover (a legume) was one source of organic residue that I studied. Legumes can make their own nitrogen fertilizer when their roots associate and form nodules with rhizobial bacteria, which have the ability to fix nitrogen from the atmosphere. During on-farm research in 1996, I discovered that the C:N ratio of mowed clover living mulch residue changed over the growing season; it was highest in July and lower in August and September. So the clover C:N ratio fluctuated over the life of the crop and was highest (the plants contained the least nitrogen) when the clover was young and had not yet formed nitrogen-fixing nodules on roots. The ratio was lowest (plants had greatest nitrogen content) when rhizobial bacteria were established in the root nodules and efficient nitrogen fixation was occurring (mid-flowering stage in August). The clover C:N ratio then began to rise again (nitrogen content decreased) as the clover began to mature at the end of September. I was surprised to learn that micronutrient levels also changed in the clover over the growing season.

Table 2.1 shows the levels of nutrients in clover over time on my Montana farm, compared with "normal" levels for clover foliage. Levels of some nutrients in my clover were higher than normal while others were lower. Notice that micronutrient and potassium levels do not follow the same pattern as nitrogen, phosphorus, sulfur, and magnesium, which are highest just before or during clover bloom in August (when farmers traditionally incorporate cover crops for best nutrient addition to the soil). Potassium levels were highest when the clover began to mature and the C:N ratio was moderate. But iron and zinc were highest in young plants, when the C:N ratio was at its highest.

My continued research showed that the timing of when I incorporated or mowed and surface-applied the clover affected nutrient availability and decomposition rate. The method of organic residue application (tillage versus mowing), the timing of application (at intervals versus all at once), and the residue C:N ratio (higher- versus lower-carbon) all interact with one another and with soil microbes to either allow or hinder nutrient availability to crops. The data from table 2.1 and my yields over time showed that vegetable crops do best in my system if I combine once yearly minimal tillage with regular surface applications of both high- and low-carbon organic residues. Tillage speeds decomposition (regardless of C:N ratio). But to avoid soil microbes tying up nutrients and to maintain active decomposition throughout the growing season, it is also vital to provide a *continuous organic residue supply* rather than adding organic residues all at once. I had discovered that my clover living mulch has different C:N ratios throughout the season, hence regular mowing throughout the season provides a steady supply of diverse microbial food.

I had discovered something vital (though it would be many more years before I had the courage to make this knowledge the cornerstone of my soil management system): By timing *when* and *how* we apply organic residues, we influence their C:N ratio, which in turn determines how much and when we build carbon in the soil and what kind of soil organic matter system we create. But as with many things, the devil is in the details. In chapters 3 and 4, I cover what I've learned about the C:N ratio details of specific organic residues and what I've learned about tillage.

Managing Tillage

Minimizing soil disturbance is another relationship to consider when building a soil organic matter system. Tillage (or not tilling) is another part of the Biodesign Farm story and my learning curve, because tillage also affects the carbon-building and nutrient-release ability of organic residues and

Table 2.1. Nutrient Content of White Clover Living Mulch, Biodesign Farm, 1996

	C:N Ratio	N (%)	S (%)	P (%)	K (%)	Mg (%)	Ca (%)	Fe (ppm)	B (ppm)	Zn (ppm)
July	16	3.30	0.27	0.43	4.10	0.40	1.40	503	21	35
August	10	4.30	0.31	0.46	4.20	0.35	1.40	165	24	30
September	11	4.10	0.30	0.45	4.90	0.31	1.20	191	23	30
Normal levels of nutrients in clover*		4.26–5.50	0.21–0.40	0.26–0.50	1.71–2.50	0.26–1.00	0.36–2.00	51–350	21–55	21–50

* W. F. Bennett, *Nutrient Deficiencies and Toxicities in Crop Plants* (St. Paul: APS Press, 1993).

microbial community health and activity (which I explain later in this chapter). I set up an experiment to compare soil nitrate-nitrogen levels during the growing season in areas with different tillage treatments. One area was a 600-by-30-foot (183 × 9 m) area of untilled, remnant grass/weed pasture; the second was a 600-by-4-foot (183 × 1.2 m) untilled clover row middle, and the third was a 600-by-4-foot Brussels sprouts crop bed created by tilling a strip in an area that had been planted to clover the previous spring. While the Brussels sprouts were growing, I left them unmulched and did not add compost prior to planting. I periodically mowed the grass pasture test plot and the clover living mulch test plot and left the residues on the soil surface.

I found that the incorporated (strip-tilled) clover from the previous spring released nitrate-nitrogen into the soil of the Brussels sprouts bed much more quickly and at higher rates than surface-applied clover or grass that was simply mowed and left on the soil surface (see figure 2.4). There was a flush of nitrate-nitrogen about 3 weeks after tilling, from 16 ppm pre-tillage to 66 ppm. This jump shows why organic farmers traditionally till in cover crops rather than surface-applying them as a mulch. I set up a second test row of Brussels sprouts, in which I added compost before strip-tilling and planting. Compost addition resulted in even higher nitrate-nitrogen levels (data is not included in figure 2.4).

The surface application of mowed clover or grass did not result in a spring flush of soil nitrogen. In fact, soil nitrogen in those plots never jumped, but it did slowly increase over time. The soil in the plot with mowed clover living mulch had higher nitrogen levels than the plot with mowed grass pasture for most of the growing season. Thus the higher C:N ratio of the grass residues resulted in lower short-term levels of soil nitrogen available for crops May through August. Another factor to keep in mind for this comparison is that growing legumes such as clover don't consume as much soil nitrogen as does grass (because legumes are benefiting from the nitrogen fixed by rhizobial bacteria in their root nodules).

In the longer term I was surprised by what happened. Four months later, by September, the soil in both the mowed clover and grass plots had similar soil nitrate-nitrogen levels and both were above 25 ppm (the level I used to strive for when I

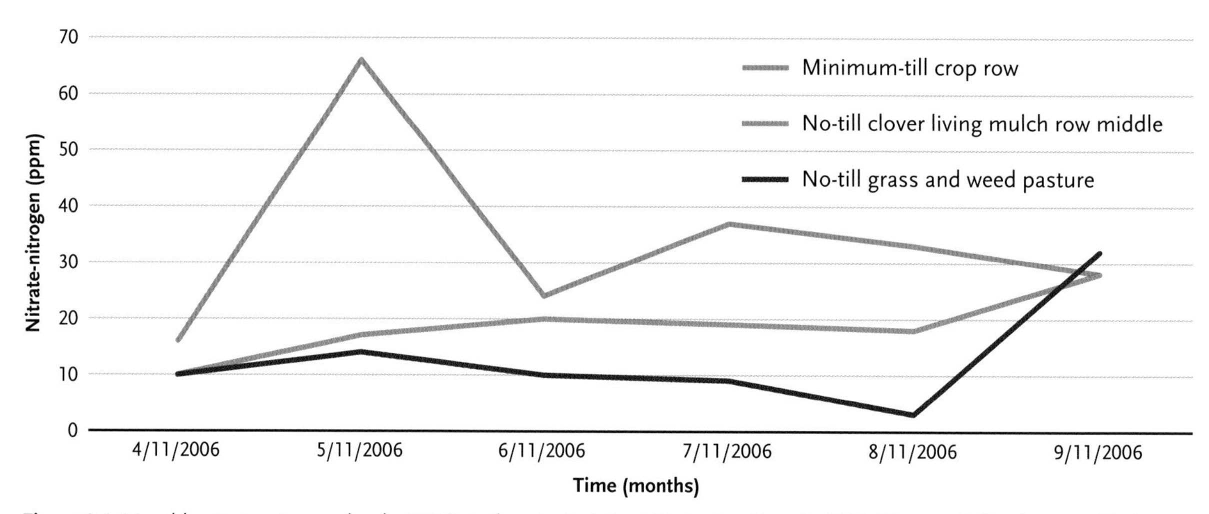

Figure 2.4. Monthly nitrate-nitrogen levels (PPM) in three test plots at Biodesign Farm in 2006. The purple line is an untilled grass/weed pasture; the green line is an untilled clover row middle (both mowed regularly). The orange line is a Brussels sprouts crop row in which a clover cover crop was strip-tilled before planting.

was a nitrogen hoarder). Plus, both had similar soil nitrogen levels when compared with the strip-tilled Brussels sprouts crop row (the sprouts had not yet been harvested). I concluded that nutrients from the surfaced-applied grasses and clovers were *eventually* released into the soil even though the residues had not been tilled in.

Here is what I learned from this 2006 experiment: As long as your soil has an active and diverse soil microbial community, incorporating legume cover crops in the spring can result in a relatively quick burst of nitrogen (in 2 to 4 weeks) and then a sustained season-long release of nitrogen as it slowly decomposes, even without compost addition. In comparison, surface-applied organic residues add nutrients much more slowly. In other words, there is a lag time for nutrient release. But over time even surface-applied clover and grass residues might be adequate fertilizers in a minimal-input soil fertility system if managed properly. (And in chapter 4, I describe in detail one way to manage these high-carbon residues "properly.") Ah-ha! After a decade of experimentation, the potential for successfully using higher-carbon, surface-applied organic residues was sinking into my nitrogen-hoarder brain.

GRADUATING TO CARBON FARMING

I remained a closet nitrogen-hoarder until I met Carl and joined in his "radical carbon-farming" methods. I was moving toward carbon farming, but Carl was already there. Unlike me, Carl did not even measure nitrate-nitrogen in his orchard's soil, or any form of nitrogen. He focused on soil organic matter instead. Carl's goal in his no-till orchard was to build a soil organic matter system that included an undisturbed soil carbon framework to support what he called his fungal food webs. I called it the soil microbial community (because it includes more than just fungi, as I explain later in this chapter).

Carl had also developed his own strategy of *mineral balancing* (which I discuss in the fruit tree management section in chapter 12). On his California farm, Carl did not include nitrogen in his fertilizer and soil health calculations. In fact, there is some evidence to support Carl's idea that fruit trees need less fertilizer (including nitrogen) than I had been taught, because the trees can recycle nitrogen. In one study with cherry trees, researchers found that three months after bloom 45 to 50 percent of the total nitrogen in the tree was derived from internal plant recycling. Carl was right not to focus on nitrogen fertilizer in his orchard! He nudged me toward another of our ecological farming principles:

> **Focus on carbon fertilizers.** Prioritize slow-release, plant-based carbon fertilizers rather than fast-release nitrogen fertilizers.

As we changed our farming methods, we began to wonder if there were better ways to measure soil fertility levels than by looking only at soluble, easily available nutrients (such as nitrate-nitrogen) and the traditional way of measuring soil carbon: percent soil organic matter. For example, to what extent (if any) were our fruit trees able to take up organic forms of nitrogen? We knew that researchers have discovered that trees with mycorrhizal root associations can take up organic nitrogen.[2] (I discuss mycorrhizal fungi in detail later in this chapter.) Further, from soil tests and microbial analysis, we knew that there were high levels of mycorrhizal fungi in our orchard soil. Could the ecological research on the uptake of organic nitrogen by plants apply to our crop plants, too? What would happen if we decreased easily available forms of nitrogen fertilizer and created soil fertility systems that more closely mimicked those in natural plant communities?[3] Regardless, we began to suspect that organic forms of nutrients may be important for the health and well-being of soil microorganisms. We were also beginning to realize that the way in which we fertilized crops would affect their ability to produce biochemicals called secondary compounds, which help protect against

Figure 2.5. There is scientific evidence that richly colored fruits like these Woodleaf Farm, California, peaches tend to contain higher levels of secondary compounds, which help naturally protect against disease and insect pests, than do light-colored fruits.

disease and insects (I discuss this in detail in chapter 5), as well as specific kinds of vitamins, minerals, and antioxidants. The plot thickens around this low-input, higher-carbon, microbial-dependent soil fertility system!

Beginning in 2016, mowed living mulch was the only fertilizer I applied to most of my crops. With vegetable crops, I discovered another benefit of maintaining a growing root: In early August 2019, living mulches expanded from the row middle into strip-tilled vegetable crop rows to act as soil cover during the winter. In the spring of 2020, these "invaders" provided a cover crop for free that I did not have to seed. The yield and quality of the 2020 onions, shallots, and squash were good with this low-input, higher-carbon fertility system (as is evident in figure 1.1). This system works because I pay attention to the management details, such as timing of living mulch applications and balancing the C:N ratios of the living mulch growth stages. I mowed the living mulch and blew it into crop rows early in the season when it had a lower C:N ratio, as well as later in the season when the C:N ratio was higher.

As we designed our new Oregon farm and evolved our thinking, we had two fundamental goals to further augment our higher-carbon fertility system: first, to keep growing roots in the soil year-round; and second, to figure out how to farm while disturbing the soil as little as possible, working with the equipment we already had. We did not want to buy new, no-till crimping/rolling equipment. We also wanted to avoid applying too much plastic to smother cover crops and weeds. Most important, we didn't want to give up the in-field habitat-building pest suppression system we had developed (and which I describe in detail in chapter 5). Finally, we also wanted to put into practice several other ecological faming principles:

Create above- and belowground diversity. Feed the soil microbial community with as much crop and ground cover, root and whole plant, above- and belowground diversity as is economically possible.

Grow your own carbon. Grow or obtain your carbon source where you are, rather than importing carbon from a distance.

Add organic residues all season. Add organic residues regularly throughout the year, not all at once in the spring.

Focus on carbon fertilizers. Prioritize slow-release, plant-based carbon fertilizers rather than fast-release nitrogen fertilizers.

Recycle rather than import nutrients. When possible, recycle minerals from on-farm sources rather than importing mined mineral nutrients from somewhere else and applying as fertilizer.

The last principle became important to us when we realized that adding minerals to our soil meant that those minerals had to be extracted from someone else's land. That did not seem sustainable or fair. We needed to know more! So we set up an experiment to figure out how to grow our own fertilizer, and I tell that story in chapter 3. Right now I want to delve into the microbial drivers that are central in a soil organic matter system.

Microbial Activity and Interconnections

My personal soil revolution takes place against the backdrop of some big developments in scientific understanding of soil ecosystem complexity and how plant roots, microbes, and other organisms interact there. In fact, so much has been discovered that hundreds of scientific articles and entire books have been written about it over the past couple of decades. Here is my simplified discussion of this topic, which is the foundation of my soil organic matter system theories.

Decomposition, release of plant-available nutrients, and retention of soil nutrients are the foundations of soil ecosystem functions within the soil organic matter system. The compost you apply to your garden or the cover crop you till into your farm field becomes the fertilizer that feeds your crops. As organic matter and organic residues decompose, plant nutrients such as nitrogen, phosphorus, and potassium are released and made available not only to plants, but also to the soil microbial community. Soil microbes are the engine that drive all these processes that release plant nutrients.[4]

No matter where you garden or farm, the soil organic matter system is built on microbial activity and interconnections. Achieving humus is not the only goal! You've probably read and heard about the importance of humus (which is simply organic matter in a very, very decomposed form). But instead of a focus on humus, we should strive for a *soil organic matter continuum*—applying a full spectrum of progressively decomposing organic residues and compounds. The soil organic matter that is constantly changing within this continuum is the *total* carbon plus other organic nutrients and microbial by-products. While maintaining this fluid concept of organic matter, it is useful to look at various forms. (I discuss these forms in detail in chapter 3.) We can think about the organic matter forms we add as plant or animal residues (like grass clippings or manure) as regular deposits we make into an optimally functioning soil organic matter system "bank account" that is rich in a continuum of decomposing, decomposed, and very decomposed organic matter. Adding a diverse continuum of organic matter to your garden and farm soil bank account feeds a diverse soil microbial community and supports microbial activity and interconnections.

Among the many players in the soil microbial community are free-living bacteria (those that are not associated with plant roots) that convert organic phosphorus within organic residues (like hay mulch) into forms of phosphorus our crops can easily use. Other free-living bacteria convert organic nitrogen from plant or animal residues into forms crop roots can take up.

Fungi are even more prevalent than bacteria in healthy soils. In agricultural soils there are approximately 25,000 known fungi species, accounting for 70 percent of total soil microbial activity. There are a few major lifestyles among fungi, including decomposer fungi and root-associated fungi. Decomposer fungi, which are free-living filamentous fungi, degrade complex compounds, such as lignin, and break down carbon. Free-living yeasts break down simpler compounds, such as sugars. Root-associated fungi, also called mycorrhizal fungi, obtain most of their carbon from plant roots rather than from breaking down soil organic matter.

The types and species of fungi that survive best in tilled farm and garden soils obtain energy from decomposing organic matter supplied as organic residues. Generally, these tillage-tolerant kinds of fungi have relatively small *hyphal networks* (the fungal webs that make up the underground nutrient-conduit superhighway). These fungi species are important for soil fertility, soil structure, and nutrient cycling, but play only a minor role in carbon storage and building the organic matter (or carbon) framework of the soil organic matter system.

Root-associated soil microbes live within the biologically active rhizosphere where plant roots, soil, and microbes commingle. If you're looking for the magic of soil, you'll surely find it in the rhizosphere. There, symbiotic mycorrhizal fungi produce glomalin, a glycoprotein that stores soil carbon and stabilizes soil particles or aggregates of many soil particles.[5] The carbon storage "heavy lifters" in soils are these mycorrhizal fungi with their larger hyphal networks. Mycorrhizal fungi account for the majority of the fungal biomass in most soils. But they are not tillage-tolerant or high-nitrogen-fertilizer-tolerant and diminish in most tilled and heavily fertilized farm and garden soils. In fact populations of mycorrhizal fungi are diminished significantly in tilled soils.[6] This is a problem because mycorrhizal fungi are so important to the soil organic matter system and to ecological farming systems. They can increase the surface area of plant root hairs by up to 80 percent.[7] Further, it is becoming clear that a significant amount of the total carbon in the soil organic matter system may be a direct result of mycorrhizal fungal activity.[8] Most crops form mycorrhizal root associations and some, like corn, legumes, and potatoes, are strongly dependent on mycorrhizal associations. The few crops that do not form mycorrhizal associations are those in the Brassicaceae family, including rape and mustard, as well as sugar beet, buckwheat, and spinach.

In addition to bacteria and fungi, the living community in the soil is made up of earthworms, insects, mites, soil-dwelling mammals and reptiles, nematodes, and protozoans. I used to manage my soil to keep earthworms happy, but these days I spend a lot more time thinking about and devising ways to also keep root-associated microorganisms, especially mycorrhizal fungi, happy. Mycorrhizal fungi have become my ecological indicator species in the soil microbe community. Kind of like the soil version of the canary in the coal mine. (For an overview of the living community in the soil, see table 3.1.) Mycorrhizae are turning out to be an indicator species not only for a healthy soil microbial community, but also for soils that grow healthy human food. There are many examples of this in current research. One example is a study from the University of British Columbia in which tomatoes inoculated with mycorrhizae had higher levels of fruit minerals, carotenoids, and other antioxidants than did fruit from un-inoculated plants.[9]

SUPPORTING THE MICROBIAL COMMUNITY

This brings me back to the farming epiphany I had in Montana: My job as a farmer and gardener is to create habitat for an effective, balanced, and active soil microbial community that leads to optimal organic residue decomposition and nutrient cycling. This epiphany is reflected in this farming principle:

> **Create above- and belowground diversity.** Feed the soil microbial community with as much crop and ground cover, root and whole plant, above- and belowground diversity as is economically possible.

There is a good reason that Carl and I put this first in our list of ecological farming principles, and why so many others who care about soil health have adopted it as well. The goal of managing a soil organic matter system is to add diverse forms of soil organic matter and enhance soil microbial activity, rather than to rely on quick-release fertilizers and composts that supply the "big-three" plant nutrients (nitrogen, phosphorus, and potassium) to crops directly. These soil organic matter systems readily process plant or animal residues. The decomposition process yields nutrients in forms that are available to plants, and over a longer span of time, contributes to the reservoir of stored nutrients and soil carbon. In other words, the *soil organic matter system* releases plant nutrients as they're needed through the intermediary of the soil microbial community. All we have to do is manage soil organic matter as if it is a system, rather than just a soil health parameter. Let's move on to how to do this.

Carl's and my third ecological farming principle is the key here:

> **Maintain growing roots year-round.** Avoid bare soil; instead, maintain plants with living roots to keep the soil covered year-round and to feed the rhizosphere a steady, balanced diet of carbon and nutrients.

Some soil scientists believe that maintaining living plant roots in the soil is the single most important way to enhance soil microbial populations and hence build and maintain soil organic matter. Why are growing roots so vital to soil health and fertility? Living plants secrete 30 to 60 percent of the atmospheric carbon they capture through

Figure 2.6. In this Woodleaf Oregon vegetable field in August, diverse year-round soil cover between crop rows is thriving—cycling and recycling nutrients and providing undisturbed habitat and root exudates for the microbial community.

photosynthesis back into the soil (in the form of carbon-rich compounds called exudates) to support mutualistic relationships with the soil microbial community in the rhizosphere. Carbon that is integrated into the soil system by growing plants and processed though the microbial community (as new microbially created substances) remains in the soil longer and thus creates a stronger soil organic matter skeleton or framework.[10] This is why I am adamant about not feeding the soil microbial community too many processed foods (like alfalfa meal, soybean meal, or pelletized organic fertilizers). The microbial community does best with a diet of nutrients that come intertwined with (attached to) more carbon. Just as humans do best (and feed their internal microbiome) when their diets supply essential nutrients intertwined with or attached to fiber.

Soil microbial biomass and activity are diminished by soil disturbance. They are increased and regulated by the quantity and quality of both stable carbon in the soil organic matter bank account and by regular inputs of carbon-containing organic residues and nutrients. Fungal growth is most dependent on carbon availability. If the soil has low carbon levels and we do not regularly add carbon inputs, our soils will have low levels of mycorrhizal fungi. Research has also shown that total carbon content and quality determine how organic residues affect total microbial biomass.[11]

Carbon quality refers to the ease and speed at which organic residue and organic matter decompose in soil. As I learned on my Montana farm, materials with a high C:N ratio tend to decompose more slowly, thus releasing nutrients at a slow, regular rate over the entire growing season. This is not the best if your goal is to be the first one to market with ripe tomatoes, but slow also means steady and balanced. Steady, balanced, and diverse nutrient release appears to support the most diverse and abundant microbial community. Researchers compared microbial inoculum collected both from forest soil (where nutrient cycling occurs slowly) and from tilled agricultural fields with added fast-release fertilizers. The forest soil inoculum improved tomato plant growth and nutrient assimilation much better than did inoculum from agricultural soils, which had fewer microbial species and total microorganism numbers. The researchers suggest that this might be explained partly because the forest soils had higher organic matter and carbon content (and thus a more abundant and diverse microbial community) relative to the agricultural soils.

In summary, this is why we need to focus on more than just rapid-release fertilizers and composts: For optimal activity, soil microorganisms require high levels of soil organic matter (soil carbon) and *regularly* added organic residues feeding the soil organic matter system. Our job is to keep them happy, because soil carbon and microbial community health are key to soil health, farm or garden productivity, and nutrient-rich food.

DELVING DEEPER

Managing the soil organic matter system for happy microbes has cascading effects on the whole garden or farm agroecosystem. One gram (or ¼ teaspoon) of healthy garden soil with a balanced and active soil microbial community can contain up to 1 billion microorganisms, including: 100 million to 1 billion bacteria, 10 million to 100 million actinomycetes, and 100,000 to 1 million fungi. In native plant communities this is easy to maintain. The problem is that most of our gardening and farming tasks undermine the health of a diverse microbial community. Some farmers add microbes raised in labs to their microbe-deficient soils. This is a good emergency intervention, but for an agroecosystem approach, we want to create habitat so soil microbes can stay, reproduce, and thrive.

I used to think adding compost to soil was the best way to encourage microbes. Compost is indeed a good way to enhance certain soil microbes and an excellent way to build soil carbon. Austrian farmers

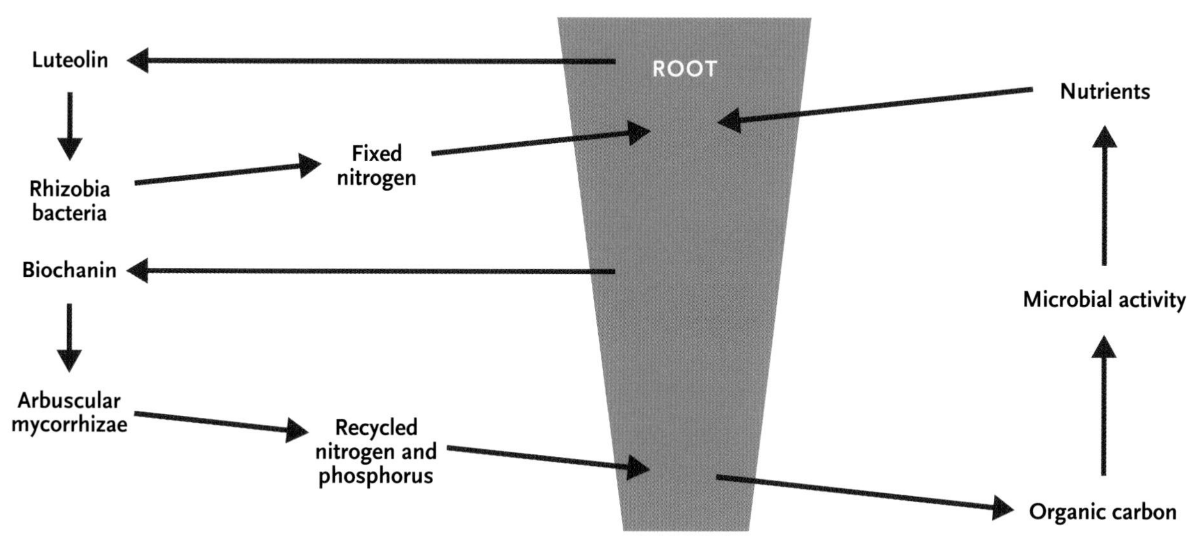

Figure 2.7. Up to 60 percent of the net carbon captured through photosynthesis ends up in plant roots. Some of that carbon (as well as other biochemicals) is released as part of the many kinds of root exudates that both feed and signal soil microorganisms, which in turn make nutrients available to plant roots and form root-microbe associations.

and composting experts Siegfried and Uta Luebke suggest that the humus crumb (or particle), with its many layers and air pockets, perfectly meets the needs of soil microorganisms. And researchers at the University of California have found that microbial biomass is positively related to the number of water-stable aggregates of particles in a soil. Thus, the better the soil aggregation, or tilth, the greater the number of total soil microbes. Any organic residue added to soil should improve soil tilth and aggregation. But we can do even more for the microbial community as a whole.

At the risk of repeating myself, I want to emphasize again that most soil microbes do best when growing roots are present and secreting carbon and plant exudates into the soil year-round. This provides good habitat for microbes in the rhizosphere. Many of the most important soil microorganisms require root associations, which are destroyed by soil disturbances such as tillage and weed cultivation. In a study comparing farming systems that maintained growing roots with other systems that did not, the systems that maintained roots showed overall better soil health and soil microbial communities. Researchers attributed this to root exudates and carbon inputs from continuous vegetation within row middles. The farming systems that maintained growing roots had greater soil particle aggregate stability, soil respiration (a measure of microbial activity), organic matter, phosphorus, and potassium, as well as the greatest number of soil invertebrates, such as collembolans and mites.[12]

To further wonderfully complicate matters, root-associated microbes can be plant-species-specific, preferring to associate with roots of particular plant species. Certain microbes also prefer the root exudates of some species more than others. Thus a diversity of plant species that provide a year-round underground smorgasbord provides habitat for the most diverse microbial community. Figure 2.7 shows a small portion of the many microbial relationships and connections provided by a growing root. The growing root-soil interface is full of biochemical and biological activity. For example, luteolin and biochanin are biochemicals that plants release as signals for specific microorganisms that their roots are ready and waiting for symbiosis and connection. Organic carbon is another root exudate that soil microorganisms use as a food source. Not all root associations are positive. There are also biochemical root exudates that send signals to plant disease microorganisms.

TESTING METHODS FOR FEEDING MICROBES

At Biodesign Farm, I tried different management methods for crop row middles with a goal of balancing the needs of crops with the needs of soil microbes. I compared the effects of tilling between rows of sweet peppers or mowing or lightly disturbing a clover living mulch between the crop rows with those of leaving an undisturbed living mulch (which served as the control). As I took data and pondered the results, another of the 10 ecological farming principles began to evolve:

> **Minimize soil disturbance.** Disturb the soil as little as possible; create year-round undisturbed refuges for above- and belowground natural enemies (biological control organisms).

At this point in my thinking, I was surprised to find that yield and fruit quality were the same for all treatments. I had expected the living mulch to rob the crop of nutrients and moisture and reduce yields. But in light of what we know today about the importance of growing roots for soil and microbial health, it seems a no-brainer that the living mulch did not harm the crop. The extra work involved in planting and managing an annual living mulch was worthwhile, and about the same amount of effort as the weeding and tilling needed to maintain bare soil between crop rows.

The data showed that soil organic matter levels and the microbial community were negatively affected by disturbing the soil to maintain vegetation-free crop row middles. By August the tilled treatment plots also had terrible tilth and compacted soil. Earthworm numbers were highest in treatments that were mowed monthly, and lowest in the tilled areas.

I measured total microbial biomass (TMB), which is a measure of bacterial and fungal mass in the soil sample. TMB was lower in tilled plots from the first tillage in July through the last in early September, and lowest in September. The data looked like a series of ascending/descending peaks and valleys, as the microbial community recovered slightly after monthly tillage disturbances, then dropped again when the microbes were decimated by the next tillage. Microbes were significantly higher in mowed treatments throughout the growing season, reaching a high of more than double the biomass measured in tilled plots.

Except in the tilled treatment, TMB was highest in all treatments in August, which is when soil temperature and moisture conditions are optimal for microbial growth in western Montana. Row middles covered by living mulch generally increased in TMB until soil temperatures dropped toward the end of the growing season. The unmowed, undisturbed living mulch (the control treatment) grew pretty tall and wild. However, it did not have the

Figure 2.8. These sweet red peppers from Biodesign Farm all are of equal high quality, even though they were grown with different management methods to test the effects of living mulch management methods between crop rows.

highest total microbial biomass of all treatments. TMB numbers were higher in the treatments that were mowed monthly. This surprised me, because I thought the soil microorganisms would be happier in the completely undisturbed control treatment. It turns out, however, that along with optimal soil moisture and temperatures, the soil microbial community needs regular carbon and nutrient additions for optimal growth. Mowing the living mulch periodically throughout the growing season provided the raw materials (plant residues) to supply those nutrients and carbon.

After seeing these results, I continued mowing the row middles in my fields monthly to feed the soil microbial community and cycle nutrients. Over time, I also adjusted my management to a system of no-till permanent living mulch (rather than annually tilled clover) in row middles, and I strip-tilled the crop rows annually before planting, as shown in figure 2.9. In this view, the perennial red clover row middles are just starting to green up, grow, and provide root exudates, long before any crops are planted. I began to wonder whether I could also reduce soil disturbance *in the crop row itself* without reducing yield too much. I had begun to realize how vital mycorrhizae are to the soil organic matter system and how lacking they are in most tilled vegetable gardens and farm fields.

Figure 2.9. Biodesign Farm 2006: Permanent perennial red clover living mulch just starting to green up in early spring as new crop rows are being strip-tilled.

I designed a study comparing several methods for managing weeds in the crop row, including a no-till approach, paper mulch, and clean cultivation. (I describe this experiment in detail in chapter 8.) Through this experiment, I discovered the beginnings of a soil health/crop yield compromise. Minimal-tillage plots provided an economically adequate yield and still retained the second highest level of my ecological indicator species, the mycorrhizae that are usually absent in tilled and cultivated gardens and farm fields. These experiments helped me to develop methods to further expand, diversify, and encourage my soil microbial community.

All of these experiments and my evolving methods to create a soil organic matter system resulted in a long-term, sustained increase in percent soil organic matter (SOM). Figure 2.10 shows this increase with long-term soil test results and a comparison of soil organic matter trends in the Old Field and the New Field. The large jump in SOM in 1994 is likely due to tilling established pasture and adding manure-based compost. Variation in the upward trend in 2001 and 2002 may be related to decreasing and stopping compost application, but SOM levels increased to the highest levels in 2006, even though I stopped applying compost altogether in 2003. In the New Field, I applied compost at very low rates in 2006 and 2007, but SOM levels continued on an upward trend through 2010 without compost addition.

Based on my on-farm experiments and the research of several microbial ecologists, I developed better methods to create habitat for a healthy, diverse soil microbial community. I describe these in detail in chapter 3.

MORE BENEFITS OF SOIL BUILDING

There are collateral benefits as we create a system in which the deposits we make into the soil bank

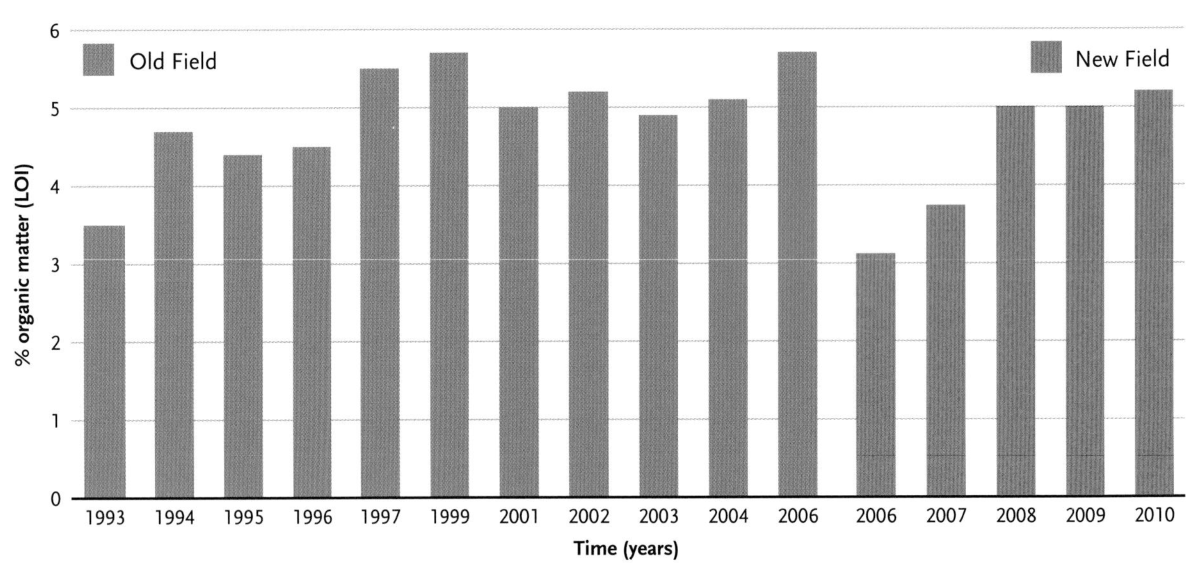

Figure 2.10. Soil organic matter levels generally increased over time in the Old Field at Biodesign Farm due to application of compost (until 2002) and mowed or tilled living mulch. In the New Field the main fertilizer was mowed living mulch.

account (such as cover crops) build soil and multitask as habitat for beneficial organisms in the farm or garden agroecosystem. Our greatest success at Biodesign and Woodleaf Farms relates to nutrient cycling and the way it works synergistically with an effective biological insect and disease suppression system (which I describe in chapter 5). For example, hay mulches, which slowly release nutrients into the soil as they decompose, also create an excellent physical environment for soil microorganisms and ground-dwelling predators. Mulch enhances the fungi that trap and consume harmful nematodes. Studies indicate that predator spider populations are significantly increased by mulching. Mulches may further prevent disease by providing a mechanical barrier between old, disease-infected leaf litter on the ground or soilborne pathogens and the non-infected foliage of growing crops.

Growing cover crops and living mulches and allowing non-crop vegetation areas to provide year-round growing roots makes sense in a system designed for multiple benefits. The space that non-crop vegetation takes up in our crop fields is worth it if it is also building the soil microbial community and creating habitat for natural enemies that provide biological control and reduce or eliminate brought-in inputs. To provide habitat, grow-your-own fertilizers (such as living mulches) need to remain as undisturbed cover for as long as possible. They also need to be allowed to bloom for an extended period to provide pollen, nectar, and even seed for beneficial insects and birds.

Soil scientist Dr. Christine Jones best sums up the importance of growing roots when she suggests that green plants are the most important drivers for building soil.[13] Plants with growing roots provide what Dr. Jones calls "soluble carbon" and habitat for the microbial community, including our main carbon storage drivers, the mycorrhizal fungi. In the next chapter I explore more fully what Dr. Jones calls the "plant-microbe bridge" as well as the myriad ecological relationships arising from that fragile bridge that can be so easily diminished by soil disturbance and excessive fertilizer additions. Then I describe in detail how diverse inputs and year-round living plants can be managed as part of the soil organic matter system to feed and enhance the soil microbial community.

CHAPTER 3

Soil System Management Principles for Farmers and Gardeners

When your goal is to build a soil organic matter system driven by microbial activity, there are a lot of details to keep track of, including how to manage carbon-to-nitrogen (C:N) ratios, optimal speed of decomposition, and nutrient cycling of cover crops, mulches, and composts. Managing ecological relationships in farm and garden soils can seem a bit overwhelming at first, but fortunately you don't have to figure everything out at once!

Native forests and prairies naturally maintain a soil organic matter system characterized by minimal soil disturbance, the presence of year-round growing roots, and regular addition of diverse organic residues with various C:N ratios. Hence most of the plant nutrients in forest soils are present in organic forms. For example, nitrogen is present primarily as amides and amines, but also as nitrogen molecules associated with polyphenols, polysaccharides, lignin residues, lipids, and other degradation products of plant and microbial origin. Unfortunately, farmers and gardeners have to work harder; to maintain soil fertility, we need to regularly add and properly time applications of organic residues like grass clippings, hay mulch, a cover crop, or compost. And we have to do this while maintaining rapid enough decomposition and nutrient cycling to meet crop nutrient demands.

It may seem easier to rely on tried-and-true organic fertilizers such as simply adding compost each spring. High-quality finished compost is a wonderful soil amendment that is much better for the soil microbial community's long-term diet than chemical fertilizer or overly processed organic fertilizers (like pelletized fertilizers or seed meals). But when we limit diversity by applying just one type of organic residue, such as a nitrogen-rich compost or seed meal, to the soil all at once in the spring, our soil microbial community feels the way we do after eating a huge cheese pizza: a bit sluggish. If the nutrient flush resulting from an organic fertilizer is too dramatic, some microbes might remain sluggish all season. High-nitrogen and -phosphorus fertilizers decrease some microbial populations, such as mycorrhizae. In fact, research at the USDA-ARS North Central Agricultural Research Laboratory in South Dakota has shown that decreasing nutrient inputs and increasing use of organic residue forms of fertilizer leads to an increase in mycorrhizal fungi. It also leads to

Table 3.1. How to Encourage a Diverse Soil Community

Type of Organism	Functions	Methods to Encourage
Plant growth-promoting rhizobacteria (PGPR)	Disease suppression, plant growth hormone synthesis, enhancement of plant nutrients.	Regularly add succulent, green organic residue; reduce tillage. Maintain growing roots.
Symbiotic nitrogen-fixing bacteria: *Rhizobium, Sinorhizobium, Mesorhizobium, Bradyrhizobium*	Fixation of atmospheric nitrogen in symbiosis with legume roots.	Use row cover or shade cloth to keep soil temperatures between 55–70°F (13–21°C). Maintain soil pH of 6.2 to 7.5. Provide moisture, soil aeration, and regular organic residue additions. Maintain legume roots year-round.
Free-living nitrogen-fixing bacteria: *Azospirillum, Herbaspirillum, Acetobacter*	Nitrogen fixation and biosynthesis of plant growth hormones (such as auxin) that stimulate plant root growth.	Minimize soil disturbance; encourage soil moisture and aeration; regularly add organic residues. Maintain growing grass and grain roots year-round.
Nitrifying bacteria	Conversion of organic forms of nitrogen into ammonium-nitrogen and nitrate-nitrogen.	Keep soil temperatures above 55°F (13°C) with row cover or removable plastic tarp in spring. Minimize tillage; add succulent green materials.
Phosphate-solubilizing bacteria	Conversion of phosphorus from fixed and/or organic forms into plant-available forms.	Minimize tillage; maintain growing roots year-round; regularly add low-carbon organic residues, like alfalfa hay mulch.
Nitrogen-fixing actinomycetes (such as *Frankia*)	Fixation of atmospheric nitrogen in forms available to non-legume woody plants.	Regularly add organic residues; minimize tillage; maintain growing roots year-round.
Mycorrhizal fungi	Formation of association with plant roots to provide plants with water and nutrients (esp. phosphorus and zinc); plant disease suppression.	Maintain growing roots. Add high-carbon organic residues such as composted wood chips or grass hay. Keep soil disturbance minimal and soil temperatures above 50°F (10°C) with row cover or removable plastic.
Disease-causing fungi and bacteria	Causal agents of root, leaf, and fruit diseases.	In this case, the goal is to discourage, not encourage, by avoiding the following: poor soil aeration; wet conditions; low organic matter; lack of crop diversity and rotation; excessive tillage, nutrients, and fertilizer.
Protozoans	Disease suppression (by consumption of bacteria and fungi).	Additions of high-carbon residues, such as straw or hay mulch; minimal tillage.
Earthworms	Mix and aerate the soil; enhance the soil structure; encourage the breakdown of raw residues; improve phosphorus availability.	Regularly add fresh or raw organic residues; keep the soil covered and keep tillage minimal; ensure adequate soil moisture; maintain soil temperatures of 50–70°F (10–21°C); mulch to encourage worms.

a healthy population of soil microbes that can eventually reduce the need for fertilizer inputs.

Further, adding only one type of organic residue as your fertilizer tends to encourage a more limited and specific microbial community. For optimal nutrition, the soil microbial community requires a varied diet of root exudates rich in carbon compounds, including sugars and vitamins, organic acids, and minerals. Our challenge is to maintain growing roots, support a diverse microbial communiy, and add a diversity of organic residues that have different C:N ratios, root exudate biochemisty, and decomposition speeds throughout the season rather than all at once in the spring.

Table 3.1 lists the main players in the soil microbial community and the "jobs" they perform. It also summarizes strategies that gardeners and farmers can implement to encourage a diverse soil community, including recommendations of types of organic residues to add and how.

PROVIDING ORGANIC RESIDUE DIVERSITY

The first goal of ecological gardeners and farmers is to provide their soil organic matter system and soil microbial community a diversity of raw, decomposing, and decomposed organic residues. Raw organic matter consists of fresh plant and animal residues, like freshly mowed lawn clippings or a succulent green clover cover crop. It encourages specific soil microorganisms. For example, in one study 75 percent of the bacteria present were found foraging near fresh, raw particulate organic matter.[1] Fresh, raw, green, and succulent organic residues are a relatively rapid-release source of plant nutrients. They result in a quick flush of microbial activity, especially bacterial activity. In my on-farm studies, green legume cover crop incorporated into my microbially active soil could release nutrients in a plant-available form in 2 to 4 weeks.

Decomposed or decomposing organic matter includes the decomposition products of fresh organic residues as well as soil microbe by-products (some call these by-products microbe manure) and soil microbe remains. Rotting leaves and unfinished composts are farther along the decomposition spectrum. And at the end of the spectrum, humus is made up of carbon and nitrogen (that originally came from the atmosphere) combined with a range of minerals from the soil and microbial by-products. This *humified carbon* derives from both soluble carbon (from plant root exudates) and the carbon from plant and animal organic residues that we normally think of as the precursors of humus. Organic matter in the humus state is not as biologically and chemically active. In other words, humus contributes less to the *immediate* availability of plant nutrients for crops than do raw or decomposing organic residues. Humus is rich in slower-release forms of organic nitrogen, phosphorus, and sulfur; it may also be lower in calcium, magnesium, and potassium than raw organic residues. Humus

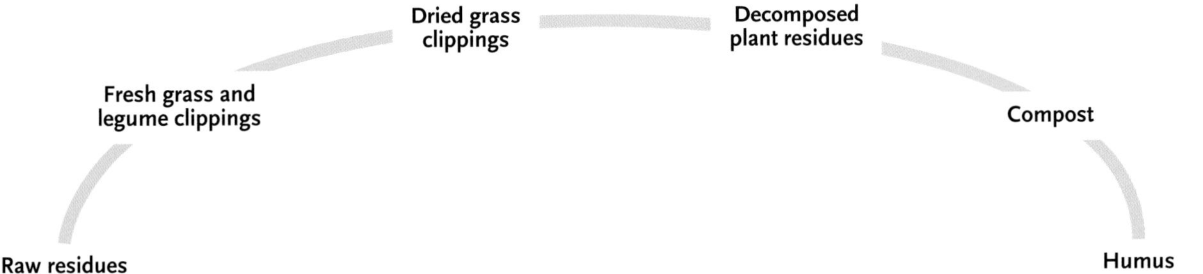

Figure 3.1. The organic matter continuum ranges from raw residues to humus. To support a thriving microbial community add a diversity of residues from across this continuum to your soil at regular intervals (rather than all at once).

produces a more durable soil structure, with a higher *buffering capability* (the ability of a soil to resist changes in pH), and also guards against nutrient loss through leaching (nutrients lost into groundwater). However, all the various forms of organic matter along the organic matter continuum are necessary (including carbon-rich root exudates and microbial by-products). Figure 3.1 illustrates where some common organic residues fall along the continuum.

Researchers in California compared various organic residue applications for their effect on soil carbon, nitrogen, and microbial community enhancement over a period of 6 to 8 years. They found that annual additions of composted yard waste (chipped branches and leaves) plus infrequent cover crops (once every 4 years) resulted in higher soil carbon levels than annual additions of cover crops without the woody compost. However, annual cover crop addition resulted in higher soil nitrate-nitrogen levels and greater microbial biomass carbon and microbial biomass nitrogen.[2] Again, adding compost or non-living organic residues as fertilizer feeds most of the soil microorganisms, but the root-microbe connection adds another dimension.

In summary, humus can be thought of as the banked, smoldering coals of a once fiercely burning fire, predictably producing a small but balanced and dependable quantity of heat. Raw organic matter during active decomposition, on the other hand, is an unpredictable hot flame that provides lots of heat but can sometimes get out of control (burning tender, young plants and leaching nutrients away from our garden soil). Another bonus of decomposing residues is that they attract and chelate (bind with) heavy metals, locking up excess amounts of toxic metals and making them less available to plant roots.

PROVIDING CARBON QUALITY DIVERSITY

Fertilizing for diversity also means attention to the carbon quality of organic residues. Here I focus on the carbon-to-nitrogen ratio as a measure of quality, but carbon-to-phosphorus content may also be important to consider for some soils, climates, and crop situations. Generally C:N ratio determines an organic residue's speed of decomposition, the way in which it stimulates microbial activity and growth, and the rate at which carbon and nutrients are released into the soil and made available to plants. As explained in chapter 2, we have some control over C:N ratio and lignin content by choosing the stage of development of the plant residues we apply. Generally, green, succulent, younger growth has "light" carbon quality and a relatively low C:N ratio. Brownish green, mature, senescent, end-of-the-year growth contains "heavy" carbon and usually needs time to decompose before it provides the quantity of nutrients to grow good crops. (Table 3.2 lists C:N ratios, nutrient contents, and decomposition rates of common organic residues and fertilizers for gardens and farms.) There may be times and farming situations for which slower decomposition

C:N Ratio and Decomposition Time

A C:N ratio of 20:1 means there are 20 units of carbon for each unit of nitrogen in the substance. There is always more carbon than nitrogen in organic residues, no matter how succulent the material is. Soil microorganisms need a good balance of carbon and nitrogen, ranging from 25 to 35 parts carbon to 1 part nitrogen, in order to decompose that residue in a reasonable amount of time. A ratio of 35:1 and higher results in slower decomposition rates and a longer time before nutrients become available to crop roots.

Table 3.2. C:N Ratio, Nutrient Content, and Decomposition Rates of Common Organic Residues and Fertilizers

Material	C:N Ratio	N*	P*	K*	S*	Ca*	Mg*	Comments
Alfalfa hay	16:1	2.6	0.6	2.2	0.3	1.0	0.02	Well balanced, faster-release; contains plant growth hormones, stimulants, and micronutrients.
Clover hay	16–19:1	2.1	0.8	1.5	0.3	1.0	0.02	Balanced, faster-release; adds micronutrients. Less nitrogen loss and builds humus better than straw.
Grass hay	25–35:1	2.1	0.7	2.4	0.1	0.3	0.1	Balanced, slow-release; adds micronutrients; less nitrogen loss and builds humus better than straw.
Lawn clippings (fresh)	18–20:1	1.0–2.0	0.3	0.8	0.1	0.2	0.1	Balanced, faster-release with micronutrients. Tends to mat and become anaerobic if used alone.
Cowpeas (dry)	16:1	3.0	0.8	2.3				Well balanced, slower release.
Weeds (fresh)	17–20:1	2.4–3.0	0.8	1.5–2.5	0.1	0.3	0.1	Balanced, faster-release with micronutrients (see chapter 7 for nutrient content of specific weeds).
Wood chips (deciduous)	100–130:1		0.2–1.0					Very slow-release; may take more than 1 year to decompose. Source of micronutrients; suppresses root rot diseases.
Wood chips (pine)	100–130:1		0.1					See wood chips (deciduous). Pine wood chips reportedly enhance soil tilth better than do hardwood chips. Composted pine bark suppresses root rot diseases.
Cornstalks (dry)	42:1	1.2	0.4	1.6				Improves soil structure, but has high C:N ratio and decomposes slowly, with slow nutrient release.
Leaves (deciduous tree, dry)	40–80:1	0.8	0.4	0.1				Slow-release. Contains good phosphorus, manganese, and other micronutrients; improves soil structure.
Worm castings	Variable	0.5	0.5					Excellent for improving soil structure and micronutrients. May increase soil pH very slightly.
Alfalfa meal	15:1	2.7	0.5	2.8	0.2	1.0	0.02	More rapidly available than alfalfa hay; contains growth stimulants and micronutrients.
Cottonseed meal	7:1	6.0	2.0	2.0				Rapidly available nitrogen source; may contain pesticide residues.
Soybean meal	7:1	6.0	1.0	2.0				More rapidly available nitrogen plus micronutrients.
Coffee grounds	20	2.0	0.1	0.6	0.1	0.1	0.1	Releases nitrogen rapidly. Adds micronutrients, including boron, iron, and zinc. pH is 5.5–6.8; can slightly lower soil pH.
Epsom salts					13		10.0	Rapidly available magnesium and sulfur source.
Granite meal			4.0					Very slow-release phosphorus source; builds soil reserves.
Rock phosphate (hard rock)			30			33		Slowly available phosphorus, calcium, and micronutrients; may increase pH.

continued on page 42

Table 3.2 *continued*

Material	C:N Ratio	N*	P*	K*	S*	Ca*	Mg*	Comments
Greensand				7.0				Very slow-release source of potassium.
Gypsum					17	22		Slowly available sulfur and calcium; does not affect pH; improves alkaline soil structure.
Seaweed (kelp)	19:1	1.0	0.2	2.0	3	0.1	0.1	Slow-release potassium and micronutrients; contains growth stimulants.
Limestone (dolomitic)						51	30–35	Slow-release calcium and magnesium; will increase pH.
Limestone (high calcium)						65–75	1.0–5.0	Slow-release calcium; used to increase pH.
Wood ash			1.6	4.0		35		High potassium, calcium, and some micronutrients. Increases pH. Can injure soil microbes, young plants, and germinating seeds.
Sulfur					99			Used to lower high soil pH and as a plant nutrient. Works slowly.

* Average pounds provided per 100 pounds (45 kg) of material.

of organic residues is desirable—for example, when we want to suppress weed growth in crop row middles or on crop bed edges.

Growers in hot, humid climates should manage their system to aim for slower residue decomposition, because heat and humidity favor rapid decomposition and discourage organic matter accumulation. Research suggests that residue decomposition can be slowed by allowing crops to grow to a more mature state before incorporating them. In a 2021 study researchers grew *Desmodium* and *Pueraria* (tropical kudzu) cover crops and evaluated decomposition rates for 12 and 18 months. The younger (12-month), succulent cover crops decomposed faster; *Desmodium* and kudzu had only 24 percent and 16 percent dry matter content, respectively, remaining in the soil after 30 weeks. But when these cover crops were grown for 18 months, it took longer for them to decompose. After 30 weeks, remaining dry matter in the soil was at least double for the more mature plants: 53 percent for *Desmodium* and 32 percent for kudzu. Furthermore, growing these leguminous cover crops for only 12 months before incorporating them resulted in release of plant nutrients into the soil much faster.[3]

MATCHING RESIDUES TO SOIL TYPE AND CLIMATE

The best choice of residues differs, depending on the particular soil type. For example, sandy soils are most quickly improved by adding both partially decomposed and well-composted organic residues and cover crops, particularly fibrous rooted grass and legume mixes. But to improve a clay soil, fresher residues are best. Raw forms of organic matter are coarser and open up the soil, so aeration occurs. The water-stable soil aggregates that help open heavier clay soils seem to form more rapidly by adding manures and grass clippings (raw organic matter) than by adding compost (stabilized organic matter). Cover crops with deep taproots (like alfalfa, forage radish, or even dandelions) can also aerate and open up heavy, clay soils. Later in this chapter I present more specific recommendations on cover crops for particular soil types and conditions.

Clearly, organic residues are a vital part of the soil organic matter and nutrient cycling system. But can *too much* residue be added to soil? Yes! An excess of residues (especially fresh residues) added to soil all at once can create problems. For example, temporarily toxic levels of soluble salts that injure plants may be released. Also, some of the initial by-products from fresh residue decay can be toxic to tender, germinating seedlings. Certain insect pests and diseases may be temporarily favored by excessive addition of fresh residue to soils. Nutrient imbalances can occur and high C:N ratios can result in temporary nitrogen deficiency while nitrogen is tied up or *immobilized* by soil microorganisms—see "Why C:N Ratios and Organic Nitrogen Matter" on page 23. For example, sawdust is sometimes applied to streams to stimulate organisms that tie up excess nitrate-nitrogen. This keeps nitrates out of the water supply. Unfortunately, sawdust or even high C:N ratio straw can have the same effect in gardens where we normally do not want nitrogen immobilized by soil microbes.

How much is too much? The answer depends on soil moisture levels, the microbial health of your soil, and your climate. Whatever amount of residue exceeds what microbes can break down in 10 weeks is too much. Another red flag that you have overloaded microbes with residues is foliar symptoms of nitrogen or phosphorus deficiencies that persist for more than 2 or 3 weeks on your plants.

Several conditions promote or diminish organic residue decomposition. Organic matter does not decompose well in soils that are too dry or too wet. Soil microorganism activity is optimal when 50 to 60 percent of soil pore space is filled with water and the soil feels moist, but not saturated. Prolonged drought will depress soil microorganism populations. Warm, wet climates promote residue decomposition. Cold, dry climates slow residue breakdown.

Making Residue Management Decisions

Which mulches, living mulches, cover crops, and composts are best to use when, and in what combination? That's a key question, and the answer will be unique to your situation, because it depends on your soil type, texture, pH, and nutrient levels, as well as precipitation, irrigation options, and climate. It also depends on how fast you want your harvest—if early harvest is a high priority, you need to ensure that nutrients are quickly available to your crops. Another consideration is what kind of soil organic matter system you want to build overall. Some gardeners and homesteaders I have worked with are comfortable managing a carbon-focused system. They are content to wait for woody, heavy-carbon materials that have high C:N ratios and high lignin content to slowly decompose as they create water-retentive planting beds with plenty of stable organic matter. Other gardeners are frustrated by the slow pace of woody material decomposition and the less-than-spectacular crop growth they get the first year or two. And most organic farmers are financially unwilling to invest in the time required for heavy carbon to decompose. But a few, including Carl and me, learned over time (and by making mistakes along the way) how to combine and manage a combination of heavy- and light-carbon amendment applications in ways that facilitate and optimize microbial activity and encourage decomposition, while still retaining growing roots in the soil year-round and good yields.

The devil is definitely in the details as we consider multiple forms of organic residue amendments and how different application methods and timing can affect soils and plant growth. Here I compare in detail the possible uses and effects of different kinds of organic residue amendments.

PLANT-BASED VERSUS ANIMAL-BASED RESIDUES

Compared with plant-based amendments such as cover crops and living mulches, manures and manure-based composts generally provide nutrients more rapidly. In one long-term study in Illinois, soil organic matter in manure-amended soil was the most active, providing quicker nutrient release into the soil. But soil organic matter in cover-cropped soil was the most stable and higher in *total* carbon and nitrogen.[4] In long-term plots at the University of California, researchers found that composted poultry manure alone was unable to support commercial yields of corn and tomatoes. A mixture of cover crops and organic mineral fertilizer (such as rock phosphate) provided the highest yields.[5]

Another study in central California compared tomato fields on multiple organic farms. Each farm used different soil amendments. Some used mostly manure, while others used mainly plant-based, composted yard waste. Yields were similar on all farms. However, manure application resulted in increased phosphorus, increased soil bacteria abundance, and decreased soil fungi abundance and diversity.[6] High phosphorus is not necessarily always a good thing; it can decrease soil fungi abundance and activity, including mycorrhizal fungi.[7] Recall from chapter 2 how important mycorrhizal fungi are to the soil organic matter system and the negative effects of high fertilizer application on mycorrhizae. Carl stopped using manure compost in his orchard in the early 1990s and switched to composted yard waste because he believed that manure compost discouraged his "fungal food webs" and ramial chipped wood increased them (for more about

Figure 3.2. Carl cutting grass, clover, weed hay mulch in late August for direct application to vegetable crop fields with the crop chopper and trailer he rigged up for the task.

ramial chipped wood, see "Wood Chips and Bark" on page 67).

Some farmers, ranchers, and soil scientists believe that animals and their manure are necessary for sustainable soil fertility. However, others suggest that plant-based soil fertility is just as sustainable for fruit and vegetable production with less potential for nitrogen and phosphorus leaching into ground water. Manure application may also lead to excessive buildup of potassium in soils and increased pH (especially if uncomposted manures are applied).

Dr. Amir Kassam (visiting professor in the School of Agriculture, Policy and Development at the University of Reading, UK, and former technical officer at the United Nations Food and Agriculture Organization) explains that manure is just plant biomass pushed through a ruminant's gut. Dr. Kassam suggests that if you return plant biomass directly onto the soil and maintain it as surface mulch, rather than having it pass through the animal first, the soil microbial community and earthworms will readily ingest the biomass and incorporate it into the soil.[8]

At Woodleaf Farm in Oregon, Carl and I discovered the efficacy of applying plant residues directly to the soil rather than feeding them to animals and spreading manure, and I am perfecting this approach in my soil organic matter system. Rather than feeding hay to animals, collecting their manure, making compost, and then spreading the compost onto my fields, I simply apply mowed hay directly to crop fields. The direct hay application saves labor and reduces diesel use on my farm by eliminating the composting and compost application steps.

Figure 3.3. A vegetable crop bed prepared for the following year with 4 inches (10 cm) of hay mulch surface-applied in September.

Figure 3.4. Woodleaf Farm, Oregon, perennial living mulches mowed and blown into sweet pepper rows in August.

Plant-Based Residue Options

Plant-based organic residue amendment options include:

Cover crops/green manures. These are plant residues in which the whole plant is incorporated or tilled into the soil.

Living mulches. These can be planted either before or along with a main crop and maintained as a living ground cover throughout the growing season. They are not incorporated, and thus maintain actively growing roots even after the annual cash crop is harvested and dies.

Mulches. Mulches are simply surface-applied plant residues. They are not alive and growing and do not maintain a growing root in the soil.

Woody materials. These include composted chipped branch wood and wood bark (often available as municipal yard waste compost or ramial chipped wood compost).

Processed seed meals. These include ground soybeans and leaf meals, such as ground alfalfa hay.

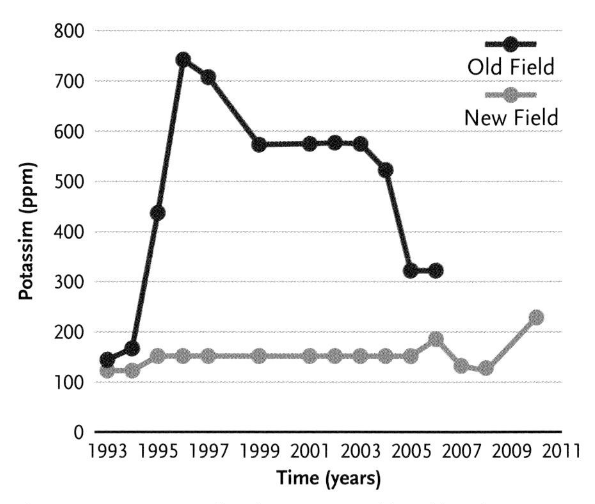

Figure 3.5. Potassium levels (ppm) in Old Field and New Field at Biodesign Farm 1993–2010.

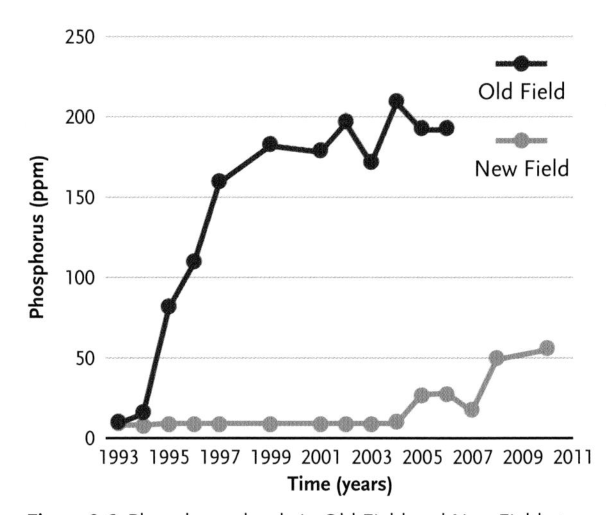

Figure 3.6. Phosphorus levels in Old Field and New Field at Biodesign Farm 1993–2010.

More about Manure: Biodesign Long-Term Records

In chapter 2, I told the sad tale of how I created excessive soil nitrogen levels in my Old Field at Biodesign Farm. Soil tests were showing alarming increases of soil nitrogen, phosphorus, potassium, and pH. In addition, I was beginning to notice problems including blossom-end rot on tomatoes and sunscald on bell peppers. Because of this, I stopped using manure in the early 2000s. Figure 3.5 shows that when I stopped manure-based compost applications, excessive potassium levels in the Old Field dropped dramatically.

On the other hand, phosphorus levels remained high for another 4 years in the Old Field after I stopped manure compost applications (see figure 3.6). Neither potassium nor phosphorus levels ever became excessive in the New Field, where manure compost was applied at very low levels in the crop row for only 2 years and then stopped. Yet yields and quality in both fields were comparable.

Soil pH increased in Old Field from 6.9 in 1993 to 7.7 in 1999 (see figure 3.7). pH decreased after I stopped using manure compost, but only to 7.6. There was no soil pH increase in the New Field; soil pH remained relatively stable around 6.8.

Sometimes applying manure may be a good intervention to rapidly improve low soil nutrient levels. It can also increase soil organic matter levels

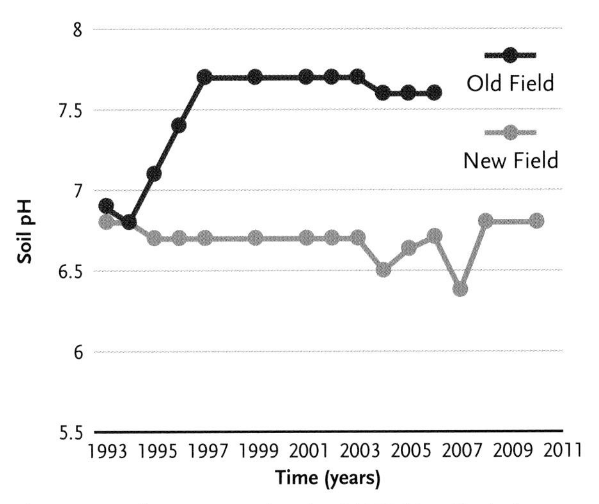

Figure 3.7. Soil pH increased in the Old Field at Biodesign Farm but remained stable in the New Field.

quickly. However, farmers and gardeners who don't have ready access to or want to use manure should not feel concerned about that—there is good evidence from my studies and on-farm experience that you can grow fruits and vegetables without making manure or manure-based compost a part of the system.

Comparing Manure- and Plant-Based Fertilizers

When I was an extension agent in Montana, my Master Gardener students and I conducted an experiment to compare common plant-based and manure-based organic residues and composts available as fertilizers in our area. These included commercially packaged composted steer manure, composted sheep manure plus straw, uncomposted (but aged) cattle manure, mushroom compost, lawn clippings, composted conifer bark, and compost made from chipped branches. My students and I ran nutrient and pH tests on these organic residues and composts and observed several generalities over 2 years of testing:

- The pH of both composted and uncomposted manures (pH 8.0–9.0) was higher than for all plant-based residues. Uncomposted manures had higher pH than composted manures.
- Composts containing wood products (composted conifer bark and chipped branches) had a low pH (pH 5.0–6.3).
- Wood product composts contained lower levels of nitrogen, phosphorus, and potassium than manure-containing fertilizers and composts, but calcium, zinc, and manganese levels were higher.
- Mushroom compost had a pH of 7.2, comparatively high levels of nitrogen, and high levels of potassium.
- Grass clippings had a pH of 6.6 and comparatively high levels of nitrogen and potassium.

Our conclusions were that grass clippings are an easy-to-obtain complete fertilizer with a good pH for vegetable crops. And for our high-pH western soils in Montana, regular manure application has the potential to raise pH levels too high for some vegetables. Finally, composts containing wood products could help to lower soil pH over time and add micronutrients such as manganese, but are not complete fertilizers because they are low in nitrogen, phosphorus, and potassium.

COMPOSTED VERSUS RAW RESIDUES

Composted plant materials tend to supply nutrients to crops more slowly than do raw, fresh plant materials, especially if fresh residues are tilled into the soil. Raw, uncomposted plant material supplies nutrients more slowly when mowed and left on the soil surface. Similarly, composted manure generally supplies nutrients to crops more slowly than does uncomposted, or raw manure. But nutrient release happens at the slowest pace of all when composted plant residues are spread over the soil surface and not incorporated.

However, use of raw manure is not recommended; it allows a greater chance to build up excessive soil nitrogen, phosphorus, and potassium, along with a high soil pH. Uncomposted manures can also result in nitrogen loss by nitrate leaching and production of ammonia gas. Raw manures may release high levels of soluble salts, which can injure plants, especially when excessive levels (more than 10 tons per acre [25 tons/ha]) are applied. (Often even composting does not reduce the salt level unless leaching occurs.)

Uncomposted manures may also contain viable weed seeds. Horse manure usually has the highest weed seed potential; sheep and goat manure have lower potential. Raw manure can contain a number of human pathogens, including *E. coli* and salmonella. These pathogens can be controlled by proper composting at temperatures of 130 to 149°F (54–65°C). If raw manure is applied to gardens and farm fields, it should be worked thoroughly into soil and crops should not be harvested for at least 60 days from the date of application.

SURFACE APPLICATION VERSUS INCORPORATION

Organic residues can be surface-applied as a mulch, such as mowed lawn clippings or hay cut and brought into the field. Or, a living mulch in row middles between crops can be periodically mowed and blown into crop rows. Living mulch provides the benefit of growing roots and an active rhizosphere near crops, whereas the mulch application does not. When a cover crop is tilled under, both foliage and roots are chopped up and incorporated into the soil. But with surface application methods, the root systems of the mulch crop are left undisturbed. This is to a grower's advantage, because though the plants may go dormant in cold-winter climates, the roots of the living mulches remain in the soil alive year-round. These root systems form important associations with soil microbes that fuel the soil organic matter system, as explained in chapter 2.

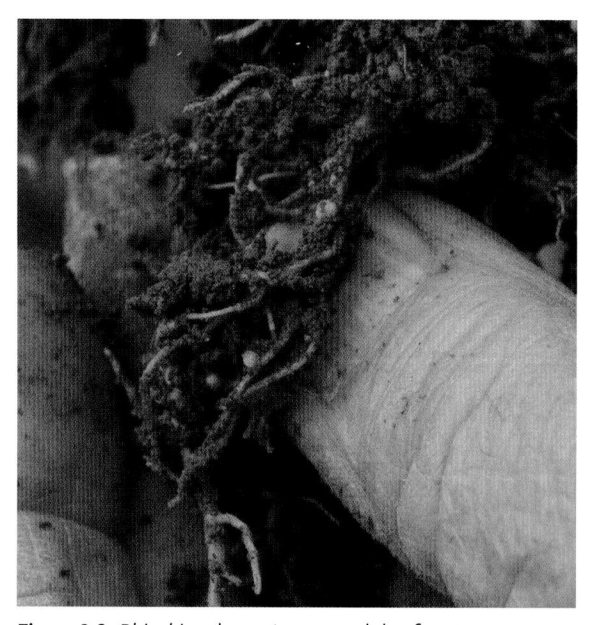

Figure 3.8. *Rhizobium* bacteria root nodules forming on a young clover plant are the result of a symbiosis between the clover plant root and the *Rhizobium* bacteria.

Cover crops supply nutrients relatively rapidly once they have been incorporated into soil and decomposed. But living mulches can temporarily absorb a large percentage of the residual soil nitrogen and sequester it in their own plant bodies. Overall, nutrients are released into the soil and to crops more slowly than when a cover crop is tilled in.

Mycorrhizae associated with living mulch and cover crop roots may absorb residual soil nitrogen and phosphorus and temporarily immobilize it in their hyphal biomass. And mycorrhizal fungi are not the only microbes tying up plant nutrients in their bodies. The process of nitrogen fixation in legume root nodules by *Rhizobium* bacteria can consume as much as 20 percent of the total plant carbon accumulated by the plant through photosynthesis. In fact, as young clover plants are growing, they have to give up some carbon, nutrients, and sugars to help the *Rhizobium* bacteria build these nodule structures on their roots. The benefit to the clover is that eventually, the *Rhizobium*-clover symbiosis results in the conversion of atmospheric nitrogen into a nitrogen form that the clover plant can use directly for its own growth. Almost all of the *Rhizobium*-fixed nitrogen goes into the legume plant while it is actively growing; not much is released into the soil for use by other plants. In fact, in perennial legumes 70 to 80 percent of the plant's nitrogen content is in the top growth rather than the roots. Hence, in one study surface-applied legumes released much less nitrogen into the soil after 40 days compared with tilled-in, incorporated legumes. In fact, the nitrogen release of surface-applied legumes was only 20 percent that of incorporated legumes. The key for gardeners and farmers is to manage for this lag time between the surface application of organic residues and the release of nutrients into the soil for crops to take up.

Managing for Timely Nutrient Release

Figuring out how long it takes for mowed, surface-applied plant residue to release nutrients for crops was an important part of my farm research at Biodesign Farm. I found that mowed legume living mulches could tie up soil nitrate-nitrogen for 2 months or longer as soil microbes processed the mowed residues (see figure 3.9). Soil microbes released the nitrate-nitrogen eventually, and over time (years) the mowed living mulch built up a strong organic nutrient reserve. Once that shift had occurred, immobilization of nutrients was not a problem as long as I managed organic residues well and added them regularly. Actually, I am still learning to manage organic residues well!

I'm still learning because the ecological relationships are complex in the case of surface-applied organic residues. As they grow, living mulches take up and cycle nutrients. Soil microorganisms, both associated and not associated with living mulch roots, also take up and cycle nutrients. The weather, the soil type, the microbial community health and diversity, and the ways in which we farmers and gardeners apply organic residues further affect the interactions among plant roots, microbes, and nutrients.

A farm trial I carried out in 2006 showed that soil nitrate-nitrogen levels were lower beneath mowed and surface-applied living mulch compared with unmowed living mulch left growing undisturbed. Levels were lowest in August (see figure 3.9). This made sense to me because other trials had shown that total microbial biomass was highest in the soil beneath mowed living mulch treatments and peaked in August. Hence my theory is that when I mow and surface apply living mulches to the soil, active soil microbes are using most of the soluble, available nitrate-nitrogen (especially when their population peaks in August in Montana). But when soil microbe activity slows by the end of September, soil nitrate-nitrogen levels beneath mowed (and unmowed) living mulches increase. There are also reports that roots of legume living mulch can "leak" nutrients into the soil in their immediate vicinity as they grow. Thus some nitrogen can be released directly to the soil. That may be part of the reason that testing showed higher nitrate-nitrogen in the soil beneath the unmowed clover living mulch in my 2006 study. However, based on my earlier resarch data, I suspect that the microbial community was not as large or active beneath the unmowed legume living mulch and thus wasn't taking up as much nitrate from the soil.

Whether you incorporate or surface-apply organic residues, there is a lag time during which nitrogen and phosphorus will be tied up in the microbial decomposition wheel. When organic residues are incorporated, the lag time is at least 2 weeks, and possibly up to 4 weeks. The lag time is longer for surface-applied residues (as discussed earlier). Plan for this and wait 2 to 4 weeks after mowing or tilling a cover crop or mulch before planting main crops. Waiting at least 2 weeks to plant will also reduce the chance of increased disease organisms, which may have been temporarily favored by an addition of fresh (particularly succulent and green) residues.

Moisture Effects

Climate (specifically humidity) and temperature also play a part in the fate of nutrients recycled

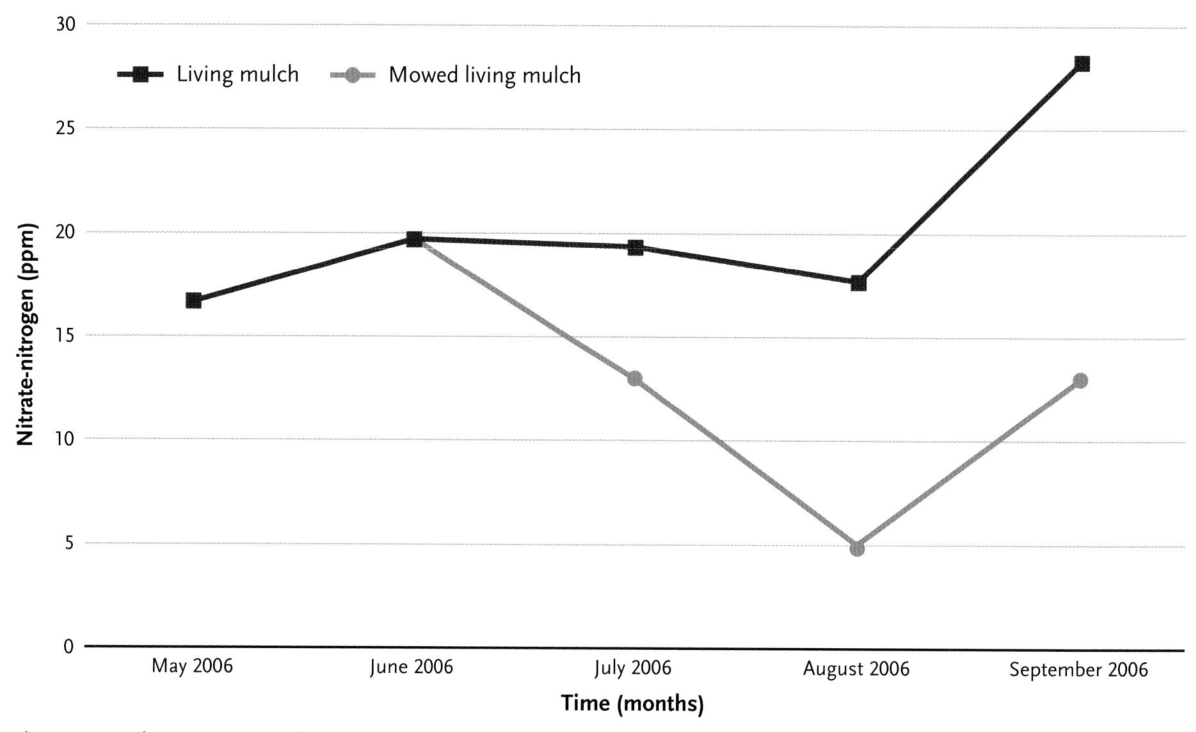

Figure 3.9. Soil nitrate-nitrogen levels in mowed and unmowed treatment plots at Biodesign Farm, 2006. Generally, soil nitrate-nitrogen levels were lower in plots where living mulch was mowed and surface-applied monthly compared with unmowed living mulch. Due to the small sample size in this study, only in August was there a statistically significant difference.

from surface-applied organic residues. Higher humidly equals more rapid decomposition and nutrient release—that is to say, the more moist the weather, the more quickly residues will release their nutrients. Studies at North Carolina State University reported that 75 percent of the nitrogen in some legume cover crops was released within 7 to 10 weeks of mowing if residues were left on the surface. But if the residues were tilled under, nitrogen release was quicker and occurred within 4 to 8 weeks. In cooler weather nitrogen release can take much longer, but it is always faster with soil-incorporated organic residues. In my Montana farm studies, nitrogen release in my microbially active soil began within 3 to 4 weeks of spring tillage of legume cover crops or strip-tilled living mulches. However, as already noted, several researchers have shown that tillage decreases soil microorganisms, specifically soil fungi.[9]

Leaving a living mulch untilled can also affect crop microclimates. The presence of the dense living mulch may decrease soil temperature and daily soil temperature extremes compared with areas where cover crops have been tilled under—such tillage usually raises soil temperature. In my Oregon orchard there were record drought conditions and high temperatures above 100°F (38°C) for weeks in July and August during 2021 and 2022. I maintained tall living mulches between crop rows to lower soil temperatures and increase humidity within the orchard. Both years the fruit yields and quality were excellent.

Application method also affects soil moisture, which is *greater* when dead mulch is applied or soil is left fallow. Soil moisture is *lower* when actively growing grass and legume living mulches are using some of it for their own growth. However, soil moisture is *lowest* in farming systems using continuous cultivation to maintain weed-free bare soil among crops. Further, living mulches can improve soil and thus also enable soil water retention when they are not actively growing, transpiring, and using water. Studies with corn crops report an initial decrease in soil moisture when a corn crop was planted into mowed hairy vetch. But by 2 to 4 weeks after the corn was planted, soil moisture levels were the same as in bare soil treatments and surpassed bare soil moisture levels 4 weeks after corn planting.[10] Mowing living mulches can also help to decrease their water use.

HIGH-CARBON VERSUS HIGH-NITROGEN RESIDUES

High-carbon and high-nitrogen residues decompose at different rates, by different biochemical pathways, and with different soil microorganisms taking the lead roles. As noted in chapter 2, fresh, succulent residues (like newly cut grass clippings) decompose rapidly. Mature, dried-out residues decompose more slowly. Generally legumes provide more nitrogen than grasses and much more than woody materials do. And compared with leaves, studies show that roots add more carbon when they are incorporated into the soil. Foliage generally adds more nitrogen.

Age and growth stage of residues also affect decomposition rate and C:N ratio. In one study young and mature grass and straw were compared. Young grass had the highest nitrogen level and lowest C:N ratio, mature grass was medium, and straw had the lowest nitrogen content, the highest carbon content, and therefore the highest C:N ratio. Not surprisingly, straw had a high lignin content; mature grass, medium; and young grass was low in lignin. After being composted for 180 days and applied, the young grass formed the most soil humus. Straw formed the least humus and mature grass was intermediate. But young grass also lost the most nitrogen during decomposition (about 50 percent).[11]

Legumes

Legume cover crops or living mulches decompose faster than grass due to lower C:N ratios. Generally, legumes:

- Provide 2 to 4 percent nitrogen.
- Provide a C:N ratio from 12:1 to 20:1.
- Release from 1 to 25 pounds of nitrogen into the soil per 1,000 square feet (5–122 kg/1,000 m^2).
- Release approximately 100 pounds of plant-available nitrogen per acre (112 kg/ha).

Legume species and even varieties vary in their capacity to fix nitrogen. Legumes must form effective root nodules to fix nitrogen and should be inoculated with the appropriate *Rhizobium* bacteria species at seeding if your soil is not microbially active and legumes have not been growing in it recently. I do not inoculate, but I have been building microbial habitat for many years and my soil is well supplied with active bacteria, including *Rhizobium*. Remember that legumes first use *available soil nitrogen* for growth before they begin to fix atmospheric nitrogen and release it into the soil.

On average, legumes contain more nutrients than grass cover crops, including four times as much calcium, twice as much phosphorus, and twice as much nitrogen.

Grasses

Grasses provide:

- Less than 1 percent nitrogen.
- A C:N ratio of 50:1.
- Little or no soil nitrogen and can temporarily immobilize up to 50 pounds of plant-available nitrogen per acre (56 kg/ha).

Grasses release nutrients into the soil more slowly than legumes because they have a higher C:N ratio and contain a greater proportion of lignin (a heavy form of carbon). The high C:N ratio of some grass cover crops may lead to a decrease in the yield of following crops because so much plant-available nitrogen (up to 50 pounds per acre [56 kg/ha]) can be temporarily tied up as the grass decomposes.[12] However, I use a mixture of grasses, legumes, and weeds as mulch and living mulch fertilizer and it works well as long as I manage application timing and C:N ratios. (I explain the nuts and bolts of this type of management in the next chapter.)

The whole question of plant biomass and how it contributes to a soil organic matter system has a new dimension when the significant contributions of living roots in the soil are fully understood and appreciated. In fact, according to Dr. Christine Jones, root exudates of carbon can be a primary mechanism for soil organic matter building. Jones suggests that root exudates, with their wide variety of chemical compounds, are able to regulate the soil microbial community in the rhizosphere, signal beneficial symbioses, and change the chemical and physical properties of the soil.[13] This statement elegantly sums up many of the benefits of building a soil organic matter system.

CHAPTER 4

Minimizing Tillage and Growing Your Own Fertilizer

There are many creative ways to minimize soil disturbance as you focus on the details of ecological relationships. We need to manage organic residue diversity and carbon-to-nitrogen (C:N) ratios, as well as the decomposition speeds and nutrient cycling of specific cover crops, mulches, and composts in different soil types and climates. The most exciting aspect of managing a soil organic matter system is that as you gain expertise in growing your own fertilizer, you may be able to phase out the use of purchased fertilizers in your farm fields or garden beds.

In this chapter I first cover practical ways to minimize tillage. (Reducing tillage is the foundation for building a soil organic matter system, although a certain amount of soil disturbance is necessary to incorporate cover crops to feed soil microbes.) The main focus of the chapter is the nuts and bolts of how to grow your own fertilizer by adding and managing organic residues, such as mulches, living mulches, and cover crops. For completeness, I also include a section on composting and the use of fertilizers that you don't grow on your farm or in your garden.

Managing Tillage

How to manage tillage has been one of the most difficult things to figure out and keep in balance as I learn to effectively grow my own fertilizer. I till mainly to feed soil microbes and cycle nutrients. I do not till or otherwise disturb the soil to cultivate weeds unless it is to incorporate a cover crop, living mulch, hay mulch, or previous crop plus weeds in need of intervention. Tillage creates too great a disturbance to the soil organic matter system to justify unless it serves the goal of feeding the soil microbial community with organic residues. Just as crops are rotated in the garden each year, keeping soil microbes happy requires that we also annually rotate the areas that will be tilled and leave some areas untilled every season. In this section I explain my methods as well as other gardeners' minimal-soil-disturbance techniques to grow a living root and keep the soil covered year-round in order to feed the rhizosphere a steady, balanced diet of carbon and nutrients.

There are plenty of good reasons to decrease tillage:

- Tillage decreases soil organic matter and microbial diversity and abundance. Several long-term studies report decreased soil organic matter in tilled compared with no-till plots.
- As mentioned in chapter 2, soil disturbance can impede or kill many types of soil fungi that are important in nutrient cycling and soil building.

- Repeated tillage, especially with a rototiller, destroys soil particle aggregation, or good soil tilth.

What you use to till makes a difference. In three separate 10-year trials in Texas, soil organic carbon levels were highest in no-till systems and lowest in moldboard plow tillage systems that turned the soil completely over. But tillage with a chisel plow (see figure 4.8) resulted in soil organic carbon levels intermediate between no-till and moldboard plow management.

On the positive side, tillage increases soil aeration (oxygen for roots and soil microbes!) and results in a flush of biological activity. It incorporates organic residues to feed the soil microbial community, helps to manage weeds, warms cool soils, and dries out wet soils. Deep digging can break up compaction and mix soil layers, improve drainage, and extend crop-rooting depth, thus increasing the range of nutrients available to plant roots.

Figure 4.1. In this midsummer view of my home garden, there is virtually no exposed bare soil and the garden is starting to look a bit wild. Tall sunflowers, the clover row middles, and the clover cover crop will remain all winter as undisturbed, untilled areas.

WHEN TO TILL AND WHEN NOT TO

If you must till, do so at the right time. Fall tillage is more damaging to soil microbial communities than spring tillage. Tillage may compact soil if done when soil is too wet. If dry or saturated, tilling will damage soil structure and in the long run negatively affect plant growth. Summary: Don't beat the organic matter out of your soil with primary tillage and weed cultivation! Practice tillage with intention.

Here are the factors I consider when I make my decisions about when to till and when not to:

- Till only to incorporate organic residues that will feed the soil microbial community, add oxygen, and speed nutrient cycling.
- Till only to incorporate weeds into the soil when preparing seedbeds to suppress crop competition (see chapter 8).
- When tilling, also design into your garden or farm nearby undisturbed spaces in which to keep untilled growing roots in the soil year-round.
- Rotate tillage and no-tillage areas each year, or every few years, and leave some areas untilled.

HOW TO MINIMIZE SOIL DISTURBANCE

One way to limit soil disturbance and avoid rototilling is to leave "islands" of undisturbed, untilled areas when preparing spring crop beds or fields.

Another option is to set up permanent bed systems. Gardens and fields can be divided into a set of beds so that foot or tractor traffic is restricted to the between-bed area. These crop row middle pathways can be managed with mulches, living mulches, cover crops, or even weed mat on a small garden scale (although weed mat does not provide living roots).

Strip tillage targets soil disturbance to the planting zone only. This helps to retain surface organic residues, preserve soil moisture, build soil structure, and reduce erosion. It shifts the balance toward annual crop transplants and germinating

Figure 4.2. Weeds, clover living mulch, and hay mulch covering soil between garden crop beds.

Figure 4.3. Strip tillage bed preparation with a 3-foot (1 m) rototiller at Woodleaf Farm, Oregon.

Figure 4.4. Wheel hoe suppressing weeds and mixing mowed and blown living mulch into the top 1 to 2 inches (2.5–5 cm) of soil in between corn rows.

seeds by warming soils, forming a good seedbed, stimulating nutrient release, incorporating organic residue fertilizers, and suppressing weeds, but without disturbing all of the microbial community habitat. Strip tillage allows some of the rapid nutrient cycling benefits of tillage while retaining some of the soil health benefits of no-tillage. For example, in Washington State experiments soil moisture was significantly greater in strip-till compared with full-till plots at days 7 and 56 after incorporating a vetch cover crop. Shallow strip tillage has proved successful in my vegetable production fields at Woodleaf Oregon, especially during very hot, dry summers.

When you do till, aim for shallow surface tillage, only 1 to 3 inches (2.5–7.5 cm) deep. With shallow tillage you can achieve a more than 50 percent reduction in the amount of soil volume disturbed. Shallow tillage is best in combination with deep tillage every few years. I employ annual shallow strip tillage with a targeted deep ripping to 12 inches (30 cm) every 3 to 5 years, using a single shank chisel plow (also called a ripper). There are many new and old, farm- and garden-scale minimal-tillage equipment options available now, such as digging forks and manual wheel hoes (see figure 4.4). At the farm scale, spaders are a good alternative for rototillers and plows. The motion of the spader works to eliminate compaction or disruption to the soil profile within a single pass. However, spaders will invert the soil much like a rototiller with repeated passes.

Once Carl and I found ways to minimize soil disturbance, we were ready to experiment further with the most effective ways to minimize the off-farm fertilizer inputs I once thought were necessary to grow crops.

An Ongoing Experiment in Breaking the Rules

At the new Woodleaf Farm in Oregon, Carl and I had the luxury of experimenting and finding practical ways to achieve our intellectual goal of stopping all off-farm fertilizer inputs. It was the ultimate test of our two most elusive ecological principles:

Grow your own carbon. Grow or obtain your carbon source where you are, rather than importing carbon from a distance.

Figure 4.5. Test plots at the start of our fertility system comparison, September 2017. The rich green plots are Fertility System 2, with newly added hay mulch applied over previously mown and blown living mulch (cut from the row middles).

Recycle rather than import nutrients. When possible, recycle minerals from on-farm sources rather than importing mined mineral nutrients from somewhere else and applying as fertilizer.

We did not want to create sustainable organic gardens and orchards on our land by mining or taking resources from someone else's land. We wanted both a closed system *and* good yields. How hard could that be? It was not too difficult for the perennial orchard.

To find out if it was possible for annual vegetable crops, we set up research plots for a 4-year experiment in which we learned how to break many of the soil science rules I had been taught in college. This vegetable experiment was conducted in a 50-year-old grass, clover, alfalfa, and weed pasture. We compared two plant-based soil fertility systems using materials grown on-farm. There were eight replications per soil fertility system treatment, and data is presented as an average of eight samples per soil fertility system. We also compared two soil test approaches to see which best predicted soil fertility in an organic production system that used only hay mulch and living mulches for fertility.

In Fertility System 1 we mowed the clover/grass/weed living mulch row middles three times per growing season and blew it into the crop rows. We strip-tilled the living mulch to form 4-foot-wide (1.2 m) crop plots each spring. We considered beneficial insect/pest insect abundance and interactions in deciding when to carry out each mowing.

Fertility System 2 was the same as the first, plus we applied 4 inches (10 cm) of clover/grass/weed hay mulch in late summer or fall and strip-tilled the mulch into 4-foot-wide (1.2 m) plots each spring. The hay mulch was cut from another field on the farm.

OVERALL RESULTS

The results were encouraging and have influenced the choices I make today in managing crop production at Woodleaf. Here are the main takeaways from the study:

Figure 4.6. In 2018 red cabbage yield and quality (especially taste) from both of the experimental soil fertility systems trialed at Woodleaf Oregon were very good.

- Crop yields were satisfying and economically sustainable all 4 years; excellent in 2020.
- Yield improvement in the second and third years was probably due to more hours spent on weed suppression.
- Crop yields for dry beans and cabbage were slightly, but not statistically, higher in Fertility System 2 than Fertility System 1. Pepper and onion yields were significantly higher in Fertility System 2.
- Labor, equipment, and fuel costs were higher in Fertility System 2 due to cutting, transporting, and application of the hay mulch.
- There were no insect or disease pest problems on any crops, except for relatively minor cabbage aphid infestations on cabbage outer leaves.

In 2022, I farmed the experiment rows, but did not collect data other than pest levels and yield. Cabbage and onion yields were lower, probably due to excessive heat in June and July (several

Table 4.1. March 2018 Microbial Activity Comparison

Microbial Indicator	Fertility System 2	Fertility System 1
Total living microbial biomass (PLFA)	5,941 excellent	5,849 excellent
Functional group diversity index	1.52 very good	1.43 good
Fungi-to-bacteria ratio	0.22 good	0.17 average
Predator-to-prey ratio	0.02 very good	0.01 above average

Table 4.2. March 2021 Microbial Activity Comparison

Microbial Indicator	Fertility System 2	Fertility System 1
Total living microbial biomass (PLFA)	6,261 excellent	5,989 excellent
Functional group diversity index	1.71 excellent	1.59 very good
Fungi-to-bacteria ratio	0.32 very good	0.26 good
Predator-to-prey ratio	0.03 excellent	0.018 very good

weeks above 100°F [38°C]). Pepper and tomato yields were good, but lower. In 2022 there was more cabbage worm damage than I had seen before on this farm, some bordering on economically significant. I also observed some flea beetle and root maggot damage to brassica crops that I had never seen before on my farm, but almost no cabbage aphid damage. My theory is that the very hot weather and drought affected my system in new ways I don't yet understand.

SOIL TEST DATA

During the experiment, Carl and I sent soil samples before and after strip-tilling to A&L Laboratories for our standard test and to Ward Laboratories, Inc., for the Haney soil test. The Haney test uses traditional and unique soil extraction methods to determine what quantities of soil nutrients are available for plants in both inorganic (nitrate-nitrogen and phosphate-phosphorus) and organic forms (organic nitrogen and phosphorus). The results also report on soil health indicators such as water-extractable organic carbon (carbon that is most readily available for soil microbes) and soil respiration carbon. Soil respiration carbon is a measure of the carbon dioxide released from soil due to decomposition of soil organic matter and organic residues by soil microbes; generally, the more carbon dioxide a soil produces, the higher the microbial activity.

PLFA (phospholipid fatty acid analysis) is used to quantify total microbial biomass and provide a general profile of the microbial community. The microbial activity measured by Ward Labs in Woodleaf Oregon's experimental plots by the PLFA method was good in 2018, 7 months after living mulch and hay mulch surface-application. PLFA results improved as the experiment continued (see the change over time as shown in tables 4.1 and 4.2). The plots in Fertility System 2 showed the highest microbial populations and species diversity. Some measurements are in the highest, "excellent" range, even after spring strip-tilling in crop rows.

As I continue to experiment with these grow-our-own-fertilizer systems and reduced-tillage farming methods, I have yet to completely figure out how much microbial activity is necessary for my agroecosystem to function well, and how to predict that amount from these tests. For example, I tilled less in crop rows in 2022. With the decreased spring tillage, organic residues did not break down quickly and crop growth was slow (in a cool spring that suddenly turned very hot). But over time, it seems clear that my soil microorganisms (and some crops) appreciate the extra 4 inches (10 cm) of hay mulch in addition to the mowed living mulch. The only microbial health parameter that I would like to improve now in my

Table 4.3. Nutrient Content of Grass/Weed/Clover Living Mulch and Hay Mulch, Woodleaf Farm (OR), 2017

	% N	% P	% K	% Mg	% Ca	Fe ppm	Mn ppm	B ppm
Hay mulch	2.54	0.23	1.91	0.34	1.24	142	63	29
Living mulch	2.24	0.26	2.17	0.30	1.21	195	56	25
Normal grass-legume mix levels*	3.2–4.2	0.2–0.4	2.6–3.5	0.1–0.3	0.5–0.9	50–200	50–150	8–15

* W. F. Bennett, *Nutrient Deficiencies and Toxicities in Crop Plants* (St. Paul: APS Press, 1993).

Table 4.4. A&L Labs Results of Standard Soil Tests, After Strip-Tillage April 2018

Treatment	Organic matter (%)	Nitrate-nitrogen (ppm)	Phosphate-phosphorus (ppm)	Potassium (ppm)
Fertility System 2	4.3	51.6	8.8	288
Fertility System 1	4.3	24.3	12.1	218

vegetable fields is the fungi-to-bacteria ratio. Test results still show a higher number of bacteria and fewer fungi, including mycorrhizal fungi than I would like to see. This is probably due to spring strip-tillage, which though minimal and shallow, tends to encourage bacteria over fungi. I am very pleased to see that the microbial predator-to-prey ratio has increased over time from 2018 to 2021, even with minimal spring tillage. That seems to coincide well with the lack of pest problems observed in my annual vegetable, dry bean, and grain agroecosystem.

Comparing the results of the two types of soil tests, the Haney test seemed generally the better predictor of soil fertility and yield in our trials. This is not surprising because grow-our-own-fertilizer systems rely on microbial activity and organic forms of nutrients from decomposing plant residues cycling into plant-available forms. We ran standard soil tests three weeks after tilling and incorporating the organic residues applied to both soil fertility systems (see table 4.4). After tillage, inorganic nutrient levels increased to an adequate range, except phosphorus. However, even before tilling in the hay mulch and living mulch fertilizer, Haney test results predicted the "total available nitrogen for the next crop" as adequate for both fertility systems. This is because the Haney test measured not only nitrate-nitrogen, but also ammonium, organic nitrogen, and total nitrogen. Figure 4.6 is an example of the additional forms of carbon measured by the Haney soil test compared with the standard (A&L) soil test. Data is presented as an average of eight samples per soil fertility system.

Yield and crop quality results seem to support the notion that microbial activity and organic forms of nutrients such as nitrogen and phosphorus are important to measure, especially in systems that are high in organic residues. Despite low phosphorus levels and lower nitrate-nitrogen in Fertility System 1, as measured by the standard soil tests, yields and quality of all crops were good, and higher only for peppers and onions in Fertility System 2. The good yields observed despite low phosphorus (and lower nitrate-nitrogen levels in the System 1 plots) are probably related to an active soil microbial community and a good amount of all the total soil carbon forms (see figure 4.6) and organic nutrients. What do I mean by "good amount"? The answer is a bit

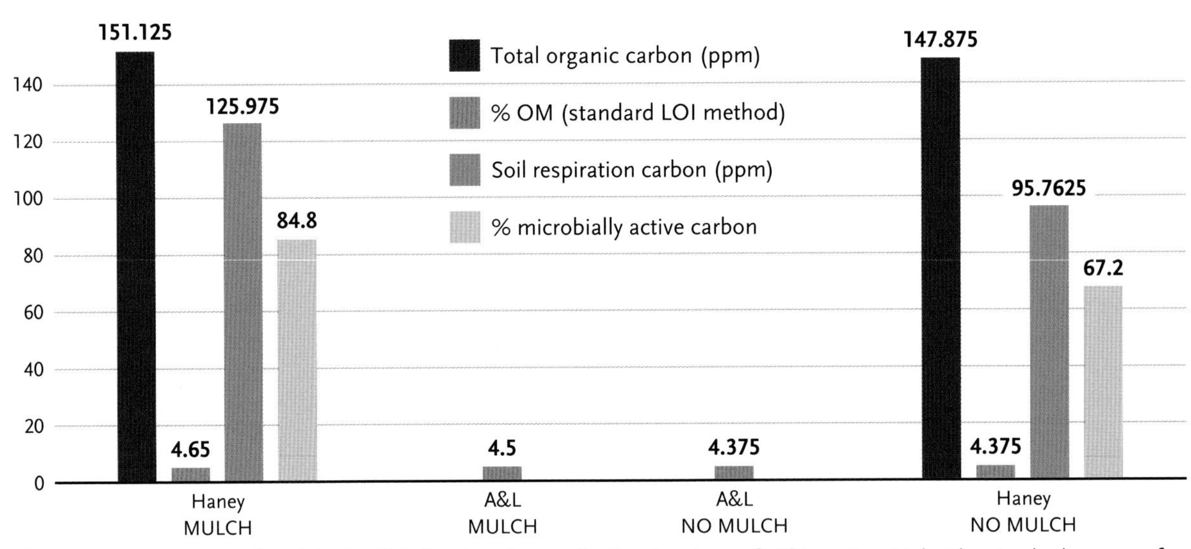

Figure 4.7. Measurements from March 2018 of various forms of soil carbon in our fertility system trials. The standard measure for soil carbon is percent soil organic matter loss on ignition (percent OM). Microbially active carbon and soil respiration carbon are indicators of the level of activity of the soil microbial community.

complex, but key to the ecological functioning of my system. According to Ward Labs' interpretation of the soil respiration carbon and organic nitrogen test results, Fertility System 2 averaged a high potential for microbial activity, with a moderate to high nitrogen credit from available soil organic nitrogen pools. Hence, Fertility System 2 had the potential for a substantial nitrogen fertilizer reduction because high microbial activity fueled by plenty of available carbon would take full advantage of the pool of nitrogen in the soil. Fertility System 1 averaged good potential for microbial activity and a moderate nitrogen credit, and it also showed the potential for a reduction in nitrogen fertilizer. This is a different way of thinking about soil fertility and the potential for using surface-applied organic residues as fertilizer!

PRACTICAL APPLICATIONS

As I write this manuscript, the grow-your-own-fertilizer experiment is ongoing, but already the results have informed how Woodleaf Farm, Oregon, grows its own fertilizer and enhances nutrient cycling and recycling. Generally, the wide no-till living mulch row middles provide a regular, slow-release crop fertilizer and both nutrition and habitat for biological control organisms and a diversity of soil microorganisms. From the experiment results, I learned that heavy-feeding crops, such as sweet peppers, yielded better when 4 inches (10 cm) of hay mulch was added once a year to the living mulch that was regularly mowed and blown into the crop row. So now, informed by the experiment results, I usually give my heavy-feeding crops, such as red peppers, onions, and sweet corn, extra hay mulch as well as mowed living mulch. For crops that are not in the ground long, like lettuce and peas, and lighter-feeding crops, like cabbage and dry beans, I generally do not provide any added hay mulch. They receive mowed living mulch only. Because crop beds are rotated each year, every bed receives an added 4 to 6 inches (10–15 cm) of hay mulch every 2 to 3 years. I was not able to add the hay mulch to any beds at the end of 2021, so in 2022 all beds received only mowed living mulch, and red pepper and sweet corn yields were lower than in the previous 4 years. The 2022 yields seem to reinforce what I learned from the experiment.

The Woodleaf Farm Soil Organic Matter System

For the past several years, I have been modifying my grow-your-own-fertilizer methods based on both what I learned from the fertility systems experiment and on what I continue to observe. Here are the month-by-month details of how bed preparation and fertilization occurs in my evolving soil organic matter system at Woodleaf Farm, Oregon.

SPRING

Spring is the time for preparing beds. Just as cottonwood tree buds swell in March, I strip-till organic residues into the soil of the crop beds. Some is newly mowed-and-blown fresh, green living mulch residues, some is living mulch (with both roots and shoots) that grew into the beds the previous summer, and some is the extra 4 to 6 inches (10–15 cm) of hay mulch applied green the previous fall, but now dry after sitting on the soil surface all fall and winter. This provides a mix of both low C:N ratio and high C:N ratio materials, some fresh, some decomposing, some almost fully decomposed, and some with both aboveground shoots and belowground living roots. This is my attempt to apply organic residues from all along the organic matter continuum shown in figure 3.1.

During bed preparation, I use shallow (2 inches [5 cm] deep) tillage because it minimizes soil volume disturbance. As mentioned earlier in this chapter, I sometimes use a single shank chisel plow for deep tillage to provide aerated habitat for soil organisms without turning over the soil or exposing soil organisms to excessive drying conditions.

Figure 4.8. After shallowly tilling the crop beds, I sometimes use a single shank chisel plow to open up the soil, add oxygen for soil microbes, and help incorporate residues.

Figure 4.9. I strip-till one to three times to incorporate residues of new, green, mowed and growing living mulch into the crop beds.

In April I again mow the succulent, greening-up grass and clover living mulch (which includes flowering weeds, such as dandelion) and blow the residue into the crop beds. I then strip-till a second time to incorporate this low C:N ratio residue that will feed microorganisms and, in turn, my crops, and also to suppress annual weeds (see figure 4.9). The timing of the second tilling, only when I am adding organic residues to the soil, is consistent with the rules for tillage summarized earlier in this chapter. Even as I mow and blow the living mulch adjacent to crops into the crop rows, I also leave some early-blooming weeds undisturbed in the center of the row middle. This helps to encourage predators and parasites back into the agroecosystem.

Figure 4.11. By June the living mulch in the row middles is wild and a bit untidy; some is left undisturbed and provides habitat for birds, snakes, and beneficial insect predators and parasites.

LATE SPRING AND EARLY SUMMER

The focus in late spring and early summer is balancing nutrient cycling/recycling with habitat management and with harvest. I leave at least 30 percent of the total row middle area unmowed and undisturbed at any one time, as shown in

Figure 4.10. In May mowed-and-blown green mulch provides good surface cover in this bed of young onions.

Figure 4.12. It also provides a barrier to suppress soilborne diseases of these squash fruits.

Figure 4.13. A strip of undisturbed living mulch in late August includes blooming red and white clover, alfalfa, chicory, common plantain, and grasses.

Figure 4.14. The mowed-and-blown living mulch covers the remnants of finished spring salad mix and pea plants.

figure 4.10. Mowed living mulch blown into crop rows forms a weed-suppressive mulch as it slowly decomposes and very slowly cycles nutrients. Repeated mowing of the living mulch every 4 to 6 weeks provides regular additions of organic residue throughout the season for slow nutrient cycling as well as a uniform cover that suppresses weeds (see figure 4.12). Repeated application helps to overcome the fact that release of nutrients from surface-applied organic residues is a slow process, requiring 2 to 4 months until the nutrients cycle into plant-available forms in the soil. This is why I generally always blow the mowed living mulch into the crop rows, unless I am about to harvest and the residues will hinder that, as they might with salad mix, where the mowed bits of living mulch are a nuisance if they mix into the harvested crop (see figure 4.11).

SUMMER

Mowing continues through summer on a regular basis, but the focus changes and I mow more selectively. I decide what to mow and when to balance the competing ecological needs: adding organic residues to maintain nutrient cycling and suppress weeds, and maintaining year-round growing roots and season-long, sequential bloom to supply pollen, nectar, and cover for biological control organisms. Once I finish harvesting a spring crop (see figure 4.14), I could strip-till the bed and replant it with a light-feeding crop such as spinach for fall harvest, or I could leave it undisturbed for the rest of the growing season, overwinter, and then strip-till and plant it the following spring.

FALL

Fall is a busy season of harvesting and bed preparation for the following year, including applying 4 to 6 inches (10–15 cm) of hay mulch to beds where heavy-feeding crops will be planted the following year (see figure 4.15). This kind of system requires careful planning of crop rotations, which I discuss in the "Rotation" section on page 136. Harvest

Figure 4.15. Hay mulch spread over a bed after harvest. The adjacent tomatoes and sweet corn are thriving in beds that received extra hay mulch the previous summer/fall.

Figure 4.16. The yield and quality of these field-grown Woodleaf Farm heirloom tomatoes is good, even though they were exposed during the growing season to all the things tomatoes hate, including strong sun, wind, and temperature extremes.

Figure 4.17. In the first week of October, these eggplants are remarkably vigorous, with plentiful glossy fruit, despite the low-nitrogen inputs of my organic grow-your-own-fertilizer system.

Figure 4.18. In late fall after the first hard frosts, my soil is covered with weed and living mulch foliage aboveground and growing roots below to maintain beneficial organism habitat all winter.

season is long in eastern Oregon: We harvest the last of the field-grown sweet peppers and heirloom tomatoes in the first week of October.

By late fall when the cottonwood tree leaves are falling, weeds and living mulch have grown into and covered the crop rows. So I mow the crop beds and mow and blow the living mulch from row middles into all the crop rows to provide winter cover for soil microorganisms and natural enemies. In this way, the rows are covered with both a surface "dead" mulch and a living mulch with active roots providing root exudates and carbon for my hard-working microbial community. It also saves me seed and labor costs because I don't have to oversow or seed a winter cover crop.

Designing Your Own Fertility System

Now that I have described my soil organic matter system, I want to cover the wide range of organic residue choices available to gardeners and farmers. As you design your own system, you can make choices that best suit your scale of operation, equipment, growing season, climate, and soils. Many of the mulch materials described below can also be used to make compost, and more detail about them is available in table 3.2 and in the "Common Composting Materials" section on page 80. Here I include information on how and when to use various mulches and cover crops, keeping in mind all of the ecological management issues discussed earlier in this chapter and in chapter 2, including C:N ratios, speed of decomposition, and effects on the soil microbial community.

MULCH OPTIONS

The hay mulch and living mulch I use are only two of the many choices of mulch. Besides adding nutrients to the soil, mulches help control weeds and maintain consistent soil moisture and temperatures. Some mulches, however, can keep the soil too cool for certain warm-season crops, particularly in the spring or in wet, cool climates. This problem can be mitigated by using reusable weed mat in the crop row and applying organic mulch in row middles. On the negative side, certain mulches may release natural chemicals toxic to seeds and seedlings. If you're not sure, apply a new mulch only after crop plants are well established. Some mulches may encourage certain pest insects (like slugs) and diseases if they remain wet and cool for an extended period of time, and if air movement is poor. As they decompose with the help of soil microbes, some mulches can also immobilize nitrogen and phosphorus if the mulch materials have too high a carbon-to-nitrogen (C:N) and/or carbon-to-phosphorus ratio. For example, corncobs, sawdust, wood chips, or wheat straw can tie up nutrients temporarily as soil microbes use the limited nitrogen they provide during decomposition.

Mulch choices include grass or legume hay, straw, leaves, lawn clippings (if not treated with herbicides), wood chips, ramial wood chips, and weeds, before they flower or form seed heads (for information on the plant nutrients available in specific weeds, see "Benefits of Weeds" on page 160).

The coarser the mulch, the less well it holds in soil moisture. Coarse mulches (straw, corncobs, or wood chips) are better on cool, wet, or heavy soils in rainy climates. Sandy, light, or well-drained soils in dry climates usually do better with fine-textured mulches (like grass or hay mulch) that hold moisture. Some mulches are best when partially decomposed (wood chips), to avoid production of toxic natural chemicals or competition with crops for nutrients as they are decomposed by soil microorganisms.

Grass Clippings

Mowed grass generally provides a balanced nitrogen, phosphorus, and potassium fertilizer. If grass clippings are still green, the C:N ratio is relatively low. If they have dried and browned, the C:N ratio will be higher. Grass clipping are great to add as mulches and to composts, with the following cautions.

- When turf has been treated with herbicides, the collected grass clippings may cause injury to susceptible crop plants.
- Under wet, warm conditions, newly cut grass clippings used as mulch can begin rapid anaerobic decomposition, which may burn tender transplants or seedlings.
- Thick applications can reduce water penetration to roots, since grass clippings tend to mat.

Leaves

As leaves decompose there is generally an increase in pH, which may slowly affect soil around mulched plants. Types of tree leaves that break down with an alkaline reaction include cottonwood, aspen, linden, willow, and green ash. Pine, spruce, and fir needles usually do not affect soil pH as they decompose. Leaf mold, unlike undecomposed leaves, may sometimes help to reduce soil pH. Leaf mold is found beneath forest or native trees on farm/garden edges. Decomposing or decomposed leaves are also an excellent garden mulch. You can leave a pile of leaves in a corner of the garden to break down over a few years on its own, or first chop the leaves using a lawn mower and then pile or compost them.

Shredded leaves break down more rapidly than whole leaves, and there is less risk of matting and resultant anaerobic decomposition. If leaves remain dry they seem to be better able to suppress weed seed germination. Wet leaves, especially if not shredded, may compact and reduce water penetration. Natural chemicals that inhibit seedling growth may be present in the leaves of certain woody (and some herbaceous) plant species at different developmental stages. Walnut leaves are a good example. Because of this, don't mulch seedlings with leaves.

Sawdust

A layer of sawdust at least 2 inches (5 cm) thick may suppress annual weeds if applied to the soil surface before weed seed germination. There are some undocumented reports that sawdust mulch may help to keep soil pH down in perennial plantings of acid-loving plants such as blueberry. However, there are many reports that sawdust mulch may have detrimental effects on plant growth and development and hence should be used with great caution on young plants. For example, in a study of newly planted trees, hardwood sawdust mulch (aged for 2 years) decreased both trunk diameter and number of permanent lateral roots. In contrast, corncob mulch treatments in the same study increased young tree survival and trunk diameter.[1] Be careful using sawdust. Thick applications of sawdust can reduce water penetration to roots and may contain natural decomposition chemicals that injure young plants. Sawdust also ties up nitrogen and phosphorus and causes crop nutrient deficiencies. Sawdust should be used only very sparingly, and probably only for perennial crops like blueberries and established fruit trees.

Straw

Straw includes wheat, rye, oat, and cornstalks, and each type has unique properties. In general, straw mulches decrease soil moisture evaporation, increase soil water storage, and keep soil temperatures cooler in the spring. Straw mulch appears to increase the levels of available potassium in soil and in crop plants.[2] Another reason to consider using straw mulch is for its beneficial effect on an important member of the soil organic matter system: earthworms. For example, one study found that straw mulches with low C:N ratios, low lignin, and low polyphenol concentrations supported 54 percent more earthworms than the unmulched control. However, mulches with high lignin, polyphenols, and C:N ratio, such as some undecomposed wood chips, increased earthworm density by only 15 percent.[3] Straw mulches have also been shown to affect pest management. Straw mulch reduced the amount of Colorado potato beetle damage in a New England study, and researchers found greater numbers of beetle-attacking soil predators in mulched compared with bare soil plots.

Specific straw mulches may be better for certain purposes. In one study wheat straw mulch (when compared with cornstalk and soybean straw mulch) increased soil particle aggregation and favored the formation of larger soil particle aggregates. This would be good for both lighter, sandy soils and heavy, wet soils. Cornstalk and soybean straw improved soil particle aggregation better than unmulched bare soil, but did not improve soil tilth nearly as much as wheat straw mulch. Mulching with straw may also affect soil temperatures. When light- and dark-colored straw mulches were compared with bare soil, light-colored straw reflected the most light and warmed the soil least. Bare soil was warmed the most. The soil with dark straw mulch was intermediate in its effect on light reflection and soil warming. Rye mulch is good for weed suppression. Oat straw that has already seeded may also cause germination of unwanted baby grain plants in gardens and fields. Clean oat straw with few or no leftover seeds is usually not a problem, but I have seen weedy disasters using oat straw from a poor combining job, especially on perennial crops like strawberries. Choose your straw carefully.

Wood Chips and Bark

Many species of trees can be a valuable source of wood chips—shredded, chipped, or ground-up pieces of wood that sometimes include bark, branches, and leaves. Individual pieces are usually small and do not pack down, thus readily allowing movement of water into the soil and water vapor out of the soil. Bark chips are usually more decomposed than wood chips. Use fresh chips only as a surface mulch or compost them before incorporating them into soil to avoid nutrient immobilization. In general, wood chips help to conserve water, but in a Texas study trees mulched with pine bark lost more water through evapotranspiration than unmulched plants, possibly because surface temperatures of the pine mulch in the study were higher than those of the lighter-colored unmulched soil.[4]

Ramial chipped wood is made solely from young, small- to medium-sized shredded greenwood branches. Young tree branches are the repository of most (up to 75 percent) of the minerals, amino acids, proteins, plant hormones, and enzymes contained in a tree overall. Fresh, young branches, twigs, and leaves have lower C:N and C:P ratios. Hence, when they are used to make chips, there is less immobilization of nitrogen and phosphorus than when bigger, older branches and woody material are used. Ramial chipped wood supports the growth of white-rot fungi, which are capable of degrading all parts of woody materials, including lignin. It also supports and feeds mycorrhizal fungi.

Organic/veganic farmer Iain Tolhurst has been experimenting with compost made from ramial chipped wood on his vegetable farm in Oxfordshire, England, for more than 20 years. He has had great success building soil organic matter and providing crop nutrients by applying coppiced branch, twig, and leaf wood chip compost over the top of a mowed legume cover crop. Tolhurst lets the composted chipped ramial branch wood sit over the winter on top of the lower C:N ratio cover crop, then tills in the cover crop/branch wood compost mix the following spring. That way his soil receives both the heavy carbon and micronutrient benefits of woody materials and the higher-nitrogen benefits of the legume cover crop.

There are some reports that fresh, uncomposted wood chips used as mulch may encourage plant disease. However, studies at Ohio State University indicate that composted wood bark provides significant disease suppression when used as part of growing media as described in "Starting Seeds in Microbially Amended Soil Mixes" on page 183. Other reports indicate that certain bark materials may contain toxins, such as phenolics. These toxins are water-soluble and can injure young plants that are susceptible to these natural chemicals. Bark chips can also harbor earwigs, termites, and sowbugs. But they also provide habitat for beneficial carabid beetles and spiders. The trick with high C:N ratio, high-lignin wood products is to balance their application with applications of a lower C:N ratio, higher-nitrogen material. Adding woody material as your sole fertilizer or soil amendment is not a good idea.

LIVING MULCH AND COVER CROP OPTIONS

Because they add living roots to the soil organic matter system, cover crops and living mulches should be the foundation of or at least included in every soil fertility farm and garden plan. Picking the appropriate species for your soil, environmental, climate, cropping, and space conditions is the key to success. Cover crops and living mulches can be the same plant species, but cover crops are incorporated into the soil, whereas living mulches grow alongside the cash crop and may or may not be incorporated into the soil when the cropping season is over. They can be a challenge to manage in smaller areas, so choice is especially important in small gardens.

Cover crops and living mulches provide protection against drought and erosion as well as weed control. Cover crops acquire and retain nutrients, they increase soil organic matter, and they conserve soil water by improvement of soil structure, which increases infiltration and water-holding capacity. They can both inhibit and encourage pests depending on how they are managed and what crops they are being grown with. I provide specific details about cover crops and diseases in chapter 7 and about cover crops and weeds in chapter 8.

Diversity matters for cover crops. Mixes of cover crops with different C:N ratios are best. For example, planting cereal rye or annual ryegrass (which has a high C:N ratio) alone can temporarily tie up nitrogen and possibly phosphorus. A solo planting of a legume like vetch or clover releases nitrogen quickly, but can result in nitrogen leaching into groundwater if the crop is tilled into a soil with low microbial activity. But a grass/legume mix, such as rye with vetch, provides slow-release nitrogen, phosphorus, and potassium, with less potential for nutrient leaching than a legume alone and less nutrient immobilization potential than a grass alone.

When you consider which cover crop is best, it's important to know how you will eventually kill the cover crop. Options include severe mowing (to the ground), tillage, rolling, and solarization with clear or black plastic. Flaming suppresses annual cover crops, like berseem clover, but not perennials, like red clover. It is hard on ground beetles and spiders.

Choosing a Cover Crop or Living Mulch

Legumes are good candidates for cover cropping or as living mulch, but not all legumes fix nitrogen at equal rates, or under all conditions. For example, clovers, sweet clovers, medics, and vetch provide 0.1 to 2.5 pounds nitrogen per 100 square feet (0.5–12.2 kg/100 m^2) while alfalfa provides 6 pounds per 100 square feet (29.3 kg/100 m^2). Legumes also contribute phosphorus, potassium, calcium, magnesium, sulfur, and micronutrients to the soil. If you have not grown legumes in your soil previously, it's a good idea to inoculate the seed of legume cover crops. Inoculants consist of species-specific bacteria that associate with legume roots to fix atmospheric nitrogen. Use the correct inoculant for the cover crop species you want to grow. Alfalfa and yellow and white sweet clovers share the same inoculant; true clovers share another; peas and hairy vetch share a third; garden beans and field beans share a fourth. Purchase inoculants when purchasing seed.

Native grasses and legumes may be better adapted to local climates and require fewer inputs when employed as a cover crop in some cases. For

Cover Crops for Different Soil Conditions

Your soil type, pH, and conditions will dictate to a large extent what cover crop choices are best. Here is a general list of cover crops that do well under specific soil conditions as well as a list of those that establish rapidly for when you need to cover the ground quickly.

Low fertility	Sunn hemp, pearl millet, cowpea, buckwheat, kura (Caucasian) clover
Low soil organic matter	High-biomass, fibrous-rooted grasses, including cereal grains, millets, and Sudan grass, alfalfa, vetch, and red clover
Low microbial activity	Diverse grass/legume/forb mixes, including mycorrhizal hosts, such as a spring mix of barley, oats, and field peas (spring types) or a summer mix of millet, buckwheat, Sudan grass, and field peas (winter or spring types) or a mix of winter wheat, clover, and vetch; mustard and radish are non-mycorrhizal, but can be added
Acidic	Oats, cereal rye, vetch, cowpea, sunn hemp, buckwheat, clovers (alsike, rose, crimson, and especially gland clover that tolerates pH down to 4.5)
Alkaline	Barley, mustards (or plants in the cabbage family), sunn hemp, sweet clovers, and clovers such as berseem, alsike, Persian, and crimson
Cold, wet	Japanese millet, oats, annual ryegrass, clovers (berseem, white, and Persian)
Dry, low irrigation	Medics, barley, pearl millet, black lentil, cowpea, lacy or tansy phacelia, yellow sweet clover, Braco white mustard, field pea, Sudan grass, Russian wildrye
Heavy, with a hardpan	Deep-rooted, vigorous crops, including radish, canola, alfalfa, yellow sweet clover, pearl millet, Sudan grass
For rapid establishment	Buckwheat, annual ryegrass, cowpea, tillage radish, cereal grains, Sudan grass

example, perennial grasses native to climates that typically experience long, dry summers go dormant when it is hot and dry, thus providing a growing (but dormant) root without risk of competition with the crop. Local mycorrhizal fungal populations may also associate preferentially with native plants over exotic grasses. But when irrigation is available, fast-growing exotic cover crops are nearly always better for biomass production, rapid soil cover, and weed suppression. I experimented for many years hoping to find a native legume living mulch that performs as well as the non-native species, but no luck so far.

Cover Crop Planting Time

Timing of planting is important because the rate of nutrient release varies among species and with stage of plant maturity. Deep-rooting, slow-to-establish, perennial cover crops take longer to start contributing nutrients to the soil. In general, a fast-growing annual legume will contribute nutrients faster and may fix more nitrogen during its growing period than perennial legumes, which are slower to begin fixing nitrogen. Annuals are usually easier to manage, especially in small garden spaces. For maximum nitrogen and carbon release, perennials should be left in place for at least a year.

On a garden scale, cover crop seed is one of the cheapest "fertilizer" inputs organic gardeners can make. So experiment! Try sowing an annual legume in early spring to grow until late spring, when you'll want the space for planting vegetables, flowers, or whatever. The only caution I offer is that because tilling the soil usually results in a decrease of soil organic matter, the tilling required to turn under a short-term cover crop may burn up reserve soil organic matter. The cover crop won't have accumulated enough biomass to offset that loss. To avoid that, instead of tilling in the annual legume, mow it short instead and cover it with a tarp, or mow it and mulch over it with hay mulch when you plant the area. I often do this as I prepare the soil in my winter high tunnel. I seed a summer buckwheat crop that I mow short and then cover with hay mulch.

If your goal is to incorporate a cover crop as a green manure, it is best to plant it in the fall or early spring so that the crop grows through the entire summer season. If planting in the fall, choose cold-hardy cover crops like perennial ryegrass. If the cover crop is to be interplanted with your crops as living mulch, it's likewise best to sow it in the fall or early spring. If utilizing the cover crop as a winter soil cover and to provide some nutrient addition after crops are harvested, a tender annual can be interplanted over a soon-to-harvest crop or planted after harvest, and then allowed to winter kill. In cold climates, sow the annual cover crop before the end of August.

Timing is also important for beneficial insect habitat and the balance between pest and beneficial insects in an agroecosystem. In several studies, mowing reduced the ability of cover crops to support predator lady beetles, big-eyed bugs, and minute pirate bugs and encouraged dispersal of pest stink bugs, tarnished plant bugs, and western flower thrips out of flowering cover crops and into crops.

Perennials

Plant the following perennials as cover crops or living mulch in late summer/early fall or in spring.

Kura clover (pH 6.0–7.0) is a spreading, winter-hardy perennial native to Eastern Europe. Kura clover can withstand poorly drained soils and soils with a high water table. It is better adapted to lower fertility and pH than other clovers and alfalfa. Kura is a good choice for areas that receive heavy foot traffic, but can be difficult to establish from seed.

Red clover (pH 5.8–7.0) is very hardy and easy to establish. It has excellent seedling vigor and grows taller than white clover. In many areas red clover can live for 2 years (and occasionally longer), but in the South it often acts as an annual. It readily reseeds itself on my Oregon farm. In areas where it is well adapted, it is the best biomass-yielding clover species. It is not tolerant of continuous short mowing, especially if combined with low or no irrigation in dry climates. Red clover requires plentiful soil moisture, but is heat- and drought-tolerant once established. It is not considered as tolerant of wet conditions as white clovers.

White clovers (pH 6.2–7.5) are hardy and easy to establish. They require consistent moisture for establishment and to maintain good growth. They prefer cooler temperatures, and germinate best at soil temperatures between 40 and 50°F (4–10°C). They are good choices for areas that receive heavy foot traffic and do relatively well with close mowing. Ladino varieties of white

Figure 4.19. Blooming red and white clover dominate my mixed-species living mulch in August in this no-till orchard-to-be. Mounds of hay mulch suppress the clovers and grasses at each spot where a fruit tree will be planted the following spring.

Figure 4.20. This alsike white clover growing as a living mulch between trellised crop rows is easy to keep short by mowing.

clover are usually most productive, but generally do not reseed well. Dutch white clover and alsike clover are vigorous and spreading. Alsike tolerates long spring flooding (7–30 days) and strongly acid soil conditions (as low as pH 5.1). White clovers were my favorite choice at Biodesign Farm in Montana because they were easier to manage than the more vigorous red clover.

Perennial ryegrass (handles a wide soil pH range) is vigorous and hardy (can be planted late in the season). Plant in combination with legumes and hairy vetch for best weed suppression. It can be difficult to manage in small garden areas and where early-spring crops are planned.

Annuals

Plant any of these annuals in the spring to serve as a cover crop or living mulch.

Arrow leaf clover (pH 5.6–8.4) is a productive annual that makes most of its growth in late spring. Arrow leaf clover requires well-drained soil. The seed can germinate at lower temperatures than most annuals. It reseeds well. Berseem and arrow leaf clovers are both preferred by predatory big-eyed bugs.

Berseem clover (pH 6.5–8.0) germinates rapidly if seed is raked in or rolled. It germinates best at soil temperatures of 40 to 60°F (4–16°C). It is fast growing, reaching 12 to 24 inches (30–60 cm) tall, and tolerates mowing. This clover requires abundant moisture and is quite tolerant of wet soils and also alkaline soils. Most cultivars are not tolerant of frost and will winterkill, but Frosty is a cold-tolerant cultivar, reported to have survived –16°F (–27°C). It doesn't do well in heavy clay soils. I used berseem clover as an annual living mulch between seeded crop rows for many years on my farm in Montana. Persian clover (pH 5.5–9.0) is an annual, prostrate or semi-erect branched clover similar to berseem but shorter.

Crimson clover (pH 6.5–8.0) requires irrigation. It germinates best at 40 to 60°F (4–16°C), doesn't tolerate high heat, and winterkills. It grows less vigorously than berseem clover in wet and cool conditions. It germinates rapidly, has better seedling vigor than most clovers, and provides beautiful red flowers that attract and support pollinators and beneficial insects.

Rose clover (pH 5.0–8.3) has a growth habit similar to crimson clover, but it is very tolerant of drought and low fertility. The hardiest variety has similar winter hardiness to arrow leaf clover or crimson clover, but varieties often grown in western states are much less cold-tolerant. It is best suited to well-drained soils.

Subterranean clover (pH 5.1–8.4) is a dense, very low-growing annual that does best in areas with mild winters. It makes most of its growth in midspring and is a very good living mulch. Subterranean clover does not yield as well as arrow leaf, berseem, or crimson clovers, but is less aggressive in a small garden. Subterranean clover is more tolerant of low fertility and shade than most clovers. It is best adapted to medium- and heavy-textured soils with good moisture-holding capacity.

Nitro alfalfa (pH 7.0–8.0) is deep-rooted and vigorous. It is an annual in cold climates, but may sometimes overwinter.

Burr medic (6.5–8.0) cultivars are the best known of the annual medics. They branch profusely at the base and send out prostrate stems that grow more erect in dense stands. They grow quickly in response to rains and fix nearly as much nitrogen per acre as true clovers. Once established burr medic tolerates shade relatively well.

Snail medic (pH 6.5–8.0) germinates best at soil temperatures of 45 to 60°F (7–16°C). It does not reseed itself in cold climates and winterkills. It grows to 12 inches (30 cm) but tolerates mowing. It is drought-tolerant, but make sure seeds do not dry out during germination. Once established snail medic can tolerate high heat and dry periods. Parabinga medic is similar to snail medic but grows vinier and shorter and matures later.

Field peas (pH 5.5–7.5). There are two types and many varieties. Austrian winter peas (black peas), yellow peas, and Canadian field peas (spring peas) are cold-hardy but heat-sensitive. Southern peas—also known as cowpeas, black-eyed peas, or crowder peas—are heat-tolerant but cold-sensitive. Southern peas are susceptible to diseases if planted in cold, wet soil, so plant after soil temperatures reach 60°F (16°C). All peas are succulent and break down quickly after incorporation. They can be harvested for dry peas, too, but removing the seed also removes most of the cover crop nitrogen addition. They are quick growing and water-thrifty.

Hairy vetch (pH 6.2–7.5) handles a wide soil pH range and is very vigorous with viny growth. It has a higher nitrogen content and lower C:N ratio than crimson clover. It is more drought-resistant than other vetches. It grows taller than most of the clovers and can produce a lot of biomass that might be a challenge in small gardens. It is a favorite pollen and nectar plant of western flower thrips (a pest of a wide range of crops), but also of beneficial insects such as minute pirate bugs, lady beetles, and pollinator bumblebees. Hairy vetch can be killed by very

Figure 4.21. Hairy vetch growing between and beneath peach tree crop rows, mixed in with the grasses and weeds in our living mulch.

close mowing when it is at peak bloom. It reseeds itself well and easily if left to go to seed.

Buckwheat (pH 6.0–6.5) is a good summer cover crop due to its quick germination, rapid growth, and ability to outcompete weeds. It grows tall, to a height of 2 to 4 feet (60–120 cm). Buckwheat has a higher tolerance of soil acidity than the cereal grains do (down to 5.2 pH). It handles infertile, low-nitrogen soils. Too much nitrogen encourages weed pressure and causes excessive vegetative growth. It does not do well in compacted, dry, or excessively wet soils. Although a field of buckwheat in full flower looks like dense ground cover, each individual plant with its hollow stem is rather spindly and does not produce a lot of biomass. Buckwheat is a quick-to-flower option to encourage beneficial insects and provide ground cover; flowering begins about 3 weeks after planting and continues prolifically for several weeks. It is easy to manage in small gardens (if you don't let it go to seed) and provides excellent summer bloom for beneficial insects.

Figure 4.22. Teff growing as a living mulch and a mowed walkway in a vegetable garden.

Annual ryegrass (pH 6.0–7.0) is a very fast-growing, nonspreading bunchgrass that can be a quick soil cover and outcompete weeds. It can be overseeded in late-summer and fall crops, and its quick establishment provides a winterkilled mulch for early-spring weed suppression. Because it is a heavy nitrogen user, ryegrass can capture and take up excess nitrogen and reduce nitrate leaching over the winter in warm climates. Ryegrass helps slow-growing, fall-seeded legumes establish and overwinter in colder climates, even if the ryegrass winterkills. However, it may outcompete legumes in warm climates, although low-nitrogen fertility generally favors the legumes. Ryegrass can become a weed if allowed to set seed. It establishes faster than perennial ryegrass but is less cold-hardy.

Sudan grass (pH 5.5–7.5) is a heat- and drought-tolerant summer cover crop. It grows very tall (5 feet [1.5 m]) and vigorously with lots of plant biomass. It also has abundant root biomass, especially if mowed, since mowing encourages root growth. It suppresses root knot nematodes and inhibits weed germination if sown at higher rates. Sudan grass decomposes slowly and leaves big chunks in the soil when turned under, so it is difficult to prepare a seedbed for small-seeded crops immediately after incorporating it. Mow Sudan grass in late summer or fall for a winter-killed mulch to provide winter soil cover, then till it in early spring.

Teff (pH 6.0–7.0) is a very drought-tolerant summer cover crop. It is a warm-season grass good at suppressing summer weeds if seeded at a high rate. It can be used as a living mulch between crop rows. Teff produces fine plants and roots that don't leave big clumps in the soil when turned in. Teff does not grow too vigorously and needs minimal mowing. Unlike annual ryegrass, it generally does not produce a lot of seed, so volunteers are not a serious problem. It tolerates drought better than buckwheat or Sudan grass, but does not tolerate frost or establish well in cool soils.

Biennials

Only one species fits the category of a biennial living mulch, and that's yellow sweet clover (pH 6.5–8.0). Planted in spring, yellow sweet clover is very drought-tolerant once well established. If overwatered, sweet clover does not compete with weeds as well. It produces flowering stalks up to 4 feet (1.2 m) tall the second year. This species has become weedy in native western grasslands and along forest roadsides due to its drought tolerance.

Compost to Rebuild a Broken System

In chapter 2 I explained why I eliminated compost from my soil organic matter system. But in some situations compost may play an important part in rebuilding (or building) a soil system, along with cover crops and living mulches and mulches. Gardens and farms in the early stages of ecological system development and those that start out on challenging, poor-fertility soils often require a quick nutrient boost that compost can provide. Microbially active composts may also be the best option for small spaces in which it is impractical to grow your own fertilizer that also maintains a year-round living root in the soil. Composting is a controlled decomposition process in which plant residues and manures are converted by microbial action into chemically more complex and stable forms and, if the composting process is well done, into a humus-like material. The great thing about compost is that as long as you trust the quality of the materials in it, and as long as it is properly *finished* compost, you will rarely have problems with it burning young plants or causing nutrient immobilization (as could happen with some high C:N ratio mulches). In finished compost the original materials have fully decomposed and reached a more stable form of organic matter. There is less chance of overfertilizing with one nutrient and ending up with nutrient imbalances. Finished compost will not tie up phosphorus or nitrogen and will not produce plant toxins.

The best-made compost adds plant nutrients directly to crops and simultaneously provides food and habitat for soil microorganisms. A desirable microbially active compost has a carbon-to-nitrogen ratio of about 20–15:1. When compost is finished, it should have a uniform consistency and earthy smell, and its original materials should no longer be easily identifiable. Finished compost should be porous and have a crumb structure—what is recognized in a soil as good tilth.

Proper temperature and pH are good indicators of whether or not a compost is finished. pH of finished compost should be between 6.5 and 7.5, and temperature should be 8°F (4°C) above the seasonal soil temperature. In the final stages of decomposition, pile temperatures may still range between 86 and 120°F (30–49°C). Finished compost may feel a bit warm, but if it feels hot to the touch, it is not yet finished.

MAKING COMPOST

Composting occurs with the help and activity of certain decomposer soil microorganisms. The composter's job is to supply those microbes with adequate nutrients and environmental conditions. Compost-making basics include attention to size of the pile, to maintaining optimal decomposition conditions, and to the quality (characteristics) of the selected materials.

Figure 4.23. A windrow pile of finished compost with good crumb structure. None of the original compost materials are identifiable.

Pile Size

Too large or too small slows decomposition. For best results, build compost piles in a pyramid shape not more than 5 feet tall and 8 feet wide (1.5 × 2.4 m). If the pile is shorter than 4 feet (1.2 m), it may not heat up well. If your pile is wider to begin with, you can scrape off the sides after the first week and add it to the top. This helps to release carbon dioxide collecting on the bottom of the pile and will improve decomposition.

Figure 4.24. A compost thermometer allows you to monitor precisely what is happening in your compost pile and is a good investment for making good microbially active, finished compost.

Optimal Conditions

Moisture content, oxygen supply, and temperature should all be monitored as the pile decomposes. Water content in the pile should be 40 to 70 percent; optimum is around 55 percent. It should feel moist but not wet to the touch. To check moisture in a home compost pile, take a handful from the approximate center of the pile (reach in 20 to 24 inches [51–61 cm] from the surface of the pile, and about halfway up the pile). The sample should feel and react like a moist sponge and release no more than a drop or two of water when squeezed. Many materials are problematic to compost because they are too wet. If a material has high moisture content, consider shredding or chopping the material first, or mix small amounts of wetter materials in layers with drier materials. If a pile is too wet, turning it will help to dry it out.

Temperatures in the initial decomposition phase should be 130 to 149°F (54–65°C). By the third or fourth week of composting, temperatures should be 86 to 120°F (30–49°C). Soil scientist Robert Parnes, author of *Fertile Soil*, suggests that "cold" composting (temperatures below 100°F [38°C]) may result in compost with greater biological diversity than hot composting—particularly uncontrolled hot composting where temperatures are allowed to go above 149°F (65°C). When I made compost, I practiced hot composting but was careful not to let temperatures rise above 145°F (63°C). In general, compost piles should be kept warm enough so that when you put your hand inside the pile, it is warm to the touch. Some weed seeds may require temperatures as high as 160°F (71°C) to be killed. Resistant weed seeds can remain viable up to temperatures of 210°F (99°C). But at temperatures higher than about 160°F (71°C), beneficial microbes are also likely to be killed. Composting at temperatures lower than 100°F (38°C) often takes significantly longer. If you do compost at temperatures above 160°F (71°C), allow a curing period before using so that the compost pile cools down and is recolonized by beneficial microbes.

Mixing Materials

Mix both coarse and fine and moist and dry materials as you build a compost pile. Pay attention to the freshness of initial composting materials and to material nutrient content. Balance the C:N ratios of the materials: Start with materials that have an initial C:N ratio between 20:1 and 30:1 (see "Common Composting Materials" on page 80 for C:N ratios of specific materials). In the recommendations below, percentages are on a volume basis rather than a weight basis.

Avoid using stored materials. Materials that have been stored for even a month may have lost nutrients and become anaerobic. Researchers comparing stockpiled and fresh cattle manure found a greater level of seedling-toxic natural chemicals in the stockpiled manure.

Include fresh green materials. Fresh greens, such as clover or grass, add vitamins and enzymes to the compost. Some experts suggest that legumes are best for this purpose. Many coastal composters add fresh seaweed. Others use fast-growing materials such as comfrey. Don't use more than 15 percent lawn clippings unless you balance them with very coarse materials, as too many clippings result in excessive moisture and matting that limits oxygen.

Add up to 10 percent soil. The soil does not have to be fertile, but it is best to use a soil with a heavier, clay loam texture. Add soil to the upper layers of the compost pile, because the weight of the soil will naturally lead it to fall toward the bottom layers. If the soil is rocky, run it through a screen first.

Add up to 10 percent finished compost. Finished compost has great water-holding capacity and thus improves moisture management. Finished compost may also contribute some of the microbes involved in decomposition.

Add no more than 5 percent kitchen wastes at one time. I recommend making no more than three additions of kitchen waste to a new pile for a grand total of 15 percent. Each 5 percent addition should be several days after the last. Remember, each addition extends the time it takes to finish a particular pile.

Inoculate with microorganisms. Some compost experts recommend inoculation to speed up the composting process. Others, including myself, suggest that inoculation is unnecessary. Commercial microbial activators, which are marketed as "compost starters," can be added. Another way to add beneficial microbes to compost is to make compost tea (using high-quality finished compost) and add it as you turn the pile, so that the entire compost pile receives even coverage. Do this after the first week or so, when pile temperatures are between 100 and 140°F (38–60°C).

Aerobic Decomposition Keeps Microbes Happy

Aerobic composting can produce compost in 5 to 8 weeks as long as sufficient oxygen is available in the pile to maintain active microbial populations. Aerobic composting is achieved by mixing layers of coarse and fine materials, and moist and dry materials. A 1- to 2-foot (30–60 cm) layer of dry, coarse material (such as cornstalks or pruned branches) on the bottom helps to maintain good aeration. The most common way to assure good compost pile aeration is to turn it often. Some experts recommend turning the pile frequently (daily or every other day) for the first week after

Figure 4.25. Steam emanating from compost when turning the pile means that decomposer microorganisms are working hard. Heat is produced as a by-product of the microbial breakdown of organic materials.

mixing the materials, then one to two times per week during the later stages. There are advantages and disadvantages to frequently turning compost piles. Frequent turning, if moisture and temperature extremes are avoided, may promote greater amounts and diversity of microbial activity in the finished compost. Microbial diversity is important because various bacteria and fungi found in quality compost have been shown to have an antagonistic effect against disease organisms. There is evidence that a highly oxidative composting process ultimately results in a better conversion of nitrogen into complex organic forms that are less susceptible to leaching. Frequent turning may also contribute to hot composting in which weed seed and disease organisms are killed. The critical temperature for killing *most* weed seeds is 145°F (63°C).

The major disadvantages of frequent turning are the time it takes, the equipment needed, and the increased potential for nitrogen loss by ammonia volatilization into the atmosphere. Again, some argue that the higher temperatures associated with hot composting may kill soil microbes. If you opt for less frequent compost pile turning, though, it is important to design your composting area to provide maximum aeration.

Figure 4.26. Building a compost pile in a bin at Woodleaf Farm, Oregon, with PVC aeration pipes extending out the side of the bin. The bin contents are hay mulch and pruned branches, with 10 percent added fertile farm soil and 15 percent kitchen wastes.

I still make a small amount of compost to apply to vegetables growing in my high tunnel at Woodleaf Farm, Oregon, because the small area in the high tunnel is subjected to more intensive, less ecological production methods. I build the compost pile in a 4-by-8-foot (1.2 × 2.4 m) bin with a finished height of 5 feet (1.5 m) (see figure 4.26). It is made with a bottom layer of ½- to 1-inch-diameter (13–25 mm) pruned branches and twigs followed by alternating layers of summer-cut coarse hay mulch, soil, and kitchen wastes. I use PVC pipes with holes drilled every 3 inches (7.5 cm) along the length to provide aeration in the pile. There are a total of 12 pipes; some are 1-inch (2.5 cm) diameter and some are 2 inches (5 cm) in diameter. The PVC pipes extend from (and through) one sidewall of the bin to the other, with the 2-inch pipes at the pile bottom and the 1-inch pipes at the top.

COMMON COMPOSTING MATERIALS

As noted above, there is some overlap between the types of organic materials used for mulch and those commonly used in making compost.

Brush and Tree Trimmings

C:N Ratio 80–100:1

Shred or grind trimmings to decrease size. They need additional moisture and nitrogen added. Young trimmings with twigs and green leaves present can be a good mix (after chipping) with lawn clippings to provide oxygen. Decomposition of woody material is very slow.

Crop Refuse

C:N Ratio 20–100:1

Unharvested portions of crops add nitrogen, phosphorus, potassium, and micronutrients. The carbon-to-nitrogen ratio is variable depending on the crop. Potato tops, for example, have a C:N ratio of 25:1. Corncobs have a ratio of 80:1 (as well as a low level of phosphorus).

Grass Clippings

C:N Ratio 20–30:1

Fresh grasses lack the structure to maintain adequate aeration in a pile. They do, however, have a favorable moisture content and carbon-to-nitrogen ratio. Freshly mowed grass clippings can add nitrogen, phosphorus, potassium, sulfur, calcium, and magnesium, as well as carbon, to compost piles. The C:N ratio depends on the type of grass and its maturity. Young, succulent grass has a lower C:N ratio than dry clippings.

Grass Hay

C:N Ratio 32:1

Grass hay increases biological activity, has low nitrogen losses, and increases humus levels better than straw. It provides nitrogen, phosphorus, potassium, sulfur, calcium, magnesium, and micronutrients. Alfalfa hay provides more nitrogen and increases humus faster than grass hay.

Kitchen Wastes

C:N Ratio 5–35:1

In general, kitchen trimmings compost rapidly (within 4 to 6 weeks if plenty of oxygen is added). The carbon-to-nitrogen ratio of kitchen wastes varies, depending on what you have. Fruit wastes run about 35:1, while coffee grounds are 20:1.

Leaves

C:N Ratio 40–80:1

Leaves are especially good for providing micronutrients. They tend to be dry, however, and have a high carbon-to-nitrogen ratio. If you use leaves, plan on adding supplementary water and some type of nitrogen-rich material to the pile as well. Leaves can be shredded before adding to a compost pile. Unshredded leaves are very slow to decompose because they mat together, keeping oxygen and moisture out. Composted leaves obtained from municipal or private leaf-composting facilities usually have a pH of about 7.0–8.0 and have a percent nitrogen/phosphorus/potassium analysis of about 1.0–0.5–0.2.

Manures

As mentioned in earlier chapters, I don't use manures because they can provide excessive nutrients, toxins, and cause soil nutrient imbalances and leaching.

Poultry manure (C:N ratio 7:1) is very concentrated. Its nutrients are rapidly available, and thus the chance of leaching loss is high. It is unusually rich in nitrogen, phosphorus, and calcium but low in potassium.

Sheep manure (C:N ratio 16:1) is the most concentrated non-poultry manure. It has high levels of nitrogen and potassium but low concentrations of phosphorus and sulfur.

Horse manure (C:N ratio 22:1) heats up rapidly and is a good balanced addition to the compost pile. Levels of potassium, phosphorus, and sulfur are somewhat lower than other manures. There are often greater quantities of viable weed seeds in horse manure than in other manures.

Cow manure (C:N ratio 18:1) rots easily with fairly low nitrogen losses, and like horse manure is a more balanced additive. It also has lower levels of potassium, phosphorus, and sulfur than other manures.

Straw

C:N Ratio 47–72:1

Straw provides less nitrogen and phosphorus than hay. The nitrogen is removed when seed heads are harvested as grain. Straw usually contains fewer weed seeds than does hay. Straw also provides less potassium and sulfur than hay, but contributes similar levels of magnesium and micronutrients. Straw decomposes very slowly; it's a good choice to help open up heavy clay soils and to control excessive moisture in compost piles. Oat straw has the lowest C:N ratio at 50:1, rye straw is 65:1, and wheat straw is highest at 125:1.

Weeds

C:N Ratio 20:1

Easily available, weeds can be used to increase the amounts of certain essential major and micronutrients in your compost. The more green and succulent the weeds, the more rapidly they decompose. Do not use weeds that have begun to develop seeds. Do not use weeds that spread by rhizomes or stolons, such as quack grass.

Wood Chips

C:N Ratio 100–130:1

Nutrients are released very slowly; they may take more than a year to decompose if fresh. Composted pine bark has been shown to suppress root rots, fusarium wilts, and some nematode-caused diseases in nursery crops.

Saving the Day with "Magic Mix"

Not everyone is ready to grow their own fertilizer or manage cover crops and living mulches, especially those gardening in a small space. Even if you are working toward that goal, you may need to employ additional fertilizer options if your plants tell you they need more food. Table 3.2 compares organic residues and fertilizers for carbon-to-nitrogen ratios, the general amount of specific plant nutrients supplied, and ease of decomposition.

Another option is to use a liquid fertilizer in situations where low soil microbial, nutrient, and moisture levels as well as unfavorable weather, timing mistakes, and poor-quality transplants or seed all conspire against a smoothly working soil organic matter system. When that happens, crops will display nutrient deficiency and/or stress signs to let us know that a farmer/gardener intervention is necessary. When I face such situations at Woodleaf Farm, I rely on my well-tested "Magic Mix" quick-intervention liquid fertilizer to fix them. Below, I first describe common nutrient deficiency symptoms, and then provide the recipe for my Magic Mix.

Let your plants tell you when an intervention is necessary. Carefully observe how they are growing. How vigorous are the crop plants and weeds? If potatoes, cabbage, grasses, tomatoes, or beets are doing poorly, there might be a nitrogen deficiency. Poor growth of beans, tomatoes, or potatoes may signal low potassium levels. Struggling root crops often indicate a deficiency of both potassium and phosphorus. Potatoes, which require a lot of early nitrogen, will develop yellowish green vines if available nitrogen is insufficient, or if soil microbes are sluggish and not busily cycling nutrients in cold spring soils. Yellow and curling leaves (on all plants) can indicate both low and excessive levels of soil moisture. In both cases plants appear wilted and, when dug up, display stunted roots. Root crops are abnormally shaped if soil drainage is poor, or if soil is compacted and hard. Nutrient deficiency symptoms can be caused by (or exacerbated by) many other conditions, including herbicide injury, diseases and insects that injure root or vascular systems, waterlogged or droughty soils, cold soils, and mechanical or wind injury. For example, leaf purpling can be a misleading symptom. This purple color signifies above-normal levels of a red plant pigment (anthocyanin) that accumulates when metabolic processes are disrupted. A phosphorus deficiency, cool temperatures, or even normal plant maturation can all cause leaf purpling. So detective work is also needed to diagnose what may look like a nutrient deficiency.

Diagnosis of soil problems should not be made based on leaf symptoms alone. Also check the plants' roots. Roots that are well branched with a profusion of white root hairs signal healthy, fertile soil. Lateral-growing, long, stringy roots with few visible root hairs indicate a soil with unavailable nutrients. Dark-colored root tips or root hairs can mean insufficient oxygen. Healthy legumes have roots with *Rhizobium* bacteria symbiosis nodules, and the nodules will be pinkish inside. Sometimes a soil test is necessary to pinpoint a specific nutrient deficiency.

DECIDING WHEN A NUTRIENT INTERVENTION IS NECESSARY

When you're using low-input soil fertility system methods, it is important to know the early signs of nutrient deficiencies so that you can intervene quickly. As mentioned earlier, some crops, including potatoes, tomatoes, and cabbage, are heavy feeders that require greater amounts of nutrients for best growth, and others, such as peas and snap beans, are light feeders. Table 4.5 provides an overview of which crops need more or less of the big three plant nutrients and calcium. This table (along with crop nutrient deficiency symptoms) can help you figure out what might be needed when crops are growing badly and look weak. Soil test results also help to confirm the possible causes of poor plant growth. However, knowing the signs of nutrient deficiency can help you react rapidly and save the day!

Nitrogen Deficiency

A general yellowing of older leaves first and stunted vegetative growth indicate nitrogen deficiency. Older leaves eventually turn brown and dry (necrotic). Whole plants can be stunted. Excessive nitrogen can also be a problem too. Plants take up nitrogen preferentially and can easily get too much for balanced growth. Too much nitrogen results in decreased hardiness, late maturity, decreased disease and insect resistance, and poor (or sometimes, absent) bloom.

Phosphorus Deficiency

Phosphorus deficiency shows up as slow plant growth and late harvests. Phosphorus is important in plant reproductive growth and root development. It promotes early maturity and fruit quality. If phosphorus is deficient in plants,

Table 4.5. Relative Nutrient Needs of Specific Vegetable Crops

Crop Name	Nitrogen	Phosphorus	Potassium	Calcium
Asparagus	Low	Average	Low	Average
Beans, *dry*	Low	High	High	High
Beans, *snap*	Low	Average	Average	High
Beets	High	Low	High	Average
Broccoli	Average	Average	Average	High
Brussels sprouts	High	Average	High	High
Cabbage	High	High	High	High
Cantaloupe	Average	Average	Average	Average
Carrots	High	Average	High	Average
Cauliflower	Average	Average	Average	Average
Celery	High	High	High	High
Collards	Low	Low	Average	High
Corn, *sweet*	High	Average	High	Average
Cucumbers	Average	Average	Average	Average
Garlic	High	High	Average	Average
Kale	Average	Average	Low	High
Lettuce	Average	Average	Average	High
Mustard greens	Low	Low	Low	High
Okra	Low	Low	Average	Average
Onions	Average	Average	Average	Average
Parsley	Low	Average	Low	Average
Parsnips	Average	High	Average	Average
Peas	Low	Average	Average	Average
Peppers	High	High	Low	High
Potatoes, *white*	High	High	High	High
Potatoes, *sweet*	Average	Average	High	Average
Pumpkins	Average	Average	Average	Average
Rutabagas	Low	Average	Average	Average
Spinach	Average	Average	Average	High
Squash	Average	Average	Average	Average
Tomatoes	High	High	High	High
Turnips	High	Average	Average	Average

Figure 4.27. Early-planted Brandywine tomatoes exhibiting nitrogen deficiency signs in cool spring soils. I am applying a root drench of Helen's Magic Mix Liquid Fertilizer Intervention.

overall growth is stunted and the plant's vegetative period may be prolonged. A general symptom of deficiency in some garden plants is the purpling of leaf undersides. In most cases, phosphorus levels in soil are sufficient. The problem is that phosphorus is usually fixed, or bound, in the soil. In acid soils, phosphorus forms complexes with aluminum and iron and is held too tightly to be easily available to plants. In alkaline soils, calcium-phosphate complexes hold phosphorus. Even humus holds phosphorus. Fortunately, there are ways to ensure phosphorus's release from bondage. Keep the soil pH at around 6.5 to 7.0, and add organic residues. Not only is phosphorus available directly from residues, but organic matter addition helps to release fixed phosphorus by stimulating microbes that solubilize bound phosphorus. Think also about the quality of added residue. If a residue with a high C:P ratio (low phosphorus content) is added, phosphorus immobilization is encouraged.

Potassium Deficiency

A deficiency of potassium is rare in plants. When it does occur, it is indicated by brown scorching and curling of leaf tips as well as yellowing between leaf veins. Purple spots may also appear on the leaf undersides of some plants. Potassium is particularly important for fruit-producing crops. When fruit trees are deficient in potassium, they are more susceptible to injury by drought, frost, and soil salinity. Some plants require higher potassium than nitrogen levels. Soil microbes also need potassium, but at much lower levels than do plants. In fact, when organic residues are decomposed, potassium is often released immediately (while microbes hold on to nitrogen and phosphorus). When released into soil, potassium becomes trapped on clay and humus particles much as very fine dust ends up in the crevices of ornate furniture.

Sulfur Deficiency

Sulfur deficiency is usually indicated by interveinal yellowing and/or a pale yellow-green coloring of younger leaves. Plants are also usually small and spindly. Levels of sulfur are usually sufficient in alkaline or neutral soils if they have a high organic matter content; levels may be low in acid soils with low organic matter. And they're lower in humid climates, since sulfur is easily leached by rain. Sulfur is highly subject to leaching but can be held by clay minerals and by humus. Adding phosphorus to the soil often displaces sulfur on clay particles so that it is more available to plant roots. Sulfur availability is microbially controlled. Sulfur-loving bacteria oxidize sulfur, making it available to plants. Another way to assure sulfur availability is to add organic matter to keep soil microorganisms happy.

Cation Exchange Capacity

Cation exchange capacity (CEC) is a useful concept to understand when evaluating whether a plant might be lacking in nutrients. CEC is a measure of the total number of exchangeable sites on soil particles available for binding positively charged ions. Clays and high-SOM soils hold on to cations most effectively. CEC can be an indicator of the potential number of cations that the soil can hold. To gardeners a high cation exchange capacity means that more potassium, calcium, magnesium are potentially available for use by plants. Plants obtain the soil nutrients they need from ions (charged particles) dissolved in the water within soil pore space. The higher the CEC, the less chance of leaching nutrients not bound onto soil particles. The solubility of charged ions depends on their electrostatic attraction to water and to clay or humus particles. These particles form a kind of magnetic complex made up of plate-like particles consisting of clay, raw organic matter, and humus. These soil particles are negatively charged magnets. They attract positively charged nutrient particles or cations (potassium, magnesium, calcium, sodium, hydrogen, and aluminum). These cations are loosely held by soil particles but can move away again or be exchanged for other cations. CEC results are given in milliequivalents per 100 grams (meq/100 g) and should be in the 10–20 meq/100 g range (higher in heavier, darker soils and soils with high organic matter).

Calcium Deficiency

Tipburn of young leaves, death of the plant's growing points, blossom or bud drop, weakened stems, and water-soaked or discolored areas on fruit (for example, small, brown corky areas within apple flesh) are all signs of a calcium deficiency. Soil organisms don't need much calcium, but it is nevertheless important to the development of good soil structure. Calcium acts as a binding agent in the aggregation of soil particles. It also helps mitigate the effects of toxic materials on plants. Some evidence points to calcium as a disease preventive. Applying lime adds calcium to soils. Gypsum is the best choice to supply calcium to high-pH soils, because it does not increase pH. Many cover crops contribute calcium to the soil.

Magnesium Deficiency

Overall pale plant color is an indication of a lack of magnesium. The earliest symptom of magnesium deficiency is interveinal yellowing of older leaves. If plant leaves are pale due to magnesium deficiency, it is almost like magic to watch them green up after a magnesium treatment, such as Epsom salts. Magnesium must be balanced with the calcium, potassium, and nitrogen levels; it should make up 10 to 15 percent of the soil reservoir of cations. Often so much nitrogen and potassium are added to soil that magnesium is overwhelmed and plants may not get enough. Magnesium is often overwhelmed by potassium in soils with low organic matter content and low cation exchange capacity. It can be added in the forms of dolomitic lime or langbeinite (a naturally occurring mineral that contains 22 percent sulfur, 22 percent potash, and 11 percent magnesium). If pH is high, don't use dolomitic lime to add magnesium. If potassium levels are too high in respect to magnesium levels, don't use potassium-containing langbeinite to add magnesium. In such cases, an Epsom salt spray is a good choice for magnesium addition.

Micronutrient Deficiencies

Deficiencies in micronutrients are less common in vegetable crops, but are more common in fruit crops. They are termed *micro* because these nutrients are needed only in minute amounts by plants and soil microbes. The two most important soil factors determining micronutrient availability are organic matter content and pH. Organic matter acts as a buffering agent, seeing to it that plants do not get too much of any one micronutrient. This is important because excessive amounts of micronutrients can be toxic to plants. Iron, manganese, and zinc are often unavailable at high pH. In some western soils with high pH, iron deficiency is often seen in ornamental and garden plants. Soil tests are a good detective tool to reveal micronutrient deficiencies.

Figure 4.28. Peach leaves showing interveinal yellowing indicative of manganese deficiency during a heat wave in eastern Oregon, with 10 days over 100°F (38°C) during fruit ripening that exacerbated the deficiency.

Iron deficiency looks much like sulfur deficiency: interveinal yellowing of young leaves.

Manganese deficiency looks much like iron deficiency: yellow younger leaves with green veins and sometimes tan, sunken spots that appear in the yellow areas between the veins.

Zinc deficiency also results in yellowing of younger leaves, as well as bronzing of leaves.

Boron deficiency is usually seen at growing tips of the root or shoot. It generally includes stunting and distortion of the growing tip with tip browning and death, brittle foliage, and yellowing of lower leaf tips. Boron is the micronutrient most likely to be deficient because it leaches easily and it can be locked up in recently limed soils. Boron is usually less plentiful in subsoils, so deep-rooting plants, like apple, may show boron deficiencies, especially in dry weather when roots are digging deep to find water. Boron contributes heavily to the quality of several crops including brassicas, apple, pear, and grape. But be careful when adding boron fertilizer because boron is needed only in small amounts and too much can also cause problems in plants.

INTERVENTIONS TO ADJUST pH

Most crop plants need a soil pH between 6.0 and 7.0 for best growth, but check the specific preferences for the crops you are growing (see table 9.2 for more details). Some fertilizers, such as composted wood bark, may help to lower soil pH slightly. Soils can also be amended to raise pH if needed.

If pH is too high, adding small amounts of sulfur and mixing it well into the upper 6 to 8 inches (15–20 cm) of soil will bring down the pH over time. The amount of sulfur needed differs depending on soil type. In general, apply 1½ to 2 pounds of sulfur per 100 square feet (7–10 kg/100 m^2) in order to lower the pH one full point (use the higher rate for heavier soils). It may take up to a year to see the effects. A faster but more economically and environmentally expensive way to lower pH is to add peat to the soil. Some commercial blueberry growers with high pH soils add up to 1 gallon (4 L) of peat to each planting hole to assure the lower pH blueberries demand.

Adding lime to the soil increases pH, but if the pH is greater than 6.1, you usually do not add need to add lime. If it's less than 6.0, however, you should probably add lime. Soil type will affect how much lime you need. A sandy soil needs to be limed more frequently because it has lower capacity to hold on to both calcium and magnesium than does a soil higher in clay and organic matter. So if pH is less than 6.0 and organic matter is high, add 5 to 8 pounds per 100 square feet (24–39 kg/100 m^2). If pH is less than 5.0, use 10 pounds lime per 100 square feet (49 kg/100 m^2). Check the ratio of magnesium to calcium on your soil test in the cation saturation ratio section of the soil test report. It will tell you the percentages of magnesium, calcium, potassium, and sodium in relation to one another. Some people, like my late husband and I, believe that there are ideal ratios among these four cations (positively charged nutrient ions), including among magnesium and calcium (I explain Carl's cation and mineral balancing in more detail in chapter 12). Using this theory, if magnesium is greater than 20 percent of the total cation saturation, use calcitic lime. If magnesium is less than 10 percent, use dolomitic lime.

Adding excessive lime can result in nutrient deficiencies. Soils with a high cation exchange capacity can handle greater lime additions than soils with low CEC. Lime should always be thoroughly tilled into the soil. Fertilizers can also affect soil pH. Rock phosphate has a liming value of approximately ⅓ pound (151 g) of lime. Thus it can slightly increase pH. Other fertilizers can slightly decrease pH. Sulfur is an example of an acidifying fertilizer. One pound (0.5 kg) of sulfur will neutralize 3 pounds (1.4 kg) of lime. If a fertilizer has an acidifying effect, the label on the bag should state how much lime is required to neutralize the added acidity.

HELEN'S MAGIC MIX LIQUID FERTILIZER INTERVENTION

Once you have diagnosed a nutrient deficiency, don't wait. Intervene as rapidly as possible. Usually a liquid root drench of organic fertilizer will provide results the fastest. Foliar fertilizer application is generally not as quick. This is when I rely on organic processed fertilizers such as seed or leaf meals to make a rapid-release organic fertilizer "Magic Mix."

I usually apply my Magic Mix with a watering can. For larger plantings, I attach a hose to a sump pump placed in the bottom of a 30-gallon garbage can full of the mix and pump about 1 to 2 cups (250–500 ml) per plant onto crops. I usually apply the mix up to four times in the first 10 days, depending on the seriousness of the deficiency. I have also used a 300-gallon (1136 L) spray tank with a hose gun attachment to apply this mix. If you are applying to a whole bed, apply at 30 gallons per 500 square feet (120 L/50 m^2) of raised bed. However, this is not meant to be used as the primary fertilizer in your fertility system; I call it "magic" because it is an extra intervention used only when really needed.

Here is the recipe for the liquid root drench mix that I have been using to intervene for deficiency symptoms for more than 20 years. Note that I don't weigh ingredients for this recipe—I simply use a 1 cup scoop as an approximate measure.

4 cups alfalfa meal
1 cup molasses
2 cups Nutramin rock dust*
2 cups soybean meal
1 cup Maxicrop (dry kelp)
1 cup magnesium sulfate (Epsom salt)
½ cup manganese sulfate (31 percent manganese)
1 cup rock phosphate
1 cup micronized sulfur
1 tablespoon Solubor (boron)
Plant extracts: nettles, comfrey, or clover†

Mix all ingredients in 30 gallons (114 L) of water and let sit at 65 to 70°F (18–21°C) for 6 to 12 hours. Stir regularly or as often as possible; otherwise, the solid and powdered ingredients will settle to the bottom.

* Nutramin rock dust contains 1 percent calcium, 2.5 percent iron, 2 percent sulfur, and 30 percent silicon dioxide (commonly found in nature as quartz).

† I make the plant extracts myself by adding the plant material to water it and letting it sit for 24 to 48 hours at 65 to 70°F (18–21°C).

CHAPTER 5

Strengthening the "Immune System" of Your Farm or Garden

When gardeners and farmers tell me they want to stop using pest control sprays—even certified organic materials—I tell them that making the shift starts with soil building. That's because building habitat for soil microbes and building habitat for natural enemies are mutually reinforcing. In this chapter I delve deeper into the steps we can take to create and manage farms and gardens that suppress crop pests naturally. This pest suppression design encourages natural enemies, or biological control organisms, including birds, bats, frogs, snakes, and beneficial insects that dwell in the soil and those that live aboveground. Aboveground beneficials require pollen, nectar, sap, and seed, so providing sources of all four is an important design consideration. Other important natural enemies to encourage include soil- and leaf-dwelling microorganisms.

There are two kinds of beneficial insects: *predators* and *parasites*. Predators eat pests directly. Parasites lay their eggs inside or on pest insects. When parasite eggs hatch, the larvae eat the insect from the inside out, a kind of "invasion of the body snatchers." Most parasites are flies and species of tiny wasps. Pollinators such as honeybees and bumblebees are included in the umbrella term of *beneficial insects*, but in this chapter I focus mostly on predatory and parasitic insects (note, though, that many predators and parasites are also pollinators).

Figure 5.1 depicts the ecological pest suppression design framework that Carl and I worked out over our years of farming separately and together. This figure is an overview of how soil and habitat building are connected and influence one another. It also shows some of the main details involved in habitat building (which plants and how they are managed) and how these details in turn impact soil building. Some aspects of the strategies change both within the growing season and from year to year depending on climate, crop choices and diet needs, market opportunities, and the stage of development of an agroecosystem (that is, whether or not the system requires intervention on a regular basis or is in a stage of relative dynamic equilibrium where nature is in charge).

There are five components for building up the immune system pest suppression capability of a garden or farm. I discuss each one in detail and present examples from my own farms and from other creative farmers employing these pest suppression steps. Several of the 10 ecological management principles presented in chapter 1 also crop up in this chapter. This is not a coincidence; as I mentioned above, soil building is a part of designing for pest suppression. Here are the five components:

- Grow healthy plants by providing optimal light, temperature, and moisture conditions and by paying attention to plant competition and nutrient balance.
- Evaluate, conserve, and create habitat for natural enemies.
- Diversify crop species and genetics over time and space within the context of your farm or garden.
- Manage habitat once you have created it.
- Maintain low levels of insect pests within your farm and garden system so that natural enemies have a food source even when pests are absent from crops.

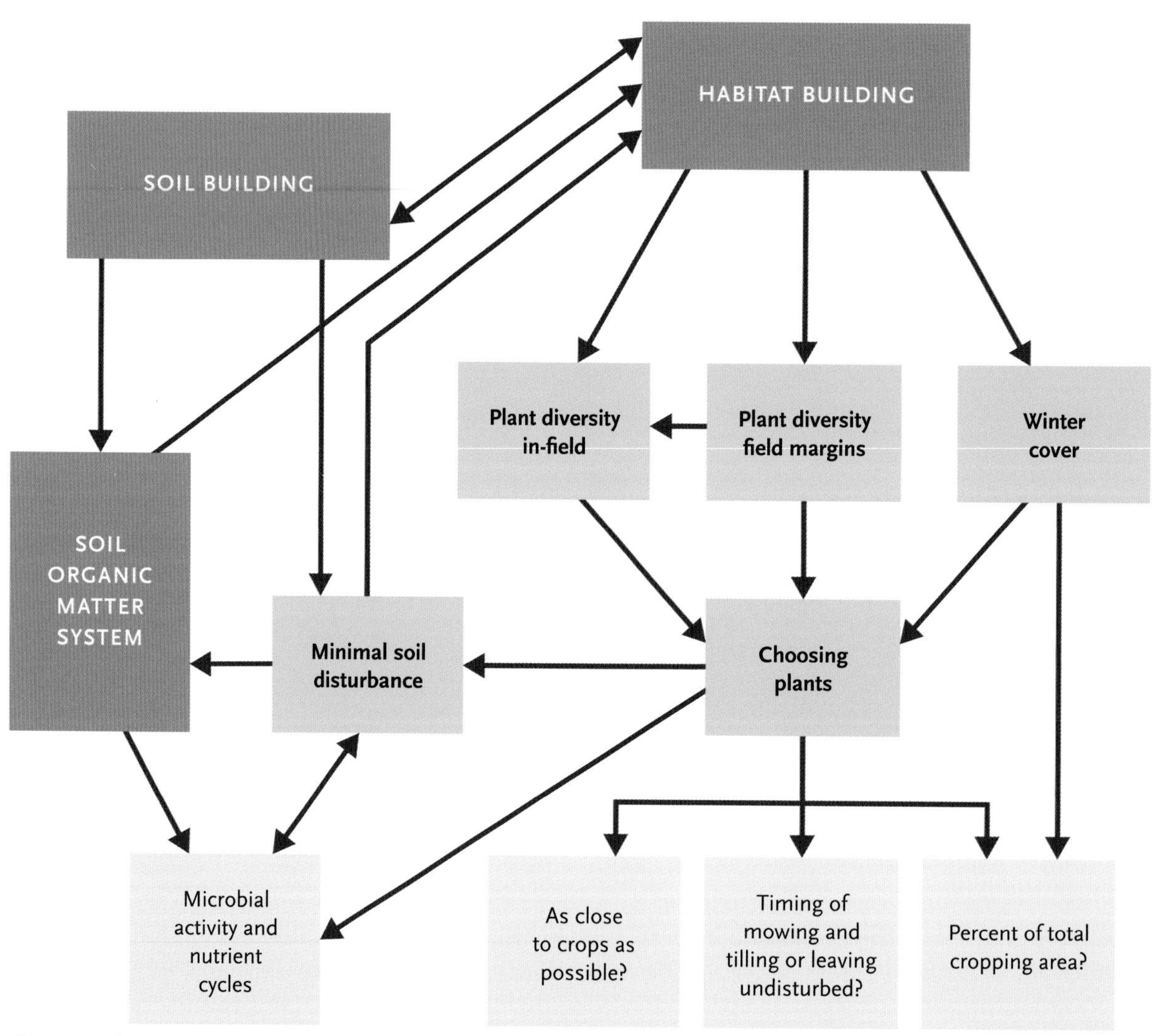

Figure 5.1. This schematic shows the ecological pest suppression design that Carl and I devised. We used it when farming jointly and I continue to refine it today. Question marks indicate decisions that need to be considered and adjusted both within the growing season and from year to year. Most interactions are reciprocal, especially those connected by double-headed arrows.

The next section describes how to apply some or all of the five components for building farm pest suppression in your garden or farm. I include details about the most important pest suppressive tools and strategies because—just as in soil building—the details matter.

Growing Healthy Plants

When you take on the challenge of farming and gardening with nature and with minimal purchased inputs, it becomes vital to pay attention to your plants' basic needs. Shifting the balance to your crops, so that they have an advantage over weeds, insects, and diseases, requires a focus on agroecosystem details.

PROVIDING OPTIMAL CONDITIONS

Start with the basics of light, temperature, and water when you are shifting the ecological balance toward your crops. Optimizing these basic factors is key to raising crops successfully without a lot of added fertilizer and pest management inputs.

Light

A grower's instinct is to focus on watering at planting time, but providing the right light conditions matters just as much. Most vegetable crops need some shade when they are establishing new roots after transplanting or as they germinate. Let your plants focus on growing roots rather than leaf and shoot growth. This is an important time to think about cropping diversity and canopy structure. Shade from adjacent taller crops, trees and shrubs in a home landscape, living mulches in crop row middles, and/or row covers will help root establishment. One of my tricks for growing tomatoes in Montana was to cover transplants in the field with hooped row cover for 4 to 5 weeks after planting to provide light shade. (Shade cloth provides heavier shade than row cover.) On the other hand, too much shade during later development of sun-loving crops can make those crops weak. When I grew annual vegetable crops in row middles in one of our California orchard fields, the shade-loving broccoli and lettuce did very well. Unfortunately, sun-loving eggplants, cucumbers, and tomatoes were weak and more susceptible to powdery mildew, slugs, and aphids (even in our dry climate). See chapters 9 and 10 for specific light requirements of particular crops, and as often as possible, work to provide the light and shade your crops prefer.

Temperature

The wrong growing temperature for a specific crop negatively affects its ability to establish and grow well. Use row covers for heat and shade cloth for cooling. For example, cold injured brassica transplants are more susceptible to cabbage worms and cool-season-loving kale and broccoli are more susceptible to cabbage aphids when excessive summer heat slows their growth. Covering young plants after transplanting into the field or garden mediates heat and cold extremes and can help prevent problems such as early flea beetle attacks or damage by root maggots on crops such as turnips. Excessive heat causes fruit crops to produce smaller fruit more susceptible to disease. I let my living mulches in the orchard grow tall to help cool my orchard when temperatures are above 100°F (38°C). Again, check each crop's preferences and provide temperature as close to optimum as you can.

Water

Humidity, rainfall, soil moisture, and irrigation—or lack thereof—can make crops more or less susceptible to pests depending on crop preferences (and the pests' preferences). For example, root knot disease of brassicas is encouraged by heavy rainfall and wet soil. Drought stress makes tree crops more susceptible to boring insects. In several studies, spider mites and rust mites feeding on drought-stressed crop plants grew faster than on non-stressed plants; mite damage was significantly higher on stressed plants.

Design Strategies and Tools for Pest Suppression

This is a list of the strategies we have used and continue to use on Biodesign and Woodleaf Farms. I have also worked with other ecological farmers who use many of these strategies and all can be used at a garden level too.

1. ***Soil building for high soil organic matter, an active soil microbial community, and an optimal balance of specific plant nutrients***
    - Reduced nitrogen-containing amendments, especially quick-release nitrogen fertilizers.

2. ***Increasing on-farm plant diversity***
    - Mixing crop families and species in the same field.
    - Allowing some crops to mature and flower.
    - Summer and winter cover crops.
    - Insectary plantings in-field and on field margins.
    - Small fields in a "patch" farm design with wild habitat on field margins.
    - Enhancing groups of generalist biological control organisms rather than introducing specialist insects.
    - Rotation of crops, cover crops, and ground covers.
    - Winter cover using hedgerows, living mulches, and winter cover crops.
    - Reduced tillage: diversification of above- and belowground plant foliage, bloom, roots.
    - Landscape complexity: maintaining wild and/or native plant habitat on field edges.
    - Woody plant hedgerows in-field and on field margins.
    - Grassy beetle banks: undisturbed perennial sod in-field and on field edges.
    - Allowing certain weeds for winter cover, early-spring bloom, and summer ground cover.
    - Interspersed diversification as close to crops as possible.
    - Annual living mulches in crop rows.
    - Perennial living mulches in crop rows.

3. ***Reduced or no spraying of certified organic insecticides to avoid killing beneficial insects***
    - Choosing insect-resistant varieties.
    - Field-scouting of pests; farm-specific pest thresholds that trigger interventions.
    - Restricting use of even minimal-impact certified organic pesticides; use only when absolutely necessary to avoid crop loss.

MANAGING PLANT COMPETITION

In an agroecosystem, the root zone is a crowded area in which crop roots interact and compete with root systems of neighboring plants and with soil microorganisms for space, water, and nutrients. Some crops handle sharing root space and nutrients well and compete well with neighboring plants for light. But others can be set back, especially when they are young and establishing roots. Poor competitors need to be considered in farm design and habitat building, and that relates to my default ecological principle regarding plant competition:

Weed selectively. Choose a ground cover mix; then identify and "chop and drop" the plants or "weeds" that are competitive with specific crops in specific growing situations and climates.

Perhaps not every gardener will want to manage a ground cover mix, especially in a small garden space. However, ground cover is part of maintaining a diversity of year-round growing roots in my low-input system with farm-grown fertilizers that multi-task as habitat for natural enemies. Weed suppression can also be a part of that system, but it requires diligent management of weed-crop interactions. I cover weed-crop interaction in detail in chapter 7.

OPTIMAL NUTRIENT BALANCE

My idea of a healthy plant has changed over the past 40 years. I once thought that the biggest, greenest, most vigorous, nitrogen-filled plants would give me the best quality and highest yield. (Thus my dedication in times past to using manure-based compost.) A healthy plant is one in which the main nutrients (such as nitrogen, phosphorus, and potassium) and the micronutrients are all balanced in relation to one another. No one nutrient should be out of a healthy range. For example, I still strive to grow vigorous plants, but now I understand that high levels of nitrogen in plant tissues don't necessarily make plants healthier. In fact, high nitrogen levels have been shown to make some crops more susceptible to insect pests such as aphids, thrips, and mites. Lower concentrations of soluble nitrogen (specifically nitrate-nitrogen) in plant tissue seem to make crops less susceptible to feeding damage by most types of insects. In several studies, applying fertilizers that rapidly release nitrate-nitrogen led to increases in sap-feeding insects. And in one review that spanned 50 years of insect and nitrogen research, 135 studies showed more plant damage and/or greater numbers of leaf-chewing insects and mites in nitrogen-fertilized crops. Only 50 studies reported less pest damage with nitrogen fertilization.[1] Just as important, fast-release nitrogen fertilizers often cause plants to reduce investments in their "immune system"—their natural chemical defenses designed to fight off insect attacks and disease. High nitrogen in the soil around plant roots can diminish the ecological efficacy of important interactions among plant roots and microbes that help to suppress insects and disease. And as I've already discussed, it is becoming clear that higher-carbon, lower-nitrogen fertilizers increase soil microbial biomass and diversity as well as root-microbe interactions. These beneficial belowground soil microorganisms help plants better tolerate stress and streamline nutrient uptake, leading to optimal plant nutrient balance. All in all, healthy plants are those that are connected to and interacting with a healthy, microbially active soil agroecosystem.

Here's the ecological principle that turns out to be a recipe for providing optimal plant nutrient balance:

Focus on carbon fertilizers. Prioritize slow-release, plant-based carbon fertilizers rather than fast-release nitrogen fertilizers.

Generally what I mean by "carbon fertilizers" are plant residues that have a higher carbon-to-nitrogen ratio (C:N ratio). Chapters 2 and 3 discuss the C:N ratios of organic fertilizers and how to manage them in detail. As I explained in chapter 2, in my on-farm research at Biodesign Farm, when I cut back on and eventually stopped adding high-nitrogen compost, soil nitrogen levels decreased. The numbers of aphids on pepper transplants also decreased (see figure 5.2). Yields went down a very small amount, but I was able to stop spraying insecticidal soap for aphids in the spring, a trade-off that I interpreted as progress. There were fewer aphids and a very high rate of parasitism by aphid parasites (70 to 80 percent). That was an indicator that the farm ecosystem was moving toward a more optimal nutrient balance. At Woodleaf Oregon now, the only "fertilizer" I apply is a living mulch or hay mulch of mixed legumes, grasses, and weeds; it provides balanced nutrients, but with less nitrogen and a higher C:N ratio. So far this nutrient balance has helped me to maintain pest suppression without using any pest sprays on fruit or vegetable crops for the past 7 years. However, during a long heat wave in 2022, some vegetable pests reached economically damaging levels on stressed crops (cabbage worms on heat- and cold-stressed kale, flea beetles on arugula, and root maggots on drought-stressed turnips). Beneficial insects eventually suppressed the cabbage worms on the kale crops, but we had to cover all successive arugula crops with row cover to manage flea beetles. This is evidence of the dynamic nature of farm and garden ecological systems discussed in chapter 1; it shows the need for farmers and gardeners to continue to monitor carefully and adjust as ecological relationships change over time.

It's not necessarily the case that the lower-carbon living mulch I use as fertilizer *always* results in a better nutrient balance. As all ecologists are fond of saying, "It depends." It depends on your soil and the state of its microbial health, the nutrient requirements of the crop you are growing, climate conditions, and how early you want to get your crop to market or to your homestead table. Some infertile, degraded soils with meager microbial activity may need a heavy shot of a nutrient-rich organic amendments. Some soils may respond best to a turbojet intervention of plowing down a rapid-release, succulent legume cover, or even an application of manure-based compost. But generally, the way to provide optimal nutrient balance is to cycle organic residue fertilizers *and carbon* through the soil microbial community. I draw the analogy to the fact that humans need fiber as well as protein and vitamins in their diet. That brings us right back to this ecological principle:

> **Maintain growing roots year-round.** Grow and maintain plants with living roots to keep the soil covered year-round and to feed the rhizosphere a steady, balanced diet of carbon and nutrients.

If we think only about crop nutrients, especially the big three of nitrogen, phosphorus, and

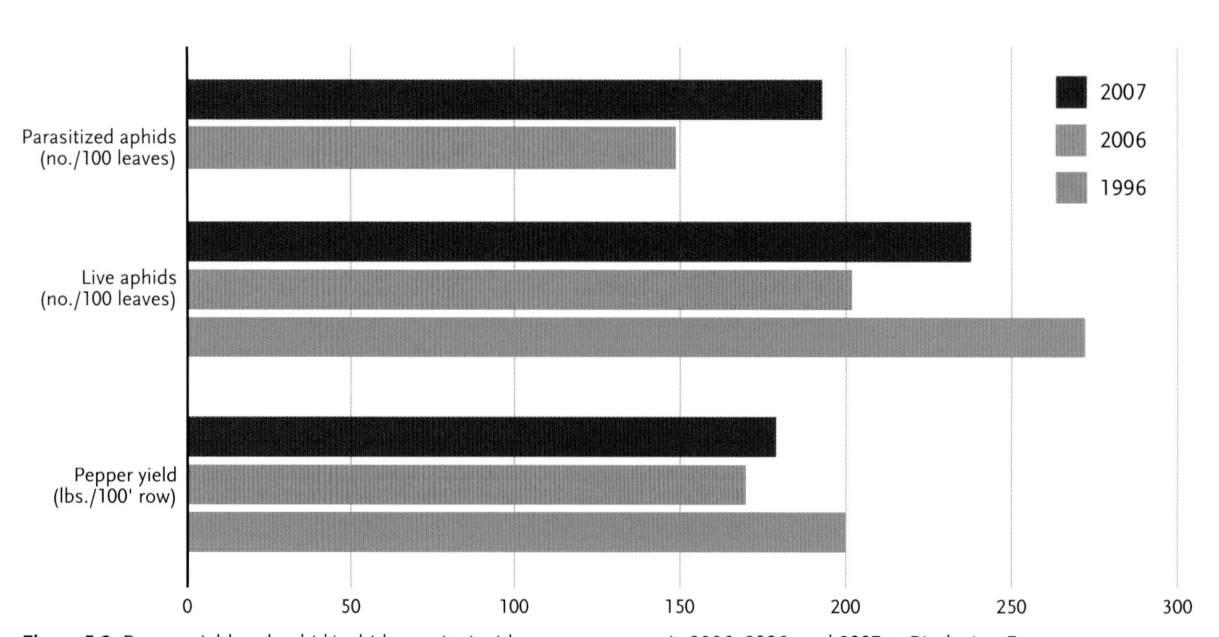

Figure 5.2. Pepper yield and aphid/aphid parasite incidence on peppers in 1996, 2006, and 2007 at Biodesign Farm.

potassium, we might forget the underground microbial labor force we are supposed to be nurturing as we build habitat for natural pest suppression.

Our goal is three-part:

- To manage the carbon-to-nitrogen ratio of organic residues added to soils.
- To manage the timing of organic residue application.
- To match fertilizer carbon quality with crop growth stage.

Crops demand different nutrient levels at different growth stages. Generally, lower C:N ratio fertilizers (higher nitrogen) work better in the early growth stages (germination and first rapid leaf growth) and higher C:N ratios are best for the later stages (fruiting or crop maturity). We can control fertilizer carbon quality through the choice of maturity stage of plant residues, as discussed in the "Providing Carbon Quality Diversity" section on page 40.

Conserve and Create Habitat

First, take a close look at your farm or garden and find the habitats for predators and parasites that already exist. Then conserve that habitat by avoiding disturbance. Disturbed soil has *significantly* fewer predators and parasites (especially spiders and ground beetles) and more pests. Several research studies report a decrease in ground-dwelling predators in fields with regular tillage and weed cultivation and an increase where minimal-tillage and no-tillage are practiced. Maintaining undisturbed habitat near the crop is key. For example, in a Michigan State University study more flea beetles were found in plots that had bare soil row middles than in plots with undisturbed rye living mulch row middles.

Next, diversify your system. But don't diversify just for diversity's sake. There should be sound ecological function reasoning behind your choices. For example, choose the best plants to encourage specific beneficial insects, communities of insect predators and parasites, and soil microorganisms. As a general guideline, habitat creation involves increasing both above- and belowground plant diversity. The specific goal, though, is to provide undisturbed, year-round plant cover (winter, spring, summer, and fall) and sequential, season-long sources of nectar, pollen, sap, and seed.

> **Create above- and belowground diversity.** Feed the soil microbial community with as much crop and ground cover, root and whole plant, above- and below-ground diversity as is economically possible.

Where we create habitat and add diversity in relation to our crops also matters. We are striving to follow this ecological principle:

> **Create habitat in the field.** Preserve wild habitat or create and maintain undisturbed habitat for natural enemies interspersed with or as close to crops as possible.

The goal is to create conditions in which natural enemies don't have to "commute"—they can live right where we want them to work. With this in mind, let's explore the difference between *interspersed* and *blocked* habitat.

INTERSPERSED HABITAT

When we create interspersed habitat, shelter and food sources for beneficial organisms are distributed within and in close proximity to crops, rather than in blocks or rows on garden or crop field edges. Interspersed habitat can be distributed randomly, such as transplanting or seeding flowering alyssum among lettuce plants within crop rows, as described in the "Phil Foster Ranch" sidebar on page 100. Or it can be a regular distribution such as row middles planted to cover crops, living mulches, or flowers.

Figure 5.3. This living mulch beneath and among peach trees at Woodleaf Farm, California, is an example of providing interspersed habitat for beneficial insects as close to the trees as possible. We mowed only the small areas around micro-sprinklers, and only in the spring, leaving 50 percent of the total orchard undisturbed throughout the season.

Figure 5.4. Sunflower, calendulas, and marigold are part of this blooming insectary block at an Oregon organic farm. Within a ½-acre (0.2 ha) planting, there were 12 small blocks of plants, which also included agastache, alyssum, orache, and fennel. GPS mapping indicated that insectary plants made up less than 1 percent of the total cropped area.

Habitat in close proximity to crops means that predators and parasites do not have to move far from plant cover and food sources into crops.

BLOCKED HABITAT

Blocked habitat is a grouping of flowering plants provided in blocks on garden or field margins or at the end of crop rows (see figure 5.4.). The disadvantage of a blocked planting is that it sometimes draws beneficial insects *out* of crop fields and into the resource-rich habitat planting. I once consulted for a large vegetable farm that wondered where all the beneficial insects could be. They were concerned because they had invested in planting insectary blocks around the farm and were expecting some serious biological control. *Insectary blocks* are blocks of plants grown to attract, feed, and shelter insect parasites and predators so as to enhance biological pest control. When I investigated the farm's fields, I found the majority of the predators and parasites gorging themselves on nectar and pollen in a ½-acre (0.2 ha) block of phacelia planted on the borders of tilled, cultivated crop fields consisting of expanses of bare soil and brassica and spinach crops (none of which were flowering). The natural enemies were not leaving their phacelia smorgasbord to forage in what to them was a food desert. Hence, cabbage worms were feeding on the brassica crops with impunity while the cabbage worm predators and parasites hung out at home in the phacelia block.

It is important to provide interspersed undisturbed and flowering habitat as close to crops as possible. Some field study evidence indicates that predators and parasites stray no more than 328 feet (100 m)—and possibly as little as 66 feet (20 m)—away from undisturbed and/or flowering habitat into crops.[2] Other research studies show similar results to what I observed at the vegetable farm with the phacelia block planting. In one study, the number of ground beetles and adult lacewings was significantly higher in crop fields that had flower strips, but the beneficial insects were present only near the flower strips.[3] In a research review, interspersed diversification increased spiders in 80 percent of the studies, while blocked diversification increased spiders in only 33 percent of studies.[4]

The moral of the story is that in theory, all diversification is good, but in practice, some methods are better than others!

Diversity Details

Diversification that leads to effective pest suppression should be based on ecological function and provide food and shelter for soil-dwelling and aboveground predators, parasites, birds, bats, and soil- and leaf-dwelling microorganisms. This kind of diversification includes:

- Flowering plants with high-volume, extended production of nutrient-rich pollen, nectar, and seed (some predators can survive on seeds when insect prey is not abundant).
- Flowers that exhibit the particular shapes and colors that best support specific predators and parasites.
- Season-long sequential bloom of pollen- and nectar-rich annual and perennial flower species.

The way you choose to diversify species and genetics over time and space in your farm or garden system will depend on your climate, crop mix, harvest plan, crop marketing plan, and scale of production. As I've described, my favorite interspersed habitat creation tool is planting perennial living mulch row middles between my annual vegetable crop rows and leaving them to flower. My living mulches consist of legumes for nitrogen fixation and a diversity of weeds and grasses for season-long flowering and production of nectar, pollen, and seeds. This works great in the arid areas of the West. But in the humid, hot, southeastern United States, living mulches can also provide habitat for slugs and disease, so they require lots of mowing to prevent them from becoming pest habitat rather than beneficials habitat. Unfortunately mowing also decreases predators such as spiders (as discussed later in this chapter).

Here is a list of options for creating diversity, beginning with methods that are easiest to implement at any scale. Moving down the list, the options become more challenging to manage at a larger market crop production scale.

Increasing landscape complexity; maintaining wild and/or native plant habitat at garden and field edges.

Arranging small crop fields or areas in a "patch" design, with habitat on field margins.

Planting mixed crop families, species, and varieties in a field or row to create a blend of varying plant genetics, plant heights, flower shapes and colors, rooting types, and root architecture.

Rotating crops and ground covers from different plant families and different ecological functional groups (producing fruits versus roots, seeds, or leaves as the crop) and having various root types

Figure 5.5. Small fields with native forest margins at Woodleaf Farm, California, create "patch design," landscape complexity, and de facto tree and shrub hedgerows at field margins. Forest patches surrounded by patches of orchard create hot spots of biodiversity within farms.

Figure 5.6. Fall crops left to flower at this organic farm in western Oregon provide early-spring pollen and nectar. Low-growing weeds and a winter cover crop provide habitat as well as soil building.

Figure 5.7. Flowering clovers in the living mulch row middles between vegetable crops provide habitat for pollinators and predators such as this crab spider at Woodleaf Farm.

(taprooted, fibrous-rooted, annual or perennial) at different times in the same space or field.

Allowing some crops to mature and flower; perhaps allowing crops to remain in place all winter.

Interseeding summer and winter cover crops into growing crops or sowing them after crops are harvested.

Tolerating certain less competitive weeds within crops because they serve as winter cover, early-spring nectar sources, and summer ground cover. Natural systems that are in ecological balance are often messy!

Establishing insectary plantings of flowering plants in fields or at field and garden margins.

Establishing tree and shrub hedgerows in-field and on garden margins.

Establishing grassy beetle banks, which are undisturbed perennial grasses in blocks in fields or at field and garden margins that provide habitat for ground beetles and spiders. Adding blooming wildflowers to beetle banks can also provide pollen and nectar for predators and parasites.

Maintaining winter cover of living mulches and winter cover crops and/or hedgerows, insectary plants, and beetle banks within or on field or garden margins.

Reducing tillage to promote diversification of undisturbed foliage, bloom, and roots.

Interspersing diverse non-crop plants as close to crops as possible, such as overseeding crops with clovers, ryegrass, or cereal grains.

Planting annuals in row middles: flowering ground covers, flower strips, or living mulches.

Establishing perennials in row middles: flowering ground covers, flower strips, or living mulches.

The more we understand about beneficial insect and natural enemy preferences, ecology, and habitat needs, the better plant diversity choices and designs we make. I have consulted with homesteaders who designed their gardens and orchards based on diversification rules using plant groupings that are not based on ecological function and do not take into consideration the needs of soil microbial communities and insect predators and parasites. Some of these homesteads ended up having to add large inputs of compost, fertilizer amendments, and pest control materials due to suboptimal natural nutrient cycling and a lack of season-long natural enemies. The owner of a 30-year-old permaculture orchard once told me that were they to start again, they would change their diversification design. For ease of pest and crop management, they would plant fruit trees of the same species or genus blocked together (rather than singly to enhance tree species diversity). Then they would create more ground cover diversity to improve nutrient cycling (with more legumes) and a mix of plants with season-long bloom. They also said that they would never again plant comfrey beneath orchard trees. They'd done so because comfrey was recommended as part of a classic permaculture guild design. However, comfrey had become a competitive monoculture beneath their trees and was not providing the season-long bloom benefits that a diversity of flowering plants in the orchard ground cover offers (such as the plant list in table 12.3).

Some plants are better choices than others for providing habitat for a more diverse and abundant community of beneficial insects. High-volume pollen and nectar producers provide the best habitat. In many studies, wildflower strips sown within crops supported greater insect abundance and diversity than cropped-only habitats, but *pollen- and nectar-rich wildflower mixtures* outperformed the general wildflower mixes.[5] Also, there is a difference between the effects of floral and extrafloral nectar on insects. Production of *extrafloral nectar* on the leaves and/or stems of some plant species can significantly prolong the duration of nectar availability for natural enemies. The primary function of these sugar-secreting extrafloral nectaries appears to be attraction of protective beneficial insects. Cowpea, passionflower, many sunflower types,

Phil Foster Ranch

Phil Foster farms on 300 acres (121 ha) near Hollister, California. His pest suppression design includes establishment of native plant woody hedgerows and grassy beetle banks on field margins and interspersed flowering habitat within crops. Phil Foster Ranch's main pest suppression strategies and tools are:

- Reduced tillage equipment, such as a spader.
- Summer and winter cover crops.
- Increased in-field plant diversity.
- Native shrub and tree hedgerows.
- Native habitat on field margins.
- Insectary plant rows: alyssum, dill, and cilantro.

Foster plants strip rows of cilantro, white dill or false Queen Anne's lace (*Ammi majus*), and yellow dill within crops rows (planted alongside crop rows at 1 to 3 percent of total acreage). When seeding lettuce transplants, he also seeds an alyssum plant every 20 plants so that when transplanting he will have interspersed diversity and blooming plants for predators and parasites. Alyssum and cilantro are well-known predator/parasite preferred plants. See the parasite habitat description information in the "Keeping Beneficial Insects and Other Natural Enemies Happy" section beginning on page 117 for other good beneficial insect-specific flowering plant choices.

Phil tries to plant a cover crop in every field once every 2 to 4 years, but drought and water shortages in California have made this a challenge. When water is short, it goes to

Figure 5.8. Newly established winter wheat cover crop.

Figure 5.10. Flowering alyssum in crop field.

Figure 5.9. Phil Foster Ranch native shrub and tree hedgerow.

crops instead of cover crops. Still, winter-, spring-, and summer-planted cover crops are an important part of the farm's habitat- and soil-building system.

Phil also has one of the oldest mature native plant hedgerows in California (see figure 5.9). It provides winter cover for birds and ground-dwelling predators and season-long, sequential bloom for predators and parasites.

Phil is part of an on-farm research study to reduce tillage, including by using rolled and crimped cover crop mulch residue. Rolling and crimping is a way to mechanically terminate cover crops. Crimpers and rollers flatten a high-biomass cover crop to produce a thick, uniform mat of mulch. They are used to kill grass cover crops (cereal rye, barley, wheat, sorghum, Sudan, pearl millet), hairy and common vetch, annual clovers, and buckwheat. Phil is also trying to minimize soil disturbance to create habitat for ground-dwelling predators. In his experiments, he has discovered that cover crop mulch residue C:N ratios are key to success or failure of his reduced-tillage habitat- and soil-building system. Experiments such as these will provide key insights on how to make the pest suppression techniques I describe in this chapter economically sustainable to midsized commercial organic farms.

partridge pea, fava beans, hairy vetch, and elderberry are examples of plants that have large extrafloral nectaries on their leaves and/or stems.

Certain flower shapes (flat umbels or tubular blossoms) and colors (yellow and/or white) attract a more diverse and abundant community of natural enemies. For example, beetle predators, some spiders, and syrphid flies show preferences for yellow and white flowers. Bees also prefer white and yellow blooms, but are attracted by purple and blue flowers, too. Flat umbels are characteristic of the Umbelliferae (carrot) family. These inflorescences are made up of hundreds of tiny flowers that small insects can easily feed from (and the adult forms of many predatory and parasitic insects are very small, as described in chapter 6). Carrot family species including Queen Anne's lace, wild parsnip, fennel, dill, cilantro (coriander), and caraway are especially attractive to small beneficial insects. Nectar accessibility is a key ecological function that makes a plant more attractive and an effective food source for specific parasites. For example, in a study at Rutgers University two wasp parasitoids of Colorado potato beetle were examined in a field experiment to see how shape and nectar availability affect parasitoid attraction and retention. The wasps remained longer on those flowers whose shape allowed the most accessible nectar. In several studies, wasps also preferred plants with short outer petals, such as buckwheat, legume, and carrot family flowers. Flower shape matters because it makes certain plants more or less attractive to the beneficial insects we are trying to attract and provide habitat for.

Timing of pollen and nectar production is also important, because season-long bloom is vital for a successful system. Choose early-blooming plants such as yellow and white flowering spring bulbs or

Figure 5.11. Blooming common buckwheat, a summer cover crop, provides excellent habitat for beneficial insects, particularly parasitic wasps. Buckwheat has flowers with easily accessible nectar.

flat, umbel-flowered sulphur buckwheat (*Eriogonum umbellatum*) as well as very late bloomers such as umbel-flowered sedums or yellow-flowered goldenrod and white-flowered asters.

In my on-farm research, parsnip, sunflowers, yarrow, buckwheat, alyssum, cilantro, dill, and certain weeds were the best plants to attract insects and provide habitat. (See "Keeping Beneficial Insects and Other Natural Enemies Happy" on page 117 for other native and non-native plants preferred by certain predatory and parasitic insects.) Many species of parasitic wasps and flies are most effective at providing biological pest control when feeding on buckwheat-species flowers, specifically common buckwheat (*Fagopyrum esculentum*) with its prolific white summer flowers.[6] Buckwheat has easily accessible nectar due to its flower shape and internal morphology; it produces a high volume of nectar that small parasitoids don't have to work hard to get. Other plant species shown to increase parasitoid life span and parasitism rates include milkweeds (*Asclepias* spp.), cornflower, common vetch, and licorice mint (*Agastache foeniculum*). A study in Washington State reported the greatest number of total beneficial insects on native wild buckwheats (*Eriogonum* spp.) and stinging nettles (*Urtica dioica*).[7] In a Maryland study, predators and parasitic wasps predominated on native plant species (such as goldenrod and milkweed) compared with weedy, non-native species, including redroot pigweed and lambsquarters.[8]

Parsnip plants that have gone to flower (see figure 5.12) are always the top predator- and parasite-attracting summer-blooming plant at Woodleaf Farm, Oregon, during my field monitoring. Cilantro and sunflowers are two other popular pollen and nectar plants at Woodleaf.

My on-farm research has also shown that some species of weeds also provide excellent habitat for pollinators, predators, and parasites. Some of the most-preferred pollen and nectar sources for syrphid flies and parasitic wasps in my studies

Figure 5.12. Flowering parsnip is a favorite source of nectar and pollen for beneficials, including a distinctive, transparent-winged lacewing and a tiny parasitic wasp (top); two bee-like Tachinid fly adults (center) (some types parasitize cutworms, codling moth, and other fruit tree pests); and several parasitic wasps (bottom), including Ichneumonid, Braconid, and Chalcid species, who parasitize the eggs and larval stages of moths, flies, grasshoppers, and beetles.

Figure 5.13. A predatory syrphid fly forages for nectar and pollen on early-spring-blooming speedwell.

Figure 5.14. Early-spring-blooming shepherd's purse provides nectar and pollen for a syrphid fly and a flea beetle.

include early-spring-blooming chickweed (*Stellaria media*), purple deadnettle (*Lamium purpureum*), henbit (*Lamium amplexicaule*), and speedwell (*Veronica* spp.). A single syrphid fly can feed upon as many as 400 aphids over the course of its lifetime.

Henbit and purple deadnettle with their tubular flowers are also a good food source for pollinators in early spring, particularly long-tongued bees such as honeybees and bumblebees, and a bee mimic called the giant bee fly. Tachinid bee flies parasitize a wide range of pests, including ground-dwelling pests such as cutworms. It is a large group with more than 1,000 species in North America. They can be recognized by the thick bristles on their abdomens.

PLANTS FOR BENEFICIALS AT BIODESIGN FARM

For 17 years in Montana I monitored which plants, in combination, were providing sequential, season-long sources of nectar, pollen, sap, and seed for beneficial organisms. My clover and weed-containing living mulch bloom sequence extended from early April, with flowering weeds such as fanweed (*Thlaspi arvense*) and shepherd's purse (*Capsella bursa-pastoris*), through late September with flowering red and white clover, weeds, and grasses. I also allowed fanweed, an annual with a relatively short life cycle, to grow in crop rows because it provides the earliest spring bloom, has easily accessible white flowers and lots of nectar for parasitic wasps, and is short in stature, with non-competitive, fibrous roots. It was especially easy to fit in fanweed among brassica crops. However, here is a caveat for developing farm and garden ecosystems that have not yet built up a large and diverse natural enemy community: Fanweed is a member of the Brassicaceae family, and thus it supports cabbage aphids (especially during flowering and seeding). Fanweed was compatible with the pest suppression system I had developed at Biodesign, where some insect pests were allowed because they fed, maintained, and built cabbage aphid predator and parasite populations. But in natural-enemy-sparse systems, allowing fanweed to grow alongside brassica crops might not work as well.

I allowed some types of more competitive weeds in the living mulch row middles only. At Biodesign, potato- and eggplant-eating flea beetles and Colorado potato beetles preferred black nightshade (*Solanum nigrum*), and so it acted as a de facto trap crop, a place where these beetles congregated first, allowing their predators to build up before there was ever a problem on crops.

PLANTS FOR BENEFICIALS AT WOODLEAF CALIFORNIA

At Woodleaf Farm in California, the living mulch provided a bloom sequence extending from early February through December. The sequence began with winter- and spring-blooming annuals such as purple deadnettle, chickweed, and Persian speedwell (*Veronica persica*) and continued through late summer with blooming perennials such as yellow

Figure 5.16. A wild row middle consisting of yellow dock, grasses, and weeds in late summer at Woodleaf Farm, California.

Figure 5.15. Chickweed and speedwell blooming in early spring at Woodleaf Farm, California.

dock (*Rumex crispus*) and yellow nutsedge (*Cyperus esculentus*), ending in late December with grasses and sedges. Woodleaf's habitat-building strategies supported a diverse group of live-in, year-round, mostly generalist predators that suppressed pest damage and allowed the farm to reduce and eventually stop spraying pest control materials altogether. Field monitoring from 2013 to 2015 revealed a diverse and abundant population of season-long ground-dwelling predators and flower-requiring predator and parasite species. More than 15 spider species were found in high numbers year-round, both in the living mulch and even in the fruit trees, including an abundance of wolf spiders who prey on codling moth and Oriental fruit moth larvae, the worms we hate to find in apples and peaches. Several species of syrphid flies were regulars all season. They seemed to prefer nectar of chickweed and Persian speedwell. Other common Woodleaf generalist predators included ground beetles (particularly carabid and rove beetles), lady beetles, soldier beetles, lacewings, assassin bugs, banded thrips, and nabid bugs. Parasites include wasps in the Aphidiidae, Braconidae, and Aphelinidae families.

Managing Habitat

At times habitat plantings begin to encroach too heavily on crops and become competitors, especially in small garden areas. If that happens, you may need to take steps to cut them back. Plantings also may need to be managed at times for ease of farm and garden operations. Once you have created cover and sequential bloom for your natural enemies, however, take care not to till, cultivate, mow, or graze it to the point that it no longer functions.

I realized just how vital it is to maintain undisturbed, blooming areas as close to crops as possible when I observed what happened over time after Carl and I sold the California Woodleaf Farm. The new owners were committed to maintaining our ecological system, but they made minor changes to facilitate farm operations. They mowed more often and kept less of the total orchard area in undisturbed (unmowed) living mulch ground cover. They also introduced some grazing chickens and goats onto the farm (though they try to keep the animals out of the orchard). By year 4 of the new management, Oriental fruit moth worms became a problem in peaches. Codling moth worm problems increased in apples and pears by year 5, and high populations of pear sawfly larvae began to feed heavily on new spring pear leaves in year 6. Also in the spring of year 6, the new owners were horrified by a large decrease in early-spring pollinators and had to bring in honeybees for pollination. I consulted with them about returning to mowing less and more selectively. In the spring of year 7, there was more early-spring bloom in unmowed areas of living mulch, and the owners did not have to bring in bees for pollination.

I've mentioned the practice of selective mowing in previous chapters, and here's a formal definition: *Selective mowing* is management of a perennial ground cover, cover crop, living mulch, or flower strip to avoid disturbance of natural enemies, especially at key pest pressure times. Some portion of the interspersed habitat within crops needs to remain completely undisturbed at all times in order for pest suppression to be effective. My goal is to keep at least 30 percent of the total habitat interspersed among my crops unmowed and undisturbed at any given time during the season. Is that percentage of undisturbed habitat area *necessary* for effective pest suppression? I don't know, but after working with several other farms that maintained a lower percentage of their total system in undisturbed habitat and still experienced pest problems, my gut feeling is that 30 percent is a sufficient level to strive for.

In the early 1990s I mowed the living mulch at Biodesign Farm regularly to facilitate farm work, reduce competition with crops, and keep weeds from flowering and going to seed. I had not yet

Figure 5.17. This spider is hunting aphids and thrips in a peach tree. Spiders were common in the orchard trees at Woodleaf California, and spiders continue to be an important component of the "in-tree pest suppression team" in the Woodleaf Oregon orchard.

Figure 5.18. Blooming weeds in the living mulch in August, such as this common plantain (*Plantago major*), provide pollen and nectar for beneficial insects like this syrphid fly.

Selective Mowing at Woodleaf Farm

Woodleaf Farm's mowing practices evolved over time as we learned to prioritize among all the ecological relationships we had to manage. In the 1990s Carl mowed the living mulch regularly for frost control and ease of farm work. When we began to monitor beneficial insects more carefully, we realized how mowing negatively affected their populations. So we began to allow a percentage of ground cover to grow wild, flower, and produce seed all season before mowing. We particularly avoided spring mowing from peach bloom to fruit set so as to avoid disturbing the predators and parasites of western flower thrips and Oriental fruit moth. Thrips populations were highest and most damaging to peach and nectarine crops at very first fruit set. During on-farm research in 2013 and 2014, higher pest thrips populations were observed in peaches immediately after mowing the living mulch. Our goal was to retain at least 30 percent of the living mulch ground cover in the orchard undisturbed and blooming throughout the season.

learned enough weed ecology details to successfully invite weeds into the garden! (Chapter 8 covers weed ecology details.) After two years of monitoring both living mulch competition with crops and beneficial insects in the living mulch, I began to let the ground cover grow taller and wilder. Beginning in 1998, I allowed the living mulch in some row middles to flower and produce seed before mowing. I particularly avoided spring mowing in order to provide wind/cold protection for young crop transplants. I also wanted to avoid disturbing predators and parasites of green peach aphid, because aphid populations were highest and most damaging to my pepper transplants in the spring. I began to manage the ground cover so that some areas (at least 50 percent) were undisturbed and blooming throughout the season. By 2000, there was no need to spray my pepper plants to manage aphids—not even insecticidal soap! Biological control and pest suppression interactions had evolved in my agroecosystem to the point that I no longer had a pest management job. As mentioned previously, I had also greatly decreased high-nitrogen compost additions by 2000, and that may also have contributed to lower aphid populations. It takes a system!

Other research supports Biodesign and Woodleaf Farms' discovery that ground cover habitat disturbance decreases beneficial insect populations and increases insect pests. USDA-ARS researcher Dave Horton found that less frequent mowing in orchards attracted more beneficial insects who prey on pear psylla, leafminers, pear sawfly, and codling moth. Horton conducted trials at three orchards and varied mowing frequency (weekly, monthly, and just once a season). With less frequent mowing, natural enemies were recorded in the ground cover in much greater numbers. They were attracted to the pollen and nectar from undisturbed flowering plants as well as to abundant prey, such as aphids and thrips in the unmowed ground cover. More lacewing larvae, spiders, lady beetles, damsel bugs, parasites, and minute pirate bugs were observed with less mowing. Some predators, especially spiders, also appeared in greater numbers in orchard trees in the less frequently mowed plots.[9]

Carl and I did not graze animals in our orchards, and I have continued to farm without grazing animals in my system. This is because grazing

Figure 5.19. The untilled living mulch in the row middles of the Woodleaf Oregon orchard includes perennial grass, clover, and weeds that provide habitat and maintain soil fertility for good crop yields.

Figure 5.20. Lots of blooming and seeding ground cover in the ungrazed and selectively mowed living mulch at Woodleaf Oregon.

decreases both cover and blooming habitat for beneficial insects. Grazing also does not allow *selective* management of interspersed habitat. Animals graze the whole area in which they are fenced. I monitored beneficial insects in an organic peach orchard near Woodleaf in California that rotationally grazed chickens under their trees (see figure 5.21). This orchard still required the spraying of certified organic pest management materials to control insect pests because pest suppression by biological control organisms was not occurring effectively. I found significantly fewer pollen- and nectar-requiring predators and virtually no

Figure 5.21. In this organic peach orchard, chickens forage with rotational grazing management. They are fenced into small areas and moved frequently, but not often enough to prevent loss of blooming ground cover necessary to shelter and feed beneficial insects.

Tolhurst Organic Farm

Iain Tolhurst farms land in Oxfordshire, England, that has been farmed for over 100 years. His pest suppression farm design includes establishment of woody hedgerows and perennial grass beetle banks on field margins. Tolhurst Organic Farm's pest suppression provides winter cover and a percentage of undisturbed habitat containing grass, legume, and wildflower cover crops somewhere on the farm each season that also provides sequential bloom of pollen and nectar-rich flowers for beneficial insects.

Tolhurst Farm has a unique rotation with 2 years in grass, legume, and wildflower cover crops in every field every 5 to 6 years. Each year, Iain sows new areas to cover crops, so that these undisturbed plots are always scattered among crop fields. The cover-cropped plots provide blooming habitat for generalist predators and parasites, birds, and soil microbes (particularly fungi that diminish with soil tillage). Iain mows the cover crop once in the first year and two or three times the second year to encourage nutrient cycling without destroying bloom or disturbing ground-dwelling predators. When tilled the third spring, the cover crop provides a balanced carbon-nitrogen fertilizer.

Iain also plants and then coppices fast-growing softwood tree species (such as willow and poplar) in farm hedgerows. He uses small, young branches to make ramial wood chips for composting. The ramial wood chip compost is spread lightly (approximately 1 inch [2.5 cm] deep) on top of mowed green cover crops in the fall of the second year of the crop's growth to better balance C:N ratios. The wood chip compost is allowed to decompose over the fall and winter and is incorporated with the cover crop in the spring. This technique works well in a climate where the ground does not freeze solid all winter and decomposition occurs into late fall and begins again early in the spring.

ground-dwelling predators (such as ground beetles and spiders) in areas where the chickens had recently grazed and were presently grazing. Chickens eat both ground beetles and spiders.

When I mentioned my findings to a friend who is a farmer and organic crop consultant, he argued that although chickens eat predatory ground beetles and parasitic insects, they also eat pests and should thus overall be ecologically beneficial. I insisted on my point of view, so he set up an on-farm experiment with two treatments as a test: rosemary production with no chicken grazing allowed compared with chicken-grazing-allowed rosemary. He found that leafhopper damage was much higher and salable rosemary yields were much lower in the chicken-grazed rosemary production area. I hope more research will be done on the effects of mowing and grazing on beneficial insect density and diversity versus pest insect populations and crop damage.

Maintaining Low Levels of Insect Pests

It's tempting to think that wiping out insect pests entirely will be the best for our crop plants. However, it turns out that it is vital to *maintain* a low level of pest insects within a balanced farm or garden ecosystem. Why? Because natural enemies need to eat, and if no pests are present, their enemies will leave the area or perish. Fortunately, undisturbed ground cover and flowering plants not only provide habitat for natural enemies, they also provide habitat for their prey—pest insects. Natural enemies need these prey as a food source when there are no pests on the crops themselves. The presence of pest insects in the habitat you intersperse through your fields or gardens allows predators and parasites to remain within the farm or garden system. Most pest outbreaks occur because there is a gap between when pests find crops and when predators and parasites show up to attack those pests. But when pests are also happily sheltering in your living mulch or insectary plantings, natural enemies stay in the system, and they can move into your crops to begin biological control as soon as any pests emerge in crops. We want to *suppress* pests, but we don't want to kill *all* the pest insects! Trying for zero insect pests ultimately causes garden and farm agroecosystem function to break down.

It has long been known that pesticides used in conventional farming, such as synthetic pyrethroids (permethrin and bifenthrin) and organophosphates (malathion and acephate), also kill natural enemies. Reducing insecticide sprays has been shown to provide more abundant and diverse communities of predator insects and birds on farms, which in turn improves biological control of farm or garden crop pest insects. More recently we are finding that even certified organic pest management materials kill too many natural enemies. I discovered this for myself during an on-farm research trial monitoring cabbage worms at Biodesign Farm. I tested three pest management approaches in two 600 feet (183 m) field rows of Brussels sprouts and the clover/weed living mulch row middles between them:

- My no-spray systems approach, providing interspersed habitat for predators/parasites and relying only on biological control organisms for suppression.
- Bimonthly rotenone-pyrethrin botanical sprays, designed to kill both cabbage worms and predators/parasites.
- *Bacillus thuringiensis* var. *kurstaki* (*Btk*), which kills only cabbage worms, sprayed whenever cabbage worms reached a threshold level, based on integrated pest management models for imported cabbage worms.

The reason I included the rotenone-pyrethrin insecticidal spray is that I wanted to test the effectiveness of predators and parasites in controlling cabbage worms. In order to have a point of comparison, I had to get rid of the "good guys" in some plots.

Some predators and most parasites are specific about what they eat, feeding only on a single prey species or type of pest. These are considered *specialists*. Other beneficial insects feed on many kinds of insects and mites. They are called *generalists*. Generalist predators feed on several alternative prey species depending on which are currently most abundant. Often specialists are parasites, and many biological control programs have traditionally revolved around specialists of particular pest insects. For example in Oregon a specialist parasitic wasp (*Ganaspis brasiliensis*) is being released to target the pest insect spotted wing drosophila. More recently, the importance of generalist predators in biological control on farms is being investigated.

During weekly sweeps over a period of 11 weeks as part of this on-farm research, I found high levels and diversity of generalist predator insects in both crop rows and living mulch row middles where nothing at all was sprayed. This no-spray treatment was designed as a measure of the natural pest control capacity of Biodesign's interspersed habitat pest suppression system. It turns out that the capacity was high! There was an abundance of predators, with different species more or less prevalent at different times of the year from spring through fall. We also discovered an ecosystem of diverse generalist predators living in the living mulch, including lady beetle adults and larvae, nabid bugs, minute pirate bugs, spiders, harvestmen, parasitic wasps, syrphid fly larvae and adults, lacewing adults and larvae, and assassin bugs (see figure 5.22 for the seasonal variation in the density of these beneficials in the field). But, we discovered no parasitism of cabbage worms. Research by others also supports this discovery. In one review of

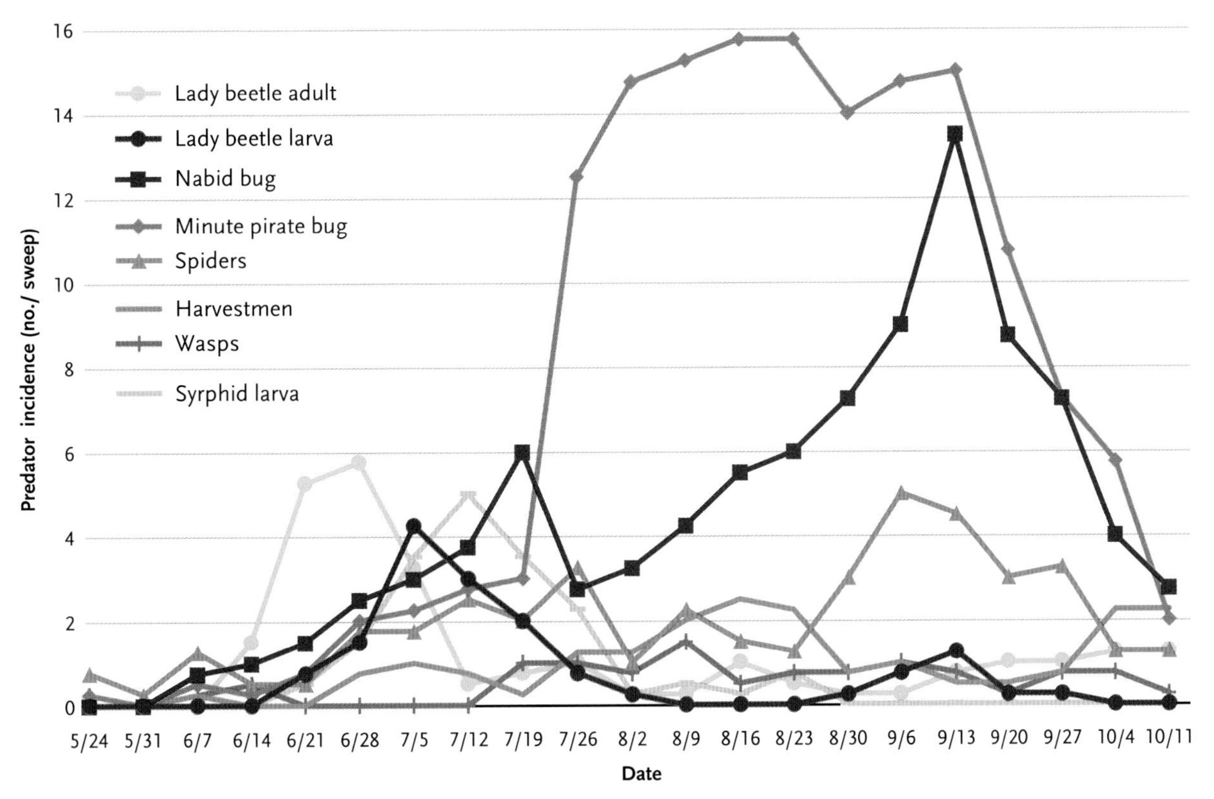

Figure 5.22. Results of a Biodesign Farm on-farm research study in 2006 to investigate the abundance and diversity of predators and parasites in living mulch in the row middles of no-spray plots of Brussels sprouts.

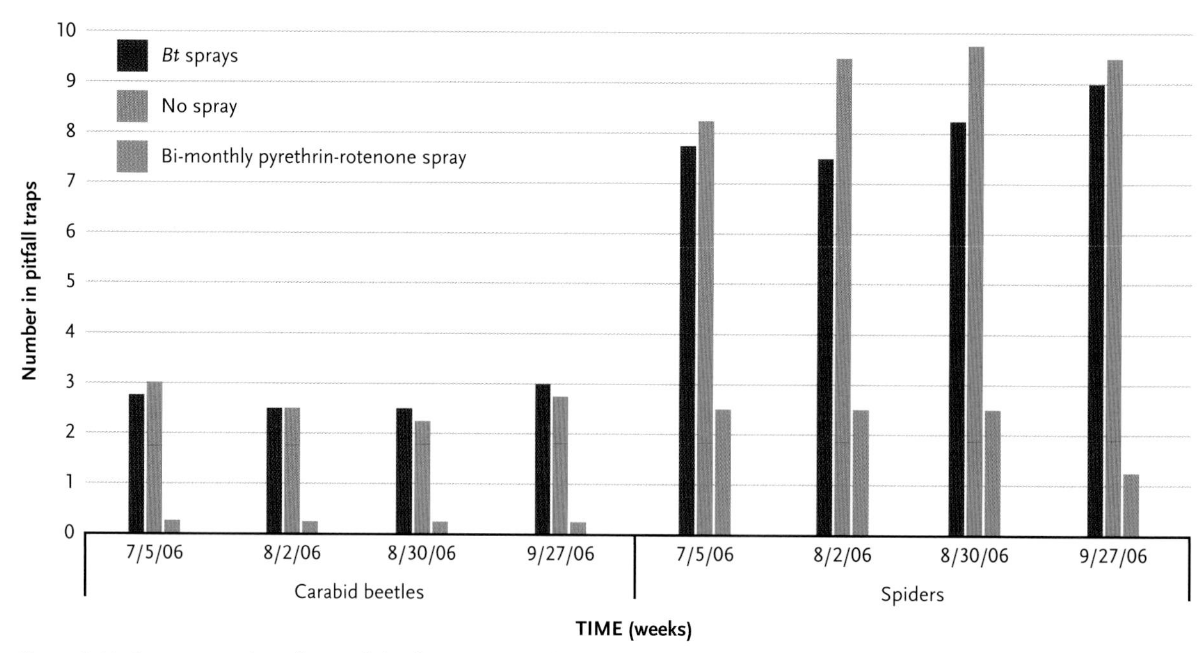

Figure 5.23. Average number of ground-dwelling predators found in pitfall traps from unsprayed plots and plots sprayed with *Bacillus thuringiensis* var. *kurstaki* and insecticidal soap or with a pyrethrin-rotenone spray.

many field experiments, in approximately 75 percent of the studies, generalist predators reduced pest numbers significantly.[10]

Predator numbers were much lower in plots where rotenone-pyrethrin botanical sprays were applied bimonthly. Syrphid flies, lady beetle larvae, and spiders almost disappeared in these plots. Natural enemy populations were reduced after each rotenone-pyrethrin spray (see figure 5.23). Even more interesting, the yield and pest damage data showed that in plots where we killed the diverse group of generalist predators, yields were lower than in plots where we relied only on our biological control system. Yield of perfect, undamaged Brussels sprouts was 88 percent in unsprayed plots and only 80 percent in rotenone-pyrethrin-sprayed plots. The best yield was in the plots sprayed with BTK. This treatment resulted in a 98 percent damage-free crop, but we sprayed it *eight* times to achieve that yield. I decided that I was willing to lose some yield in order to maintain my agroecosystem approach and save the labor and material cost of spraying my brassica crops eight times. Further, I was able to sell 40 percent of the "damaged" Brussels sprouts from this experiment at a farmers market for a reduced price. I think these results are convincing enough to persuade many gardeners and homesteaders to make the decision not to spray. For some commercial farms, however, there may be reluctance to lose any potential crop yield to pests. However, it's important to keep in mind that input costs sometimes outweigh increased yield.

Clearly, not all agroecosystem farms and gardens or pest suppression systems will look alike, or even similar! Sometimes our gardens and farms require pest-specific, minimal-impact interventions to slowly bring them back into balance with all the ecological relationships happening within and around them as the system "calibrates" itself. In chapter 6, the focus broadens to include managing insect pests as well as suppressing them using a systems approach in which we strive to manage ecological relationships rather than simply managing specific insect pests.

CHAPTER 6

Working with Pest and Beneficial Insects

Insect–plant interactions in the farm or garden ecosystem are like an intricate spiderweb. If we cut vital interconnected threads, the pest suppression systems we have designed can collapse. Some common interventions to manage insect pests (such as spraying insecticides, even those that are accepted for use on organic farms) have the effect of cutting those vital threads. As a systems approach farmer, my general rule for working with insect pests is to first understand as many of the details as possible, because if you kill something in the system, *you inherit its job*. In other words, let's say I spray a certified organic insecticide, such as spinosad, to control codling moth. Unfortunately, spinosad also kills some of the beneficial insects who help to naturally suppress codling moth on apples and pears, such as syrphid fly adults and larvae, lacewing larvae, and predator beetles. I have thus killed some of my natural suppression system, so if I want to avoid the problem of wormy fruit, I have to monitor for codling moth larvae in my orchard and continue to spray pesticides to kill those larvae, possibly several times over the course of the summer.

Let's start with a reminder of the end goal in a systems approach: Whenever possible we want to avoid pest problems by *suppressing* pests rather than *killing* them. And in this context, suppressing means inhibiting the growth or development of pests—so it's a preventive approach, not a reactive one. In most biologically diverse native plant communities, it is the role of natural enemies—such as insect predators and parasites, microorganisms, snakes, frogs, birds, and bats—to regulate plant pest populations. That does not mean there are zero pests present in a native plant community. Think about it: There has to be a basic supply of prey available at all times for the natural enemies to eat, or those enemies would die out or leave in search of food elsewhere. On the other hand, the pest populations never become large enough to destroy all the plants, or even a large amount of them, because they are always being held in check (suppressed) by their natural enemies. Likewise, in order to support healthy, diverse, and abundant predator and parasite populations ready to respond quickly when you need them, there have to be pest insects for them to eat. Therefore, any farm/garden ecosystem relying on pest suppression cannot have zero pest damage. Insect pest suppression has four working components. We need to:

- Learn about the biology and ecology of the pest and beneficial insects present in the farm or garden landscape.

- Build and manage habitat to shift the ecological balance toward beneficial insects and natural enemies and away from pest insects.
- Tolerate low levels of pests in order to support diverse, healthy, and vigorous beneficial insect and natural enemy populations.
- Use minimal-impact methods allowed for use by certified organic operations first, and only escalate to certified organic materials that have greater ecological impact when a pest's population is so high it negatively affects your farm or garden economics, market system, and ecological considerations.

In this chapter I cover in detail how to manage the balance between beneficial and pest insects in a garden or farm so that as little intervention as possible is required. I also explain the difference between minimal-impact interventions and higher-impact interventions. Knowing ecological impacts can mitigate potential detrimental ones when we decide to intervene and "manage" a pest.

A Quick Guide to Identifying Insects

It's impossible to make wise ecosystem management choices unless you can identify both the pest and beneficial insects that are present on your crops. So let's begin with a quick guide to understanding the basics of insect identification and some tips on identifying common pests and beneficials. For more ID information and photos, the internet is a great resource. Simply type in the names of specific pests and beneficials on the search engine of your choice.

TELLING THE GOOD GUYS FROM THE BAD GUYS

There are three easy ways for farmers and gardeners to identify insects at a basic level that will help you decide who is a beneficial good guy and who is a plant-eating bad guy. In some cases, this level of ID will be enough; or if not, it's a good first step that you can follow by looking for more information online or in an ID handbook.

By Mouthpart Type

Gardeners can learn to identify insects by the type of feeding injury they inflict on plants. And such damage is usually something we can easily spot on our plants.

Insects with *chewing mouthparts* make *large round or jagged-edged holes* in leaves. All beetles and caterpillars (larvae of moths and butterflies) have chewing mouthparts (see figure 6.1). So when you find chewed leaves in your garden, you know you'll generally want to look for a beetle or caterpillar as the culprit.

Feeding by insects with *piercing and sucking mouthparts* results in *tiny stipples* that make leaves look as if they have been poked with a pin (see figure 6.2). Thrips are one type of insect with piercing/sucking mouthparts. See table 6.1 for more hints as to which insect may be causing specific plant injury.

Many beneficial insects have chewing mouthparts to feed on their insect prey, but some pierce and suck out the contents of other insects.

By Life Stage

If you can find the actual pest on your plants, you can narrow down the ID a bit further. Most pest and beneficial insects develop through three or four stages in their life cycle: *egg*, *larva*, *pupa*, and *adult*. Their appearance changes, sometimes dramatically, from stage to stage. The egg and pupa are nonfeeding stages. You may never see the eggs of some pest insect species, but others are highly characteristic, such as the bright yellow eggs of Colorado potato beetle on potato leaf undersides.

Larvae and adults are the feeding stages, but what an adult insect eats is likely to be entirely different from what the larval stage feeds upon. In fact, adults of some species do not eat at all. On the other hand, the adult phase of some types of insects may cause as

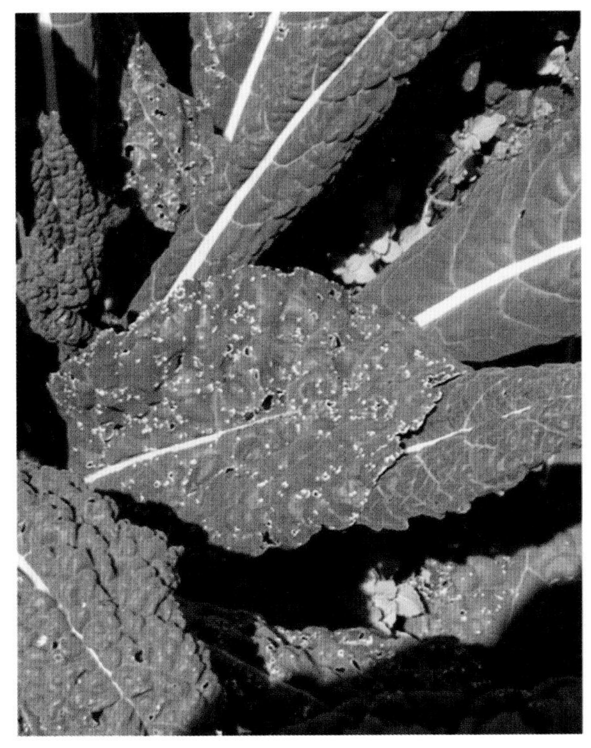

much or more damage than their larvae; this is especially the case for both beetle and true bug adults.

If you can identify what type of larvae are feeding on your plants, it helps to pin down the ID of the insect you're up against. For example, caterpillars (such as cabbage worms) are the larvae of moths or butterflies. Maggots (think of onion or cabbage root maggots) are the larvae of flies. Grubs are the larvae of beetles. See "Identifying Larval Stages of Insects" on page 116 for a discussion of the differences between types of larvae.

By Where You Find Them

Insects don't usually eat every crop in the garden (despite what it sometimes feels like!). Generally insect pests feed on specific crops or crop families. For example, cabbage aphids feed only on plants in the cabbage family (Brassicaceae); they will not attack tomatoes or lettuce. So another clue in identifying

Figure 6.1. Damage due to insects with chewing mouthparts: small holes in kale leaves due to flea beetles (top); large holes in kale caused by cabbage worms (bottom).

Figure 6.2. The whitish pinprick holes, also called stipples, in these onion leaves are caused by onion thrips, which have piercing/sucking mouthparts.

insects is to observe what species of plants they are eating. See chapters 10 and 13 for information on insect pests that specific vegetable and fruit crops host or attract. Beneficial insects are often found on crops being eaten by pest insects, or where pest insects are building up. For example, if you find groups of aphids on shoots and leaves, you will usually find lady beetles and their larvae there, too.

IDENTIFYING LARVAL STAGES OF INSECTS

Most insect identification books or websites group local or regional insect pests into these common insect orders:

Coleoptera: Beetles
Diptera: Flies
Hymenoptera: Bees, wasps, and sawflies
Lepidoptera: Moths and butterflies
Hemiptera: True bugs

You'll probably recall from high school biology that an *order* is a classification term used by scientists—orders are made up of groups of *families*, and families are made up of groups of *genera*. If you can identify an insect as belonging to one of the four orders, you can then easily look it up in a regional ID guide or on a website to pin down its identity more closely.

Being able to identify the life cycle stages of common pests is important because the larval stage is usually the best time for intervention. In most pest species, the larval stage usually is the most destructive, but it is also the most vulnerable to minimum impact intervention. If we can identify insects at the larval stage, then we can use minimal-impact options most effectively. With some minimal-impact materials, it is vital to know which order of insect pest you are dealing with, too. For example, *Bacillus thuringiensis* (*Bt*), a bacterium that causes disease in many types of pest insects, is available in various formulations that work only on specific insect orders. There is a formulation of *Bt* for beetle larvae and another for moth and butterfly larvae, but the *Bt* formulation for beetle larvae is completely ineffective on moth and butterfly larvae and vice versa.

Beetles (Coleoptera)

Coleoptera larvae, or grubs, have six legs and are usually soft-bodied (see figure 6.3). Their head is distinct from the body and their legs are somewhat enlarged compared with those of caterpillars. Some types feed on leaves, some by boring into stems and branches, and others feed inside fruits and roots. For example, *wireworms* are narrow, hard-bodied beetle larvae that are found in soil and feed mostly on roots. Adult beetles come in many sizes and shapes; *weevils* are one type known for their elongated snouts.

Beneficial beetle larvae include lady beetle larvae (see figure 6.3), which are strong generalist predators that eat insect eggs, beetle larvae (root worms and weevils), aphids, and other soft-bodied insects, such as leafhoppers, scales, mites, and mealybugs.

Figure 6.3. It's important to be able to identify generalist predators like this lady beetle larva. Note the three enlarged legs on either side of the body and distinct head. Photo by Katja Schulz.

Flies (Diptera)

Diptera larvae are mostly legless. They are called *maggots* and are small, greenish to pale yellow colored, and wormlike; they do not have a distinct head. They feed inside fruits, seeds, roots, or between leaf layers. The larvae of predatory flies feed on a wide range of insects including aphids, small caterpillars, psyllids, scales, and mites.

Bees, Wasps, and Sawflies (Hymenoptera)

Hymenopteran larvae have six legs on or near their heads and prolegs on the lower segments of their bodies. Some types, such as sawflies, are leafminers who feed between leaf surfaces. The larvae of beneficial parasitic wasps hatch from eggs laid in pests like aphids or caterpillars; as they grow, these larvae consume the pest and kill it.

Figure 6.4. This Pacific chorus frog is sitting patiently on an almost ripe sweet red pepper at Woodleaf Farm, waiting for a tasty beetle, moth, or caterpillar to happen by. Frogs also eat aphids and many other insects.

Moths and Butterflies (Lepidoptera)

Lepidoptera larvae have six legs, including smaller, partial, stumplike legs on the middle and last segments of their bodies. *Borers* are larvae that tunnel into branches, stems, and roots. *Caterpillars* feed on flowers, fruit, and leaves. As noted above, large holes in leaves and/or entire chewed leaves are characteristic of moth and butterfly larval feeding. Some types are *leafminers*, and feed between leaf layers.

For a quick guide to the signs of damage by the larvae and adults of specific insect pests, see table 6.1.

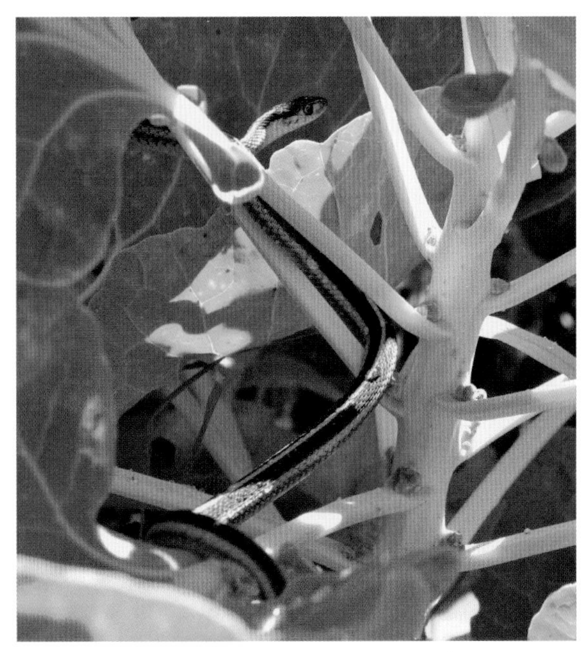

Figure 6.5. A surprising sight among the Brussels sprouts at Woodleaf Farm: a garter snake hunting for insects! Snakes eat beetles, caterpillars, and aphids, as well as slugs and mice.

Keeping Beneficial Insects and Other Natural Enemies Happy

Building and managing habitat for beneficial insects means providing food and undisturbed, sheltered areas for them to safely mate, reproduce, and overwinter (as described in chapter 5). Nectar, pollen, sap, and seeds are necessary food sources that fuel predator and parasite survival, movement, and reproduction.[1] These are the threads farmers and gardeners need to maintain in their farm and garden systems to encourage biological suppression of pest insects.

Beneficial insects can be classified as generalists or specialists. Generalists are usually the first to respond when there is a pest outbreak. If you provide the proper habitat, they will be living on-site before pests flare up. With specialists, on the other hand, there is usually a gap period between when the pest flares up and when the specialist begins to respond. But they can be very effective once they build up and become numerous. Both the generalists and specialists are important to farm and garden insect pest suppression.

Here I cover some of the more common predators and parasites of pest insects and describe the habitat, habitat management, and plant species they most prefer. The insects are grouped by order, and you'll notice that more orders are listed than for pest insects above—beneficials are very diverse! The listings also include family name. Different genera and species of these beneficial insects are present in different parts of the United States and the world. However, no matter where you live, there are likely to be native species of beneficial beetles, wasps, and other predators and parasites.

Besides beneficial insects, I also provide ideas for improving habitat for birds and bats. And keep in mind that snakes and frogs are part of the natural forces that suppress pests on ecological farms. Snakes and frogs eat a variety of insect pests, slugs, and rodents. Rubber boa snakes, which are found in most of the western states and Canadian provinces, specialize in eating mice and voles, going down the rodents' tunnels in search of a meal. I often come across snakes hunting for prey beneath the hay mulch on my farm.

BEETLES (COLEOPTERA)

There are several types of predatory beetle, including beetles that spend most of their time on the ground and species that I find hanging out in the very tops of fruit trees.

Ground Beetles

Most ground beetles (carabid and rove beetles for example) are predators, but some are parasites of

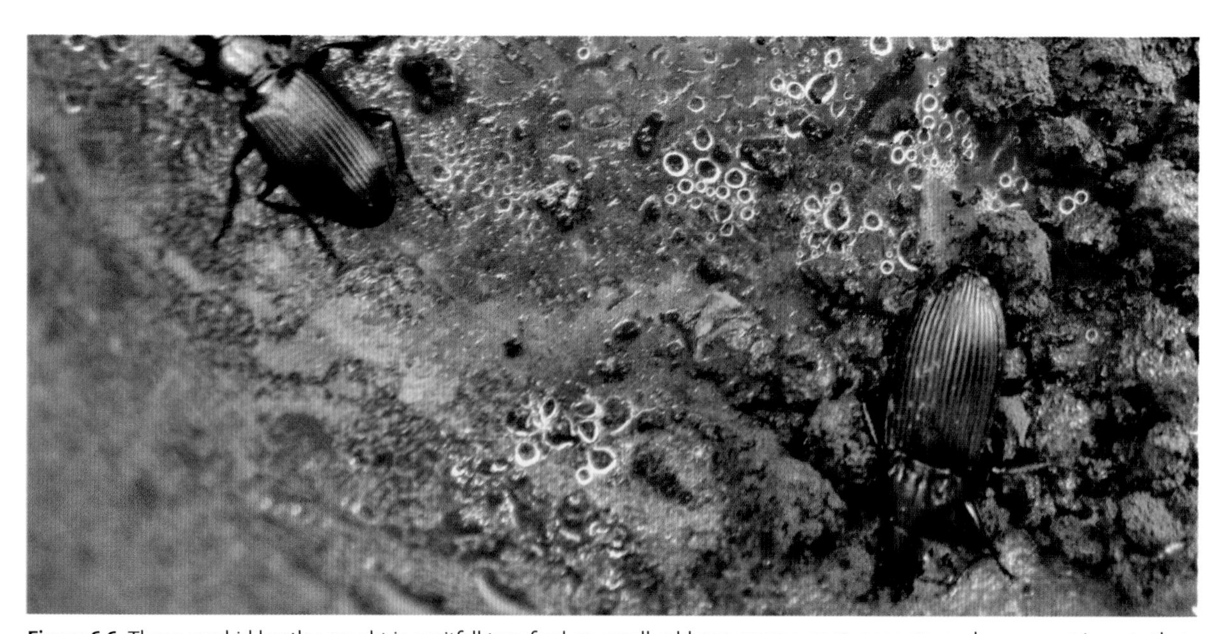

Figure 6.6. These carabid beetles caught in a pitfall trap feed on small cabbage worms, root maggots, and many pest insects that spend part of their life cycle in the soil.

soil-dwelling stages of insects, such as Colorado potato beetle pupae. There are hundreds of species of ground beetles, occurring in a wide range of colors and sizes from 1/8 to 1 inch (3–25 mm) long. Larvae are generally whitish to light brown and elongate with a large head. One generation is produced per year. Adults overwinter in undisturbed plant residue or mulch.

Habitat: Undisturbed (untilled), moist soil with thick ground cover. Ground beetles avoid dry, bare, open, cultivated soil areas. They feed on insects (including cutworms, caterpillars, beetle grubs and adults, fly maggots and pupae, slugs, and snails) and also weed seeds. In a Michigan State University study, only 73 seed-eating ground beetles were captured in a tilled soybean field, while more than 1,500 were captured in an adjoining native grass strip. Ground beetles hate soil disturbance!

Lady Beetles

The most common species of lady beetle is probably the convergent lady beetle, *Hippodamia convergens*, which has red wing covers with converging white markings just behind its head that give it its scientific name. Both larvae (see figure 6.3) and adult lady beetles (see figure 6.7) are predators. The adults are more specific and eat mostly aphids. The larvae are generalists and eat insect eggs, root-feeding beetle larvae, weevil larvae, aphids, and other soft-bodied insects, such as leafhoppers, scales, mites, and mealybugs. Lady beetles fly away whenever food is scarce, but remain in an area if sequentially blooming nectar sources are available season-long. Ants that eat the honeydew produced by aphids will protect the aphids by attacking the lady beetles.

Habitat: *Achillea* spp., *Ceanothus* spp., *Asclepias* spp., *Atriplex* spp., *Rhamnus* spp., native grasses, *Salix* spp., annual alyssum, cilantro, and sunflowers.

Soldier Beetles

Adult soldier beetles (*Chauliognathus pennsylvanicus*) are ½-inch (13 mm), narrow insects that are orange or red in color with black or brown wing covers. Larvae are velvety grubs and prefer damp ground. Larvae feed on soil-dwelling stages of insects like beetle grubs, caterpillars, maggots, and

Figure 6.7. Convergent lady beetle adults searching for aphids.

Figure 6.8. This soldier beetle on a plum tree leaf is seeking a meal of aphids.

grasshopper eggs. Adults are strong aphid predators, but also feed on other insects.

Habitat: Goldenrod and other plants in the daisy family (Asteraceae) such as sunflowers and cosmos, and weeds such as flowering and seeding yellow dock.

FLIES (DIPTERA)

Several genera and species of flies can be both predators and parasites of common insect pests.

Gall Midges

Adults gall midges (Cecidomyiidae family) are tiny (1/8 inch [3 mm]), fragile-looking, long-legged, dainty flies that resemble mosquitoes. The maggot-like larvae are tiny and bright yellow to orange in color. Gall midges are frequently seen where outbreaks of aphids or mites occur, but aren't so common where pest populations are low. They also feed on scales, whiteflies, thrips, and eggs of insects and mites. *Aphidoletes aphidimyza* is reportedly one of the most effective aphid predators in commercial greenhouses.

Habitat: Higher humidity and ground cover is important for gall midges. They need shady areas with protection from intense sunlight and high temperatures.

Figure 6.9. Two wormlike syrphid fly larvae are almost hidden by the large infestation of cabbage aphids they are feeding on. Sometimes you have to look very closely to find hard-working predators.

Parasitic Flies

Adult parasitic flies (Tachnidae family) look like large, bristly houseflies. They deposit eggs into the bodies of their prey; eggs hatch and the larvae feed internally. Parasitic flies can lay eggs in the larvae of butterflies and moths, beetles, sawflies, squash bugs, stink bugs, and crickets.

Habitat: Adults feed on pollen and nectar and are attracted to annual alyssum, cilantro, fennel, lovage, and *Achillea* spp., *Eriogonum* spp., and *Rhamnus* spp. In my on-farm research they are common on parsnip left to flower (see figure 5.12).

Figure 6.10. A syrphid fly adult (*Toxomerus marginatus*) feeding on pollen and nectar from a blooming weed, shepherd's purse.

Syrphid or Hover Flies

Syrphids (Syrphidae family) are found throughout the world and are common on many crops fed upon by aphids and other small soft-bodied pest insects. Syrphid fly adults are winged, ¼ to ½ inch (6–13 mm) long, with a yellow-and-black abdomen (resembling bees or wasps, but without a pinched-in wasp waist). The adults feed on flower nectar and pollen. The larvae are strong aphid predators and adults often seek aphid colonies and then lay their eggs nearby. Eggs hatch into headless, legless, greenish to pale yellow or pinkish maggots ⅛ to ¼ inch (3–6 mm) in length (see figure 6.9). These larvae (maggots) are important generalist predators in most crops, eating mealybugs, scale insects, whiteflies, leafhoppers, and small caterpillars in addition to aphids.

Habitat: Syrphid flies need diverse, undisturbed vegetation that provides them alternative prey, shelter from wind, sequentially blooming flowers and weeds (early to late season), and overwintering habitat. They prefer the following types: *Achillea* spp., *Ceanothus* spp., *Asclepias* spp., *Eriogonum* spp., *Prunus* spp., annual alyssum, cilantro, buckwheat, mustard, *Phacelia* spp., and fennel. In my on-farm research, *Brassica*, *Veronica*, and *Stellaria* weed species attracted the most syrphid flies. These flies are very sensitive to certain pesticides, including synthetic pyrethroids. Populations were greatly decreased in experiment plots on my farm where I applied rotenone-pyrethrin sprays, compared with no-spray plots.

PREDATORY BUGS (HEMIPTERA)

Most true bugs (order Hemiptera) are plant feeders (think of many stink bugs, for example), but a few are important predators. Predatory bugs feed on small pests when they are young, but switch to larger soft-bodied insects, especially caterpillars, as they mature. The assassin bug (*Reduvidae*) is a general predator that has long legs, a low, flat body profile, and a pointed head with long antennae. Note the triangle just below the bug's head in figure 6.11. This is an important identification hint because all true bugs have that characteristic triangle.

Figure 6.11. Assassin bugs are generalist predators that feed on a wide variety of pests ranging from aphids to caterpillars. This assassin bug is feeding on a coreopsis flower.

Damsel bugs (*Nabis*) are important predators of aphids, leafhoppers, true bugs (including squash bugs), thrips, and small caterpillars.

Minute pirate bugs (*Orius*) are tiny (1⁄16 inch [1.6 mm]), with an oval body that has a distinctive black with white pattern. They are generalists and feed on eggs (including corn earworm and grasshopper eggs), thrips, spider mites, aphids, psyllids, whiteflies, small caterpillars, and plant bugs (including squash bugs). They are especially attracted to colonies of thrips and also feed on leafhoppers.

Habitat: Most true bugs need undisturbed overwintering sites, such as untilled hedgerows, living mulches, or insectaries with season-long, sequential bloom. Weed species can provide nectar and pollen as a food source when prey is not available. Assassin bugs favor *Polygonum* spp., *Achillea* spp., and native grasses. Favored plants for minute pirate bug include *Achillea* spp., *Eriogonum* spp., annual cosmos, sunflowers, and corn as well as some shrubs, like *Baccharis* spp.

PREDATORY WASPS AND PARASITIC WASPS (HYMENOPTERA)

There are hundreds of species of parasitic wasps (called *parasitoids*), including Chalcid, Ichneumon, Braconid, and Trichogramma wasps that vary in size, shape, and color. All wasps have a pinched-in narrow waist, which is how you can tell them apart from flies. Trichogramma wasps, who are tiny, gnatlike, and yellow with red eyes, attack over 200 species of crop pests! Wasps especially like to parasitize caterpillar eggs. The wasps lay their own eggs on or inside the eggs, larvae, pupae, and adults of other insects. Their emerging larvae eat their way out, killing the host in the process. Although some wasp parasitoid species are generalists, most are specialist feeders.

There are also some species of predatory wasps, and they feed on caterpillars, aphids, mealybugs,

Figure 6.13. Several species of parasitic and predatory wasps feeding on the nectar and pollen in parsnip flowers.

Figure 6.12. Yellow jacket (Vespinae) predator wasp searching for cabbage worms on a broccoli leaf where feeding cabbage worms have chewed round holes.

Figure 6.14. Parasitized aphids (called mummies) on this leaf are tan colored and look bloated compared with the non-parasitized green aphids.

and leafhoppers. Some species feed only on particular species of caterpillars or aphids. Paper wasps such as yellow jackets are predators of caterpillars, and in a Wisconsin study were found to prey on cabbage worms.

Habitat: Wasps need pollen and nectar sources; they also prefer many plants in the Apiaceae, such as cilantro, parsnip, fennel, dill, and Queen Anne's lace, *Achillea* spp., *Helianthus* spp., and many plants in the Brassicaceae, such as annual alyssum and broccoli (when plants have bolted and are in flower). They also feed on species of native and common buckwheat (*Eriogonum*) and *Baccharis*.

LACEWINGS (NEUROPTERA)

Adult lacewings are delicate, light green to brown insects with transparent, veined wings. They are strongly attracted to the smell of aphids and even eat the honeydew secreted by aphids.

Larvae are small (1⁄8 inch [3 mm]), yellowish green creatures with brown stripes. The body resembles the shape of an alligator, with two distinct, elongate mouthparts sticking out of the head like tiny swords. Adults of most lacewing species feed only on nectar and pollen, but some also feed on insects. Lacewing larvae eat insect eggs, mites, and thrips when they are small. As they grow

Figure 6.15. Lacewing adult feeding on cilantro flowers.

Figure 6.16. Crab spider collected from apple trees.

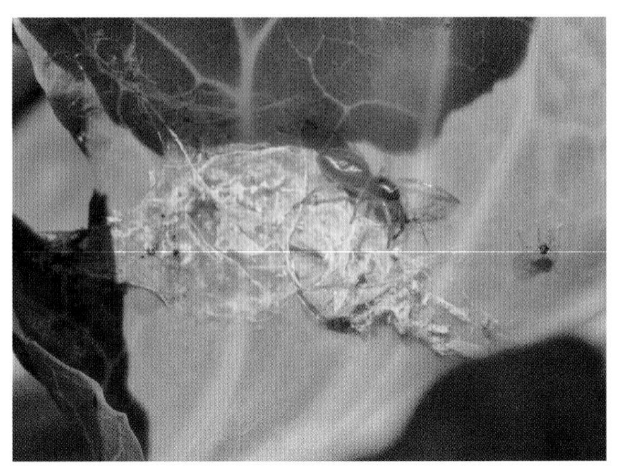

Figure 6.17. This spider is feeding on cabbage aphids.

they switch to larger soft-bodied pests like aphids, mealybugs, whiteflies, and small caterpillars.

Habitat: Lacewings are commonly found on carrot family (Apiaceae) flowers, such as cilantro, grasses, *Eriogonum* spp. and *Achillea* spp. They need a steady source of nectar, honeydew, and pollen for successful reproduction, so sequential, season-long bloom is vital. Some lacewing species prefer trees and shrubs, such as *Prunus* spp., *Ceanothus* spp., *Baccharis* spp., and *Rhamnus* spp.

SPIDERS (ARACHNIDAE)

Spiders are not classified as insects; they are in the class of living organisms called arachnids. As you probably know, an easy way to tell a spider from an insect is that all spiders have eight legs, while insects have only six. Spiders come in a multitude of sizes, colors, and shapes. Generally, spiders are of two main types: web making and hunting. Web spiders trap insects in their webs. Hunting spiders actively pursue and catch crawling insects. The most common hunting spiders are crab spiders, wolf spiders, and jumping spiders. Crab spiders are brown and/or white, ¼ to ½ inch (6–13 mm) long, with plump bodies. They move sideways like a crab when running. Wolf spiders hunt at night and are ground dwellers. They have dark, mottled colors and are ⅕ to 1 inch (5–25 mm) in size. Jumping spiders are stocky with short, stout legs, and a longer pair of legs in front. Harvestmen, also called daddy longlegs, are closely related to spiders and hunt in a similar way. Their bodies are ½ to 1 inch (13–25 mm) across and they have very long legs.

Spiders are often the most abundant predators in agroecosystems, especially in tree fruit orchards.[2] They are important generalist predators early in the season when specialist predators and parasites have not yet arrived or increased to high enough numbers for effective biological control.[3] They are highly effective predators because they control pests as part of a predator community in which different species occupy complementary niches.[4] Spiders within this predator community reportedly suppress aphids, codling moths, leafhoppers, pear psylla, mites, and plant bugs.[5] Spiders also prey on grasshoppers, cutworms, caterpillars (including overwintering codling moth larvae), aphids, scales, and insect eggs. In fact, spiders consume large numbers of insect eggs. In a Kentucky study one nocturnal spider ate 15 to 20 corn earworm eggs per day and 440 eggs during its development. Spiders even search out and attack leafminer larvae within their mines between plant leaf layers.

Habitat: Spiders prefer higher humidity and are very sensitive to physical and chemical disturbance. They are very sensitive to insecticides and are reported to be more abundant in unsprayed orchards.[6] They are also sensitive to some herbicides.[7] Populations of spiders are slow to recover from chemical applications, probably because most spider species have only one generation per year.[8] They are found in higher numbers where soil is not disturbed and pesticides (even certified organic pesticides) are not sprayed. In my own on-farm research, spider populations were decimated by spraying rotenone-pyrethrin sprays, compared with no-spray plots (see figure 5.23). The populations of ground-dwelling spiders decline with tillage and weed cultivation. Straw mulch increases spider populations and decreases pest damage.[9] Encourage spiders by leaving wild, blooming, undisturbed areas interspersed within crops and/or using living mulches between rows of perennial and annual crops.

PARASITIC NEMATODES

Many growers are familiar with the root knot nematode, a microscopic roundworm that infests and damages the roots of various vegetable crops. There are also many species of soil-dwelling parasitic nematodes, which feed on larval and pupal stages of soil-dwelling insects, or insects that overwinter as larvae within the soil, such as Colorado potato beetle. You can purchase parasitic nematodes to apply to the soil, but it is essential to use the species or strain that is adapted to your target pest.

Habitat: Nematodes require moist soil. They are ineffective in dry soil. Soil temperature is also critical. Most nematodes are inactive at soil temperatures below 47°F (8°C). Parasitic nematodes are sensitive to soil disturbance and have been found in higher numbers where tillage and weed cultivation is minimal. If you create habitat, parasitic nematodes will come, and that's the ideal approach. However, you can apply purchased parasitic nematodes as a minimal-impact intervention. Be sure to time spray applications when soil temperatures are greater than 50°F (10°C), and be sure the soil is premoistened. In outdoor field situations, high application rates (2 billion nematodes per acre [5 billion/ha]) are necessary. Repeat applications may be necessary. Nematodes usually persist for 2 to 4 weeks.

BIRDS

Research is beginning to show how important birds can be as predators in vegetable, tree fruit, vine fruit, and nut crops. They feed on caterpillars such as codling moth, imported cabbage worm, and cabbage looper. Most birds have a varied diet of insects, seeds, and fruits. Insect feeding is usually greatest during nesting and when feeding young in the spring and early summer. However, I have observed sparrows patrolling all of my spring, summer, and fall cabbage and broccoli plantings, eating large numbers of imported cabbage worms and cabbage loopers. One of the best sources of information I know of about birds and the specific crop pests they prey upon is *Supporting Beneficial Birds and Managing Pest Birds* published by the Wild Farm Alliance (www.wildfarmalliance.org).

Habitat: Generally, the most important component of bird habitat is nesting sites and roosting sites, such as snags on orchard or field borders. In dry areas, maintaining a cat-free water source also helps to attract birds. Some birds eat a diet of mainly insects; some eat insects when they are nesting and feeding young, but then turn to fruit and seeds. The resource mentioned above provides information that will help you encourage birds that eat insect pests rather than fruit-eating birds.

Tolerating Low Levels of Pests

At Biodesign and Woodleaf Farms, Carl and I tolerated 5 to 10 percent crop damage as our "tithing to ecology." In my work as a consultant for large organic vegetable and fruit farms, I have seen that significant crop loss can happen, simply due to poor timing and inattention to details. That led me to realize that accepting a small amount of crop loss from the pest insects that feed my "free ecosystem

Table 6.1. Quick Guide to Damage Caused by Common Plant Pests

Symptoms and Signs	Possible Causes
Leaves with large, round, chewed or ragged holes	Slugs and snails, animal pests Adult or larval stages of beetles (Colorado potato beetle, Mexican bean beetle) Larval stages of moths or butterflies (armyworms, cabbage worms)
Curled, twisted, puckered, deformed, or distorted leaves	Aphids, leafhoppers, tarnished plant bugs Nutrient deficiencies, viral diseases
Curled leaves with webs or sawdust-like frass present	Leafrollers, larval stages of moths or butterflies (webworms).
Leaves with many small round holes	Flea beetles Plant diseases
Whitish or yellowish leaves with stippling (tiny pinholes)	Tarnished plant bugs, spider mites, thrips
Silvery appearance of leaves or white stippling of leaves	Thrips
Small wartlike bumps, scales, or cottony growths on leaves and stems	Scales, mealybugs Plant diseases
Small blister-like bumps on leaves, often reddish or yellowish in color	Eriophyid (blister) mites
Leaves with whitish or tan spots visible on leaf surfaces and shallow tunnels between upper and lower leaf surfaces	Leafminer fly larvae, sawfly larvae
Shiny, slimy, sticky coating on leaves	Aphids, mealybugs, scales, pear psylla, spittlebugs, whiteflies Slugs, snails
Yellowing leaves chewed in a lacy pattern; stems chewed and wilted	Striped cucumber beetles, spotted cucumber beetles
Withering and wilting leaves and vines on squash plants	Squash vine borer, squash bugs Disease
Tunnels, rotted areas, and sawdust-like frass inside fruit; wormlike larvae present	Moth larvae (codling moth, European corn borer, corn earworm, armyworm), apple maggot, western cherry fruit fly larvae
Distorted, misshapen, twisted, or cat-faced fruit	Tarnished plant bug, stink bugs, western flower thrips Hail
Root or bulbs have rotted spots, chewed areas, and tunnels	Wireworms, beetle grubs, maggots (cabbage, onion, and carrot root maggots), weevil larvae (carrot)
Roots or stems with round, wartlike galls and swelling	Gall wasps, nematodes, plant diseases
Wilted leaves, hollow stems filled with sawdust-like frass and wormlike larva inside	Moth larvae (European corn, squash vine, peach twig, apple round-headed, and flat-headed borers)
Seedlings chewed off at soil level	Cutworms Animal pests.

providers" (all the beneficial organisms within my farm ecosystem) is a very smart management choice. If I can save the labor required to spray a crop, as well as the cost of spray materials, I am willing to put up with some crop damage on certain crops. And most years, most of the crops at Woodleaf are surprisingly pest-free.

Encouraging wild habitat or ground cover within crop fields can provide a reservoir of insects for natural enemies to feed on when pests are scarce on crops. But this approach does require careful, regular monitoring of crops to ensure that pest damage doesn't get out of control. I monitor my orchard and vegetable fields at least once each week, and I keep records. I use the Problem Identification Forms to keep track of what is going on with both pest and beneficial insects on my crops. They also show me whether pest levels are increasing or decreasing in the short term and give me a long-term record of pest and beneficial insect population change. Recording long-term change helps me decide whether or not my pest suppression strategies are working.

When I am considering an intervention, I fill out the Pre-Intervention Information Form to guide my decision making. These forms are in the appendix on page 338.

If pest levels rise too high, I consider intervention. I have developed farm-specific pest thresholds that trigger treatment with minimal-impact methods (See "Understanding Pest Thresholds" on page 128). I continue to monitor my crops weekly, but as my efforts at building and managing habitat have evolved, I rarely, if ever, observe pests reach a threshold high enough to cause economic damage.

Intervention When Insects Impact Crops

When things get out of balance and pest populations exceed the tolerance of your farm or garden, it's time to intervene. In this section of the chapter, I provide detailed information about how to use *minimal ecological impact*, *moderate ecological impact*, or *heaviest ecological impact* intervention materials that are approved for use by certified organic operations.

INTERVENING WITH THE LEAST POSSIBLE IMPACT

I strive to avoid killing natural enemies of crop pests, especially beneficial insects and parasites. Thus before I intervene at any level in response to a pest problem, I first apply all the components of ecological systems thinking:

- Prevention
- Pest diagnosis
- Research
- Monitoring
- Ecological decision making

These components seem like common sense, but over the years I have noticed that when even the most ecologically oriented farmers or gardeners get busy, their first response to a pest problem is: How do I kill the culprit pest? Let's consider each of these components in turn.

Prevention

Prevention is a long-term commitment, but it is the most important of the ecological thinking components. Prevention includes some of the important concepts and techniques discussed in earlier chapters, such as attending to crop health and creating a farm or garden design rich in diversity to enhance the environment for beneficial insects. Farm or garden hygiene plays a role, including removing all insect-infested or diseased plant tissue and using clean tools. Resistance is yet another key strategy. Choose resistant or tolerant varieties if possible when particular pests are a problem.

Pest Diagnosis

Be sure you identify the pest or other cause of a problem before intervening. Sometimes crop

Understanding Pest Thresholds

A *pest threshold* is the pest or crop damage level that triggers intervention to avoid significant economic damage to crops. It is a prediction about the population density of a pest that may cause an unacceptable amount of damage unless action is taken. Thresholds can be based on models available from land-grant university integrated pest management (IPM) programs and/or on records that you develop over time for your own farm or garden as you monitor both pests and the damage they cause in your agroecosystem. Ultimately, pest thresholds depend on the amount of crop damage your farm can sustain economically or you are willing to accept in your garden.

damage is not caused by an insect, or at least not the insect you are seeing on the plants. Chapters 9 and 12 list the specific symptoms of the most common pests for particular vegetable and tree fruit crops. Sometimes damage is a result of environmental conditions: too hot, too dry, too cold, wind, hail, frost. Check recent weather events as part of your diagnosis. Don't kill things unless you are sure they are responsible for the problem!

Research

Once you identify a pest on your plants, research what it looks like in its other life stages, too. Which stage does the damage? What stage is most vulnerable to minimal-impact interventions? What is the pest threshold level at which damage will be unacceptable? If you can answer these questions, you will have the ecological knowledge to suppress pests relying on the ecological interactions within the agroecosystem you have created. And you also have the information necessary to choose a minimal-ecological-impact intervention if it is truly needed.

Pest Monitoring

Develop a plan for weekly monitoring during which you check your crops and jot down notes, such as any damage you observe, how serious it is, and whether the plants are also suffering from any environmental stress. I take my hand lenses and notebook into the field and walk the same path every week. I fill out my weekly monitoring forms each time (see the forms in the appendix). These completed forms and my full notebooks (the long-term repository of the forms) over the years have taught me a good portion of what I know about pest suppression. They have helped me understand the positive and negative implications of my decision to stop spraying my crops.

Also keep track of weather conditions, especially those that affect pest development such as temperature, humidity, and precipitation. You can also use pheromone and visual traps. Pheromone traps lure male insects by using the scentlike substance (pheromones) given off by an insect of the opposite sex. Visual traps use light, bright colors and shapes to attract insects of both sexes.

You can also use degree-day models to predict pest emergence and vulnerable stages.

Degree-day models. The rate of insect development is closely related to temperature: Insects grow and develop faster in hot conditions than in cold conditions. Scientists quantify this in terms of degree days. A degree day for a particular species equals the average daily temperature minus the insect's *lower development threshold*, which is the temperature below which the insect does not grow or develop.

Let's consider the example of codling moth. Study of this insect has shown that its lower development threshold is 50°F (10°C). We can determine how many degree days of growth are available for a codling moth by first calculating the average daily temperature. If the maximum temperature on a certain day is 80°F (27°C) and the minimum is 40°F (4°C), the daily average is 60°F (16°C):

$$(80 + 40)/2 = 60°F$$

From this daily average of 60°F, we subtract the lower development threshold of 50°F.

$$60°F - 50°F = 10 \text{ degree days}$$

Hence, on that date 10 degree days of codling moth physiological development time accumulated. A degree-day model for codling moth predicts the number of degree days that must accumulate before a certain life stage of a pest occurs. The Washington State University model for codling moth predicts that moths will emerge when 175 degree days have accumulated (beginning from January 1 of the year). These models may vary from one area to another. In the United States, you can find out more about degree-day models for pests by contacting your local extension service or consulting the integrated pest management websites of your local land-grant university. It can take some research to find a degree-day model for a pest, but consulting the model will significantly help improve the timing and efficacy of your interventions. Using degree-day models is complex and a time commitment; it is probably a tool best utilized at the farm scale.

Figure 6.18. Ecological decision making means taking into consideration the health and habitat of beneficial insects, pollinators, and birds, like this yellow warbler in my apple tree, catching insects to feed his babies.

Ecological Decision Making

Develop an integrated suppression/management program for specific pests, rather than a "see and spray" program. Each program should be based on as much biological information as you can gather about the pest. Choose minimal-impact interventions first and heaviest-impact interventions only as a last resort. Any insecticide intervention should be:

- As specific as possible to the target pest. Applying nonspecific insecticides and/or materials that have heavy ecological impact often makes pest problems worse.
- Short-lived. (This refers to how long the insecticide will remain active after it has been applied. It should be only a few hours or a day, rather than several days to a week.)
- Compatible with beneficial organisms, because killing beneficials makes pest problems worse.

MINIMAL-IMPACT INTERVENTIONS

Biological controls, traps, and microbial insecticides are the methods and materials that present the least chance of injuring beneficial insects or other natural enemies. Biological control and habitat for beneficial organisms is covered in chapter 5 and in the "Keeping Beneficial Insects and Other Natural Enemies Happy" section on page 117.

Traps

Traps are a useful tool to help us suppress insects. They alert us when the first insect pests of a specific species arrive in our gardens and farms. They also help us to figure out whether a certain pest is present at a high enough level to do damage to our crops (thus perhaps requiring intervention). And some traps can help to suppress pest insect populations directly, by trapping and killing some.

Pheromone traps are usually used for monitoring only, not to try to combat an infestation. For example, codling moth traps catch only *male* moths, not females, so using such traps will not significantly lower pest populations.

Red sphere sticky traps can be used to help decrease or monitor some pests that move into fruit crops in the spring, such as western cherry fruit flies and plum curculio (see "Apple" on page 284 and "Cherry" on page 292 for more detail).

Slug and snail traps can be effective to reduce damage caused by these pests.

Yellow, white, or blue sticky traps work for trapping aphids, thrips, whiteflies, and fruit flies, both to monitor and to reduce populations. Choose the color recommended for the particular pest you want to control. Unfortunately, these traps will also trap and kill certain flying beneficial insects.

Microbial Insecticides

The "active ingredients" in these materials are living beneficial microorganisms. These organisms generally attack only one group of insects, or one *life stage* among several groups of insects. Their efficacy is very dependent on environmental conditions. Temperature, moisture, and relative humidity must be within certain limits. Timing applications properly is also key to efficacy. Take time to learn the details of how a microbial insecticide works before you try applying it.

Bacillus thuringiensis (*Bt*) is a microscopic organism that occurs naturally in the soil. There are different varieties, and each affects a narrow range of pest insects. Because *Bt* is relatively pest-specific, it is safe for most beneficial insects. The *Bt* insecticide is applied to foliage so pests consume it as they feed. Insects must ingest *Bt*, or it will not affect them. Thorough coverage of crop foliage with the insecticide is essential.

Application timing is important because the microbes survive only 1 to 2 days after application. Thus you should apply *Bt* as soon as you find newly hatched pest insects on vulnerable crops. Freshly hatched larvae are the most susceptible to *Bt*, larger larval stages are less susceptible, and adults are not affected at all. Feeding stimulants (such as molasses) help to improve *Bt* consumption by insect pests.

Note also these limitations: *Bt* degrades rapidly when exposed to the ultraviolet radiation (UV) in sunlight. Warm, moist conditions favor bacterial growth that can also decay this product. *Bt* is ineffective at temperatures below 45°F (7°C).

Dry formulations are more stable than liquids. Therefore, *Bt* should be stored in a cool, dry, dark place in a tightly sealed container. Wettable powders and wettable granulars should be mixed as needed, and used within a day or two of mixing. Spreader-stickers can be added to ensure adherence of the *Bt* to foliage and block UV rays to some degree. Granular *Bt* formulations combined with cornstarch have been shown to improve adherence to foliage and pest mortality.

Mixing *Bt* with other materials can actually decrease its efficacy. One example is neem, whose anti-feeding properties appear to reduce the consumption of *Bt* by pests.

There are three main varieties of *Bt*, and they are effective against different types of insect pests.

- *Bacillus thuringiensis israelensis* (*Bti*) is recommended for fungus gnat larvae.
- *Bacillus thuringiensis* var. *kurstaki* or *Berliner* (*Btk* or *BtB*) is formulated only for caterpillars or worms that turn into moths and butterflies. Repeat applications may be required for control of eggs that hatch several days later.
- *Bacillus thuringiensis* San Diego (*Bt*SD) is intended for use on beetle larvae only. Apply at the first sign of infestation, then again several days later.

Some insects, such as codling moth or corn earworm, feed in a way that allows only a small window during which they will be exposed to foliar *Bt* before they burrow into the fruit or corn ear. *Bt* does not work as well on these kinds of insects.

Granulosis virus (GV) and nuclear polyhedrosis virus (NPV) are two viral insecticides that are used for certain caterpillars, including codling moth and Oriental fruit moth larvae. Granulosis viruses are highly specific and generally safe for beneficial insects. The virus has slow action, a restricted host range, and low persistence in the environment. Application timing is critical. First sprays should be applied just before larvae hatch, so that the new larvae are exposed to the virus as soon as they begin feeding.

Nosema locustae is a microorganism specific for grasshoppers. This biological insecticide is applied as a bait when grasshoppers are very young. The grasshoppers die slowly over a period of several weeks. Feeding is reduced, but not eliminated. An application of *Nosema locustae* can reduce grasshopper populations for 1 to 2 years. However, grasshoppers are strong fliers and continually migrate from surrounding areas. So at a garden scale, unless a neighborhood cooperates to treat large areas, application in a single yard might not provide much control.

MODERATE-IMPACT INTERVENTIONS

These methods and materials present more of a risk of injuring beneficial insects or other natural enemies, but if they are used carefully, sparingly, and discriminately, they will not permanently disrupt the agroecosystem pest suppression you have designed.

Beauveria bassiana

Beauveria bassiana is a disease-causing fungus that grows naturally in most soils. It acts as a parasite and affects many types of insects. It can be applied to fruit, nut, and vine crops, and also to vegetable crops for insect problems such as aphids, thrips, whiteflies, mealybugs, caterpillars, and beetles. Immature stages tend to be more susceptible than adults, but *B. bassiana* is a generalist feeder and infects all life stages of leaf-feeding insects, including some beneficial ones. For example, this fungus can infect and kill several species of lady beetles. However, *B. bassiana* does not affect most beneficial insects as seriously as another parasitic fungus, *Metarhizium anisopliae,* and is generally considered to be relatively safe on most beneficial insects. *B. bassiana* works via contact with the pest, so good coverage is essential. It is not necessary for the insect to ingest this fungus.

As with all living organisms, it takes time for *Beauveria bassiana* to develop fully. It is not a quick fix. Optimal conditions for spread of this fungus are temperatures between 75 and 85°F (24–30°C) and 60 to 97 percent humidity. *Beauveria bassiana* is best applied in the late afternoon, in the evening, or on a rainy day to avoid exposure to sunlight. Some researchers have found that when mixed with *Bacillus thuringiensis* San Diego, it provides better management for Colorado potato beetle than if either is applied alone. *Beauveria bassiana* should not be applied around honeybees or near natural water sources, because it will kill both bees and some types of water-dwelling invertebrates.

Soaps

Insecticidal soaps are similar to liquid household soaps except that they are highly refined to be more effective for insect control and to prevent damage to plants. They are very low in mammalian toxicity and break down quickly in the environment; however, they are toxic to many soft-bodied predators and parasites, such as syrphid fly larvae. Repeat applications are necessary. Formulations currently available contain either soap alone or soap and pyrethrum (see "Pyrethrum Powder" on page 134). Formulations containing pyrethrum will be more effective on a wider spectrum of pests, but are also much more toxic to beneficial insects. Soaps work best for soft-bodied insects such as aphids, whiteflies, and immature scales. They also kill thrips and mites effectively, but it is difficult to get the soap spray to penetrate into the small spaces where these insects are found. Soaps have little activity against grasshoppers, beetles, flies, wasps, and caterpillars. They are not effective for organisms such as slugs and snails. If you make a "homemade" insecticidal soap spray using household soap or detergent, use it cautiously—damage to plants is common as a result of additives in the household products.

Horticultural Oils

Horticultural oils are highly refined paraffinic oils that are manufactured specifically for pest control on trees and ornamentals. They degrade rapidly through evaporation and have almost no toxicity to humans or wildlife at the rates used for pest management. But, they can be toxic to soft-bodied predators, such as syrphid fly larvae and predatory spider mites. Superior or supreme oils are quite safe for most plants when used according to directions. Superior oil doesn't contain sulfur and can be applied when plants are in leaf. Supreme oil is a highly refined product similar to superior oil (sometimes the terms are used interchangeably). Some plants that are sensitive to these materials are listed on the labels. Oils can be used at any time during the season, though some are labeled "dormant" oils.

Most oils available for pest management are petroleum-based. However, vegetable- and citrus-based horticultural oils are available also.

Horticultural oils are most effective against eggs of most insect species, larvae, and many adult species that have soft bodies. They are most effective on eggs if used close to the time of hatch, because the developing larvae inside eggs become more susceptible to suffocation the closer they get to hatching. Oils work well to suppress mites, scales, mealybugs, aphids, whiteflies, and leafhoppers, among others.

Several precautions should be followed when using oils to avoid damage to plants. Most important is to make sure the plants are not water-stressed. Water trees and ornamentals several hours before application of oils, just to be safe. Avoid spraying if humidity is less than 65 percent or if leaves are wet. Apply oils on cooler days or in early morning. Spray when temperatures are greater than 40°F but less than 75°F (4–24°C).

Minerals

Minerals have a relatively long natural persistence when compared with botanical insecticides, but they are generally low toxicity. Some minerals, such as kaolin clay, are made up of particles that are not broken down rapidly by high heat or direct sunshine.

Diatomaceous earth (DE) is made from fossilized silica shells of microscopic algae. Most insects are covered with a protective coating of wax. Diatomaceous earth absorbs this wax, causing the insect to dry out and die. The particles of powder also have sharp edges that are abrasive to an insect's protective coating, causing small cuts that are fatal. It will affect all insects, including beneficial ones. One limitation when using DE outdoors is that it loses its effectiveness when it becomes wet. Also, DE can injure earthworms if it is washed into the soil or if earthworms come into contact with it on the soil surface.

Kaolin (Surround WP) is a clay that repels insects such as leafhoppers, leafrollers, weevils, and smaller caterpillars (such as codling moth). Studies in Missouri, Washington State, and New York State suggest that Surround WP also keeps apple fruits (on the tree) cooler and may reduce sunburn and heat stress by as much as 50 percent. Good coverage of fruit and leaves to be protected is essential. The white clay coating decreases with rain, irrigation, and/or time, and repeat applications are often necessary.

HEAVIEST-IMPACT INTERVENTIONS

Some methods and materials used for pest management present a very high probability of injuring beneficial insects or other natural enemies. Although these materials are designated as acceptable for use on certified organic operations, their use will likely disrupt the agroecosystem pest suppression you have designed. They should be used only as a last resort when pest thresholds have been exceeded and unacceptable economic damage is predicted.

Plant *extracts*, also known as *botanical insecticides*, are usually, but not always, less toxic to humans, beneficial insects, and wildlife than are synthetic chemical insecticides used in conventional farming. Extracts generally break down quickly in the environment, usually in 1 to 7 days. There are many formulations and trade names of pest management materials that claim to be "natural" or "organic." Read the information on the product label about the main ingredient and other ingredients so you know what you are using! If you can't figure out what the ingredients in a "natural" insecticide are, best not to use it. Most formulations contain one or more of the botanical ingredients listed below.

Neem Oil

Neem oil is extracted from the seed of the neem tree (*Azadirachta indica*), which is native to Africa and the Middle East. Neem acts as an anti-feedant and repellent; it may also disrupt insect molting, prolong the larval period, and disrupt mating. The oil can control a wide spectrum of pests. It is effective against leafminers, whitefly, thrips, caterpillars, aphids, scale, beetles, and mealybugs. Neem is generally safe for predators, but it has been shown to negatively affect some beneficials, including the larval stages of lady beetles, the spined soldier bug (*Podisus maculiventris*), syrphid flies, and some parasitic wasps. Neem has a short activity period, only 24 to 48 hours. It is toxic to fish, so avoid spraying near water. It is *slow* to control insects, especially if populations are high. Frequent applications are recommended. Neem oil should not be applied at temperatures below 50°F (10°C) or above 80°F (27°C). Like any oil, it should not be applied to plants in direct sunlight at the height of the day, nor to water-stressed plants because of the potential for burning plant tissues. Neem foliar sprays can last 4 to 8 days, but rainfall washes the spray off leaves and sunlight degrades it.

Pyrethrin Liquid

Pyrethrin is derived from the flower of the *Chrysanthemum cinerariifolium*. Liquid pyrethrin is available under many labels and is one of the most widely used botanical insecticides. Most formulations contain about 5 percent active pyrethrin. All formulations are broad-spectrum insecticides that kill beneficial insects as well as pests. Pyrethrin is most effective for flies, ants, aphids, mosquitoes, thrips, and some kinds of caterpillars and beetles. It is less effective for grasshoppers, spider mites, mealybugs, and adult scales, or insects that burrow inside plant material or fruit. In my on-farm studies, pyrethrin was especially hard on spiders and syrphid flies, but decreased the numbers of all beneficial insects monitored. In another study, pyrethrin and pyrethrin plus neem were toxic to spined soldier bug. It is also toxic to honeybees. Pyrethrin is of relatively low toxicity to mammals, including humans, but is very toxic to fish and certain other aquatic life, so avoid application near water. Avoid formulations that contain piperonyl butoxide (PBO), which are very toxic to predators and parasites and not certified organic. *Pyrethroids* are very different from pyrethrin; they are synthetic chemical compounds that are far more toxic to insects than the natural compound. Pyrethroids have no place in organic and ecological farming or gardening.

Pyrethrum Powder

Pyrethrum (powder) is also made from dried flower heads of *C. cinerariifolium*. Pyrethrum is of lower toxicity than pyrethrin, and not as effective against pests. Pyrethrum remains active on plant leaves for 1 to 3 days, whereas pyrethrin formulations can remain active for up to 5 days. Both compounds break down in the sun and are less effective at temperatures above 80°F (27°C).

Spinosad

Spinosad is an organic compound produced by a group of filamentous soil bacteria (Actinomycetes) called *Saccaropolyspora spinosa*. It is a broad-spectrum microbial insecticide that kills lots of insects, including some beneficial ones. It has been used against caterpillars and worms, sawflies, thrips, leafminers, and foliage-feeding beetles, working best during their early developmental stages. Spinosad works both through contact and ingestion, causing rapid death by affecting the nervous system. It has a long residual activity (about 7 to 14 days) and some systemic action. Leafminer control is improved with a vegetable oil adjuvant, which helps encourage a pest management material to spread across and stick well onto leaf surfaces. Spinosad has less activity against predators than most insecticides used in conventional agriculture, but in some studies it was highly toxic to two predators—the big-eyed bug (*Geocoris punctipes*) and the spined soldier bug (*Podisus maculiventris*)—and also to parasitic wasps, such as braconid wasps and *Cotesia marginiventris*, a generalist parasitoid of cutworms, armyworms, corn earworm, and cabbage looper. In some studies, spinosad has been reported to be toxic to predatory mites, syrphid fly adults and larvae, lacewing larvae, predator beetles, and bees for up to a day after spraying. It is safe for humans, pets, and birds, but toxic to fish and honeybees. Organic farmers have been using spinosad for many years, and there are some anecdotal reports of certain insect species (imported cabbage worm and cabbage looper and codling moth) becoming resistant to it.

CHAPTER 7

Working with Beneficial and Disease-Causing Microorganisms

The fungi, bacteria, viruses, and nematodes that cause plant disease are related to but different from the microorganisms that are so important for soil health and that serve as natural enemies of plant pests. Low populations of disease-causing microorganisms are often present in the soil and on plant leaves, doing no significant harm. They balloon out of control and cause disease outbreaks only under certain conditions. And just as specific insect pests don't usually feed on every plant in the field or garden, generally, specific plant diseases attack only particular crops or crop families.

Plant disease is not inevitable. It does not happen unless four conditions are met:

- The plant or plant family must be susceptible to a specific disease. For example, peaches do not get apple scab disease and cucumbers are not susceptible to early blight of potato and tomato.
- The "bad guy" microorganism must be present. Some diseases occur in many parts of the US or the world, but others occur only in particular areas. Further, some disease-causing organisms are not present in a new garden or farm because susceptible plants that support the disease have not been grown there for years.
- The environment, temperature, and moisture conditions must be favorable for infection by the pathogen (the bad-guy microorganism). All disease microorganisms have specific temperature, moisture, humidity, light, and "nutrition" preferences. Disease pathogens do not prosper if the conditions are unfavorable.
- The garden or farm ecosystem balance must have been disturbed and disrupted.

Preventive strategies—such as using disease-resistant cultivars and rotations, avoiding the conditions that initiate infection by specific diseases, balancing plant nutrition, and balancing interactions between beneficial and disease-causing microorganisms—are the focus of disease suppression. Unlike insect pest suppression, there are fewer effective, field-tested disease management material options for organic farmers and gardeners. However, though not yet well tested in gardens and on farms, there is a lot of new research into the use of beneficial microorganisms to suppress disease pathogens. I present some of the new research in this chapter, both for those who want to try the commercial beneficial organisms being studied and to encourage all of us who are designing disease-suppressive gardens and farms based on maintaining habitat to support our native beneficial microorganisms.

Designing and Maintaining a Disease-Suppressive Ecosystem

Disease suppression is not a simple cause-and-effect phenomenon. It involves a complex of relationships within healthy plants, a healthy soil and its microbial community. There are four components for achieving ecological disease suppression in gardens and on farms: creating diversity, rotation of crops, balancing crop plant nutrition and soil pH, and enhancing disease-suppressive microorganisms.

DIVERSITY

Diversifying crops, ground covers, and in-field and field-margin plants is the first component in designing for disease suppression, as discussed in chapter 5. Polycultures that include multiple plant families, genera, and species suppress disease much better than monocultures. Diverse above- and belowground plant life supports so many more ecological functions and relationships, especially within the soil microbial community. Many studies report that plant diversity has a beneficial influence on the soil microbial community balance that is the backbone of disease suppression.[1]

ROTATION

Rotating crops is a time-honored tool to suppress disease in gardens and on farms; it is a way of adding diverse above- and belowground plant life to

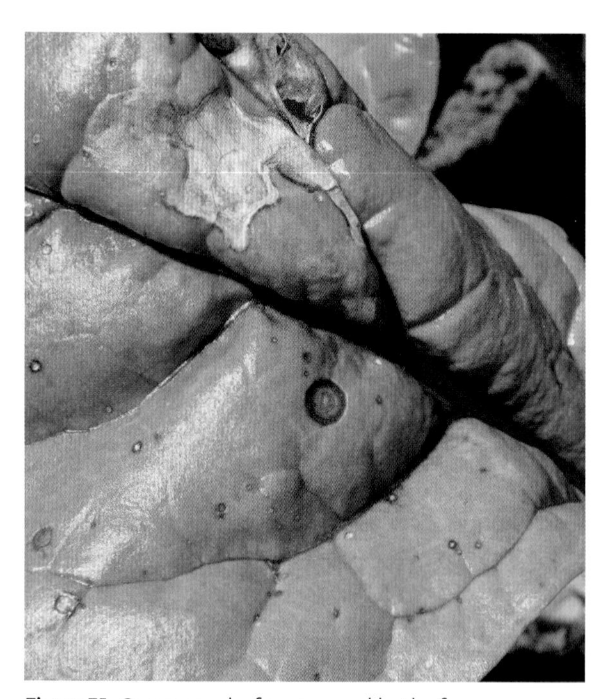

Figure 7.1. Cercospora leaf spot caused by the fungus *Cercospora beticola* shown on this Swiss chard leaf. Small round spots with a bull's-eye pattern coalesce into dead patches as the fungus grows.

Figure 7.2. Note the fungus-plant interaction pattern on this apricot leaf infected by shot-hole fungus (*Wilsonomyces carpophilus*). Spores germinated on the warm, wet leaf surface and began to grow, but the plant used its defense system to stop the fungal growth. The result is different colors and striations in the spots and the indistinct outer yellow halos.

Table 7.1. Vegetables, Herbs, and Flowers by Family

Plant Family	Family Members: Vegetables, Herbs, and Flowers
Amaryllidaceae	Asparagus, chive, garlic, leek, onion, shallot
Apiaceae	Carrot, celery, cilantro, dill, fennel, parsley, parsnip
Asteraceae	Artichoke, calendula, lettuce, marigold, sunflower, zinnia
Brassicaceae	Broccoli, Brussels sprouts, cabbage, cauliflower, daikon, horseradish, kale, kohlrabi, mustard greens, radish, rutabaga, sweet alyssum, turnip
Chenopodiaceae	Beet, spinach, Swiss chard
Convolvulaceae	Morning glory, sweet potato
Cucurbitaceae	Cantaloupe and other melons, cucumber, gourd, pumpkin, squash (winter and summer), watermelon, zucchini
Fabaceae	Lima bean, mung bean, pea, snap bean, soybean, sweet pea
Lamiaceae	Sage
Malvaceae	Mallow, okra
Poaceae	Corn
Rosaceae	Strawberry
Solanaceae	Eggplant, pepper, petunia, potato, tomato

a field or garden over time. Rotate the location in which you plant annual crops each year. To avoid diseases specific to particular plant families and species, do not grow crops or flowers from the same family in the same location for at least 1 to 3 years (see table 7.1). For example, if you planted tomatoes in a raised bed in your garden this year, wait 3 years before planting potatoes, eggplants, peppers, or tomatoes in that particular bed again.

But what if you are growing intensively on a relatively small plot of land? How far away from last year's site must a crop be rotated in order to benefit from the practice? The answer depends on how a specific disease organism overwinters and spreads. As a general rule, though, the farther away you can rotate a crop the better, especially if the problematic disease organism overwinters in the soil rather than on plant debris. Rotation distance can be reduced if you are able to take steps to prevent the disease organism from spreading. For example, if the disease spreads by way of splashing water from sprinkler irrigation, you can use drip irrigation only. If the organism splashes up onto foliage from the surface soil, you can be sure there is no bare soil around susceptible plants. If human dispersal (spreading spores via boots, clothing, or hands) is a factor, you can avoid walking in the field or garden when leaves are wet from a heavy dew, rain, or irrigation. Part 2 of this book provides more information about how specific diseases spread.

BALANCING PLANT NUTRITION AND SOIL pH

Healthy, vigorous, nutritionally balanced plants can compensate for losses of leaf and root surface area due to disease infection and damage. But plants don't fare as well when they have too much of certain nutrients or are growing in soil with pH that is too high or too low. Plants that are nutritionally imbalanced or deficient are not able to exert their natural defense systems against disease infection. These natural defenses include activating enzymes that then spur production of defense biochemicals, altering the composition of root exudates, changing soil rhizosphere pH, and promoting soil microbial activity.[2] Plant disease defense systems also include:

- Thicker epidermal cells (the outermost layer of cells covering the stem, root, leaf, flower, fruit, and seed parts of a plant) and waxy cuticles on leaf and stem tissue.
- Increased lignin, callose, or silica in plant tissue and rigid cell walls. *Callose* is a compound produced by plants to act as a temporary cell wall in response to stress and/or leaf, stem, and root damage.
- Higher production of defensive biochemicals, such as phenols and glucosinolates, and synthesis of polyphenol compounds called phytoalexins (see "Phytoalexins" on this page).

Phytoalexins

Plants synthesize the antimicrobial biochemicals known as *phytoalexins* relatively rapidly at the initial sites of disease infection. The phytoalexins act as toxins to the invading microorganisms, disrupting their metabolism or preventing reproduction. Most phytoalexins are products of specific plant biochemical pathways, such as the phenylpropanoid pathway and the mevalonic acid pathway. It's significant to us humans that many plant phytoalexins are also antioxidants that are important in human nutrition to suppress human disease. For example, *resveratrol* is a phytoalexin produced by plants in response to various stresses; it promotes disease resistance in plants. And as you probably know, resveratrol is thought to have antioxidant and anti-inflammatory properties to protect humans against diseases such as cancer, diabetes, and Alzheimer's.

The effects of nutrient balance on disease suppression are complicated. Specific mineral nutrients have been reported to help plants avoid and/or overcome disease infection (see table 7.2). In general, disease microorganisms are stimulated or suppressed by the presence of specific plant root exudates. Some root exudates act as signals to disease-causing microorganisms, while others signal beneficial microorganisms (as discussed in "Delving Deeper" on page 32).

Further, there is a relationship between plant production of disease-stimulating root exudates and certain plant mineral nutrients. For example, the interaction of low potassium and excessive nitrogen levels may stimulate some disease microorganisms or make plants more susceptible to disease microorganism infection. Hence in many cases of crop disease, low potassium and excessive nitrogen levels in plants enhance disease infection.[3] Researchers have found that increasing the nitrogen in grape leaves resulted in a decrease of the defensive biochemical responsible for downy mildew resistance. Another plant disease, *Fusarium oxysporum*, is favored by high levels of phosphorus and the ammonium form of nitrogen. Rust diseases and powdery mildew are generally encouraged by high nitrogen levels. However, nitrogen is not always a disease stimulant. Enhanced plant nitrogen levels have been reported to *discourage* diseases such as *Alternaria* (leaf spot disease), *Fusarium oxysporum* (wilt and rot disease), and *Xanthomonas* (bacterial spots and wilt). This is an excellent example of the way disease suppression always depends on more than one relationship in the ecosystem.

Other plant nutrients elicit a more uniform response. Higher potassium concentrations generally increase the resistance of many plants to fungal diseases, particularly those that infect aboveground plant parts. Higher levels of calcium in plant tissue also seem to help plants resist certain diseases. Related to calcium levels, higher pH almost always helps most plants resist specific diseases (as shown in table 7.2)

Micronutrient levels also play a role in helping plants resist disease.[4] Low levels of boron, manganese, copper, and zinc often enhance plant disease susceptibility. Carl researched micronutrient effects on brown rot disease suppression in the 1990s and developed his Mineral Mix Bloom Spray, which I still use. In fact, it is the only spray I now apply to my orchard. (See table 7.3 for details about this spray.) Over the years we found that applying the Mineral Mix Spray was more effective for suppressing disease on apple and pear trees than spraying with biological control organisms or disease control sprays (such as sulfur). This was also true for peach disease suppression. Notice in table 7.3 that the rates and numbers of applications of our Mineral Mix Spray changed each year based on climatic conditions, specifically rainfall. In years when the bloom period was rainy, we would add more sulfur and gypsum to the spray.

ENHANCING DISEASE-SUPPRESSIVE MICROORGANISMS

Ecological management focuses on enhancing microbial ecology, rather than simply killing disease-causing microbes. Beneficial soil and leaf microorganisms both have the ability to suppress plant disease microorganisms, but in this chapter I primarily cover the effects of soil microbes. The miraculous way plant roots manage to simultaneously communicate with neighboring plants, the soil, and both beneficial and disease-causing microorganisms creates a balance that keeps disease microorganisms from taking over. Some of the exudates released from plant roots initiate and manipulate interactions between roots and soil microorganisms. Beneficial microbes receive biochemical signals from plants to initiate root colonization and symbiosis. Root exudates also act as antimicrobial defense mechanisms against disease microorganisms. Soilborne diseases occur when there is disturbance of the rhizosphere and the soil microbial community balance.[5]

Table 7.2. Relationships Between Mineral Nutrients and Disease Resistance

Mineral Nutrient Addition	Increased Resistance	Decreased Resistance
Gypsum	Potato to potato scab Tomato and onion to pythium root rot	
Lime*	Cucumber, tomato, and watermelon to fusarium wilt Pea, pepper, cucumber, and bean to pythium root rot	Potato to potato scab
Potassium	Celery, melon, and tomato to fusarium wilt Tomato to verticillium wilt	
Nitrogen	Potato to verticillium wilt Tomato, cucumber, and watermelon to fusarium wilt	Eggplant and tomato to verticillium wilt
Phosphorus		Tomato and melon to fusarium wilt

* This effect may not be due directly to lime (and the calcium it adds), but also to the increase in soil pH.

Table 7.3. Mineral Mix Bloom Spray Composition and Application Rates per Acre

Material	Purpose	Pounds per Acre*		
		2013[1]	2014[2]	2015[3]
Micronized sulfur	Sulfur fertilizer, disease management	8	10	6
Maxicrop (dry kelp)	Micronutrient fertilizer, plant growth regulator	1.5	1	1.5
Azomite rock dust (calcium aluminosilicate plus 70 minerals)	Mechanical coating of bloom to create a barrier to suppress disease infection	5	10	5
Gypsum (22% calcium, 16% sulfur)	Calcium and sulfur fertilizer, mechanical coating of bloom	6	10	6
Ferrous sulfate	Iron fertilizer, pH adjustment (to reduce pH to 6.6–6.8)	5	10	5
Sulfate of potash (50% potassium, 18% sulfur)	Potassium and sulfur fertilizer, mechanical coating	5	6	5
Manganese sulfate (31% manganese)	Manganese fertilizer	5	6	5
Solubor (boron)	Boron fertilizer	1	1.5–2	1
Nutramin rock dust (1% calcium, 2.5% iron, 2% sulfur, 30% silica)	Mechanical coating of bloom to create a barrier to suppress disease infection	2	1.5	2
Activate (a humate)	Ingredient aggregator	1	1	1
Therm X70 (derived from yucca)	Spreader sticker	¼	¼	¼

* Material rates change depending on weather and rainfall. These examples are for peach bloom sprays 2013–2015. All ingredients mixed in 300 gallons (1,136 L) of water and applied from first bloom through petal fall using an air blast sprayer. 1 pound per acre is equivalent to 1.12 kg per hectare.

[1] One rain event occurred during bloom; there were three applications of Mineral Mix Spray.

[2] Two rain events occurred during bloom; there were three applications with higher rates of sulfur and gypsum to cover and protect blossoms and suppress brown rot disease microorganisms.

[3] The weather was very dry; only one application was made with a lower sulfur rate, but normal amounts of mineral and micronutrient were added.

Garden and farm soils become disease-suppressive by shifting the balance of the total soil microbial community from disease-causing to beneficial and non-harmful microorganisms. Beneficial microorganisms suppress disease-causing microorganisms by directly preying upon or outcompeting them and also by enhancing plant growth and turning on plant resistance and defense systems.[6] In other words, beneficial microorganisms inhibit the bad guys by eating them, producing antimicrobial biochemicals, and/or taking up most or all of the living space in the root and on above-ground plant parts. Further, particular plant species defend themselves against soilborne diseases through *selective* stimulation and support of beneficial microorganisms.[7]

Most researchers report that a diverse and abundant soil and plant microbial community provides the greatest disease suppression. But not all studies concur on this point and some researchers insist that specific beneficial microorganisms will suppress particular diseases best. Several types of disease-suppressing organisms and relationships are present in a healthy soil and plant microbial community.

Mycoparasites and hyperparasitic fungi. These are fungi that parasitize other fungi.

Plant growth-promoting rhizobacteria (PGPR). These bacteria colonize plant roots, improve plant growth, and suppress disease-causing microorganisms.

Consortiums of mycorrhizal fungi and PGPR. Different types of beneficial microorganisms, such as mycorrhizal fungi and PGPR, often work together to provide disease suppression. In one study, seed inoculation with a mix of PGPR, *Rhizobium* bacteria, and mycorrhizae resulted in the best yields, plant growth, nitrogen and phosphorus content, and disease suppression in dry bean crops.[8]

Specific antagonistic bacteria and fungi. Certain beneficial bacteria, such as *Pseudomonas* and *Azobacter,* and fungi, such as *Trichoderma* and *Gliocladium,* have been reported to suppress and/or prey upon disease organisms. There is also evidence that biochemicals produced by *Pseudomonas* bacteria (called siderophores, pseudobactins, and pseudomycins) may also help to suppress disease organisms.

Specific microbiomes in the root zone. Scientists suggest that a microbiome of specific plant-associated microorganisms can actually broaden the immune and defense functions within plants.[9] In fact, plants may even actively recruit their own specific microbiomes by releasing biochemicals into the soil in the zone where roots and microbes interact. These biochemicals selectively stimulate beneficial rather than harmful microorganisms.[10]

Epiphytic bacteria and fungi. These bacteria and fungi live non-parasitically on the surface of plant leaves, roots, flowers, buds, seeds, and fruits. Most do not affect the plant on which they live, but some can be either beneficial or detrimental. For example, back in 1994 a microbiologist at California State University–Chico cultured a yeast that was found on the leaves of our peach trees in the Woodleaf Farm orchard. It turned out to be *Aureobasidium pullulans,* which is a fungus now used commercially to suppress diseases of stone fruit caused by *Monilinia* spp. fungi (that cause brown rot disease) and *Botrytis cinerea* (gray mold). It is also now used as a biocontrol agent for fire blight protection in organic apples and pears.[11] However, when we sprayed the commercial formulation of *A. pullulans* (not collected from our orchard tree leaves) on our apples for fire blight in 2013 and 2014, it was not as effective as our standard Mineral Mix Bloom Spray. In chapter 12 I describe in more detail the native beneficial microorganisms discovered on peach

Hyphal Networks

Filamentous fungi are composed of *hyphae* (branching filaments that make up the vegetative part of a fungus) that grow and branch into a complicated, expanding network called the mycelium. Filamentous fungi are able to colonize most natural environments due to their ability to form an expanding interconnected network. For example, hyphal networks created by mycorrhizal fungi connect individual plants together and transfer water, carbon, nitrogen, and other nutrients. These mycorrhizal hyphal networks are also reported to be conduits for interplant signaling, which appears to influence plant defenses against plant-eating insects and fungi.

leaves at Woodleaf Farm, California, and the part they played in suppressing brown rot disease (see figures 12.10 and 12.11).

Disease suppression occurs by these diverse soil and foliar microbial communities acting against soilborne and foliar disease-causing microorganisms in multifaceted and usually interconnected ways. Beneficial soil microbes suppress disease as they interact to compete for mineral nutrients and soil and root space, feed on and parasitize fungi and bacteria, produce antimicrobial biochemicals, and promote induced systemic resistance within plants. Beneficial microorganisms have been shown to induce resistance in plants both pre- and post-disease infection.

This phenomenon is called *induced systemic resistance* (ISR) or *systemic acquired resistance* (SAR). SAR and ISR are similar, but have different biochemical signals or "signal transduction pathways" in plants. Generally, this is what occurs: Plant growth-promoting bacteria and fungi in the rhizosphere sensitize, or turn on, the plant immune system and prime the whole plant for defense against a broad range of disease microorganisms (and also pest insects). These plant defense systems can be triggered even as seeds germinate and beneficial microorganisms (including *Pseudomonas, Bacillus, Trichoderma,* and mycorrhizal species) colonize their emerging root systems. In other words, once associated with the plant's root system, these beneficial microorganisms signal the plant to turn on its own defense system. That signal is then relayed throughout the entire plant so all parts of the plant, not just the roots, can develop resistance to disease microorganisms. The beauty and complexity of the ecological relationships involved seems almost magical!

Several biocontrol products based on bacterial biochemicals produced by *Pseudomonas, Bacillus, Streptomyces,* and *Agrobacterium* strains are available commercially (see "Beneficial Microorganisms as Biological Controls" on page 146). Commercial formulations of PGPR can be applied to seeds as they are planted. These commercial PGPR have reportedly activated induced resistance in several crops and reduced the incidence of anthracnose, angular leaf spot, fusarium wilt, and cucumber mosaic virus. Results are spotty, however. To build disease suppression in the long term, system strategies to support native beneficial microbial communities are usually necessary. We stopped using commercial microbial products once we realized that it was cheaper and more effective instead to take care of and enhance the native microbial populations in our soils and on our crop leaves. Some new evidence supports our findings.

Strategies to Enhance Disease-Suppressive Microorganisms

The list of ways to enhance disease-suppressive microorganisms is not dissimilar to the ways to make soil microorganisms happy (as discussed in chapter 2). You will see a pattern emerge here!

REDUCE SOIL DISTURBANCE

Frequent tillage and weed cultivation can harm populations of beneficial microbes that suppress disease organisms. Several studies report a connection between soil disturbance and increased disease. Bacterial spot of radish was significantly lower in no-till than tilled soils. *Pythium* disease organisms were highest in cultivated forest tree nursery soils as compared with undisturbed forest soils.[12] In a long-term study, disease suppression was improved by reduced- and no-tillage farming strategies, even without the addition of crop rotation. The researchers compared conventional deep plowing (that turns the soil over) with non-inversion tillage methods (such as chisel-plowing). The non-inversion methods changed the vertical distribution of soil carbon and increased the amount of soil microbial biomass. They also resulted in carbon and microbial biomass being concentrated closer to the soil surface. Further, this long-term study found greater abundance of saprophytic (non-disease-causing) *Fusarium* species in reduced-tillage plots and lower levels of the *Fusarium* species that cause plant disease.[13] In other words, increased soil disturbance changed the composition of good versus bad fungi in the soil, shifting the balance toward disease-causing microorganisms.

Other studies have found some increased disease on no-till farms, but those are generally not ecologically managed reduced-tillage farms.

APPLY MULCHES AND COMPOSTS

Mulch and compost contain or create habitat for and feed disease-suppressing microbial populations.[14] Composted wood bark has been shown to suppress several diseases, including root rots and damping-off disease.[15] Researchers report that the duration of disease suppression is highly variable, from 1 week to 3 years. They suggest that suppression is related to biologically active organic matter, or what I refer to as unstable organic matter, that is still decomposing and providing nutrients and energy for microorganisms. Some researchers report that soils with long-term annual organic matter additions are more disease-suppressive.[16] Others suggest that the quality, decomposition speed, and carbon-to-nitrogen ratios of specific organic matter additions affects a material's disease suppression ability.[17] For example, Australian researchers compared horse, chicken, and cow manure as well as alfalfa hay and mushroom compost for disease suppression potential.[18] They found that alfalfa hay was the most suppressive of *Sclerotinia* lettuce drop disease. It turns out that different mulches and composts suppress different diseases better than others. So if a garden or farm has a serious, recurring disease problem, it might be worthwhile to read up on experiments on the effects of mulching on that specific disease. In general, most surface mulches can suppress the spread of diseases. On the other hand, some mulches have been shown to encourage disease, especially in wet climates with heavy, poorly draining soil when only one kind of mulch (like straw) is used regularly year after year.

COVER CROPS AND LIVING MULCHES

Although there are complex and numerous ecological relationships involved in disease suppression, generally if we keep living roots present in the soil and create diverse microbial habitat, then disease suppression organisms will thrive. But the management details of those living roots matter, because they can both suppress and encourage some diseases. Planting cover crops and living mulches before a garden or crop field is established or as a mowed pathway between crops encourages beneficial microbes and discourages disease-causing microbes in several ways. Chapter 4 provides detailed information about cover crops and living mulches, but there are a few important details to consider when focusing on them specifically for disease suppression.

- Choose the appropriate cover crop for your climate, soil type, crops, and specific disease history.
- Choose when to use an untilled living mulch versus an incorporated cover crop.
- Choose the timing for mowing a living mulch or incorporating a cover crop based on rainfall, humidity, soil moisture, and temperature (which affect both residue decomposition and the development of particular diseases).

Cover crops and living mulches help to suppress disease in several ways. First, they cover bare soil and act as a physical barrier against disease microbes splashing from the soil onto plant leaves and stems. An interesting study shows this principle in practice. A no-till wheat cover crop/living mulch significantly reduced splash spread of the disease microorganism *Phytophthora capsici*, and hence reduced disease severity on bell peppers.[19] Incorporated cover crops also suppress diseases in several ways. They exude substances from their roots as they begin to decompose. These decomposing plant residue biochemicals can be temporarily toxic to disease-causing microbes. For example, cover crops in the Brassicaceae family contain biochemicals called glucosinolates. When the cells of brassica cover crops are ruptured (such as when they are incorporated into the soil), glucosinolates are enzymatically converted to compounds that suppress soilborne diseases. Specific cover crops also differentially affect the levels of carbon and nutrients in the soil that both soil microbes and crops depend upon. Thus they can change microbial community dynamics, shifting the balance to beneficial or to disease-causing microorganisms.

In China, where common North American species of living mulches are used widely for soil management in apple orchards, a long-term study found that all the living mulches studied increased soil bacterial community diversity. However, white clover and crown vetch living mulches improved the activity levels of several soil enzymes while perennial ryegrass improved the activity of only one soil enzyme.[20] Soil enzyme activity is important because it increases the speed at which plant residues decompose and release plant-available nutrients. They are also a general indication of microbial abundance and activity. Other researchers studying living mulches in Chinese apple orchards found that white clover, crown vetch, and bird's-foot trefoil living mulches promoted soil microbial carbon metabolism more effectively than did orchard grass. They suggest that providing a diversity of both root exudates and plant residues from a mix of cover crop species may further improve disease suppression.[21] With that in mind, let's consider the disease suppression relationships of various species commonly used as cover crops and living mulch.

Grasses

Any living mulch or cover crop is better than bare soil in terms of disease suppression, but particular crops are better than others for suppressing particular diseases. For example, Sudan grass was the most suppressive of verticillium wilt (*Verticillium dahlia*) on potatoes when compared with other cover crops including canola and field peas in an Idaho study.[22] Sudan grass as a living mulch also reduced splash spread of anthracnose-causing microorganisms when compared with bare soil plots.[23] (As mentioned above, most cover crops that form a thick sod have the potential to decrease splash spread of diseases.) In an Ohio study, oats (spring-sown, killed, and left on the soil surface as mulch) and winter rye (fall-sown as a cover crop) both reduced pumpkin fusarium fruit rot damage compared with bare soil plots.[24]

Figure 7.3. A young Sudan grass cover crop provides a thick sod and yields a lot of high-carbon residue that helps suppress diseases like verticillium wilt.

Legumes

Disease suppression following legume cover crops can be variable, according to research and my own and other farmers' results. More research needs to be done to pinpoint the mechanisms for suppression by legume cover crops. Part of the answer may be that incorporation of legume cover crop residues generally increases bacterial biomass. For example, researchers found that an overall increase in bacterial biomass in soil following a hairy vetch cover crop may have resulted in competition with disease microorganisms for both nutrients and pumpkin root colonization. *Trichoderma harzianum*, a fungus that suppresses diseases caused by *Pythium* and *Fusarium* fungi, is able to colonize red and crimson clovers and hairy vetch. This may be part of the reason that watermelon following hairy vetch and crimson clover cover crops had reduced fusarium wilt and improved mycorrhizal colonization of the roots. Clover living mulches have been shown to enhance mycorrhizae and suppress root rot diseases in some orchards, probably by enhancing root-associated beneficial fungi populations. However, there are other reasons. For example, in another study, clover living mulches in peach orchards enhanced native populations of the beneficial microorganism *Beauveria bassiana*. Spring-sown medic maintained as a living mulch was shown to reduce pumpkin fusarium fruit rot damage compared with bare soil plots.[25] Pumpkins grown in no-till hairy vetch had a higher yield and 36 percent less plectosporium blight and 50 percent less black rot compared with conventional tilled, bare soil plots.[26] A tilled hairy vetch cover crop also reduced fusarium wilt on watermelon by 20 to 60 percent compared with non-cover-cropped plots.[27] The hairy vetch also increased the total number of bacteria in the soil.

On the flip side, in some studies disease-suppressive bacteria actually *decreased* when legumes were part of an incorporated cover crop mix. More research needs to be done to discover exactly why. Further, some organic farms have reported that certain legume cover crops seem to increase particular diseases. For example, legume cover crops may increase fusarium diseases in pea crops that follow them. Common vetch can encourage verticillium wilt caused by *Verticillium dahlia*. California research shows that the cool-season legumes 'Windsor' broad bean, bell bean, field pea, hairy vetch, common vetch, purple vetch, and 'Lana' woolypod vetch are susceptible to colonization by *Verticillium dahliae* and can spread it to susceptible crops, even though all of these infected legume cover crops showed no signs of verticillium wilt. Because there is less chance of spreading the disease to susceptible crops, the researchers suggest that bell bean and hairy vetch are the best of these legume cover crop choices if verticillium wilt is or has been present in your crops.[28]

Brassicas

Cover crops in the Brassicaceae family have shown the best disease suppression results in research trials and in practice on organic farms. For example, when pea, rapeseed, and wheat cover crops were compared, rapeseed inhibited bacterial wilt (caused by *Ralstonia solanacearum*) most effectively on tomato.[29] Soil incorporation of rutabaga lowered populations of root knot nematodes (*Meloidogyme chitwoodi*). Broccoli and Brussels sprouts have been shown to reduce incidence of verticillium wilt when incorporated into the soil.

Based on research, different species of brassica cover crops affect specific diseases differently. For example, Indian mustard was most effective against potato scab (*Streptomyces scabiei*), whereas canola and rapeseed were effective against black scurf (*Rhizoctonia solani*) on potato. Even at the cultivar level, mustards vary in their disease suppression ability. For example, 'Caliente 61' and 'Caliente 119' mustard varieties resulted in the greatest potato scab suppression during studies at the University of Vermont. White mustard grown as a fall crop/

winter cover, then turned under the following year suppressed root rot of peas. In another study, Indian mustard was most effective for reducing powdery scab and common scab potato diseases, whereas rapeseed and canola were most effective in reducing *Rhizoctonia* potato diseases.[30]

Brassicas seem to suppress disease in several ways. Disease-suppressive bacteria have been shown to increase substantially with soil incorporation of brassica plant material. As mentioned above, glucosinolates that are released from tilled or mowed brassica cover crops have been shown to suppress fungal disease. Glucosinolates from a finished crop of decaying cabbage reduced damping off and root rot disease caused by *Pythium ultimum* and southern blight caused by *Sclerotium rolfsii* during field studies in California. Diseases caused by *Pythium* and *Sclerotinia* fungi were reduced by tilling in cabbage leaves and stems.[31] Glucosinolates released from incorporated Brassicaceae family cover crops reduced potato verticillium wilt.[32]

Brassica cover crops suppress disease-causing microorganisms so well, in fact, that they should not be used regularly because they have the potential to suppress fungi *too much*. Too much fungal suppression could throw the microbial balance off in almost the same way as can disease-killing minerals like copper. Most of the Brassicaceae are non-mycorrhizal and inhibit mycorrhizae fungal spore germination due to the antifungal compounds produced by their roots. Decomposing brassica mulch or litter can dramatically decrease fungal diversity. More research on these ecological relationships needs to be done, but in one study, canola decreased fungi species, including mycorrhizae, and total microbial biomass.[33] It turns out that, generally, living brassicas may not be so broadly antifungal. However, in another study invasion of the weedy species *Alliaria petiolata* (garlic mustard) decreased levels of other disease-suppressive fungi. Another reason to be careful is that some Brassicaceae cover crops inhibit crop seed germination and growth for periods of 1 to 2 weeks after cover crop incorporation into soil. Decaying brassica residues have been shown to especially inhibit small-seeded crops like lettuce. To avoid crop injury, mow and incorporate cover crops 2 to 4 weeks before planting annual crops. So if your main goal is to suppress a specific disease in a field or a bed, incorporated brassica cover crops may be a good choice, but use them with some caution.

Intervention When Diseases Impact Crops

The ecological relationships involved in disease suppression are complex. During the unusually hot, dry summer of 2022, for example, for the first time I observed vegetable crop disease in my high-carbon, microbially rich, diverse system. Bacterial spot occurred on drought-stressed radishes. After some research, I discovered that the bacteria that cause this disease are better able to survive dry soil conditions than disease-suppressive microorganisms are. Because researchers and growers are still learning how to suppress disease using a systems approach, direct intervention to manage disease may sometimes be necessary. This section explains how to use minimal-impact, moderate-impact, or heaviest-impact intervention materials only when disease populations become high enough to be economically damaging or simply too high for you to accept in your garden. As with pest insects, I limit the discussion here to materials that are accepted for use on certified organic farms.

MINIMAL-IMPACT INTERVENTIONS

These are the methods and materials that present the least chance of adversely affecting beneficial soil or leaf microorganisms.

Beneficial Microorganisms as Biological Controls

These are sometimes called and sold as biofungicides, but I don't use that term because it removes the focus from a systems approach in which disease suppression is an integration of healthy plants, healthy soils, and balancing beneficial and disease-causing microorganisms.

Ampelomycea quisqualis is a fungal hyperparasite of powdery mildew on grapes, tree fruits, vegetables, and flowers. It has been shown to have an enhanced efficacy when mixed with horticultural oil. High relative humidity conditions are required for this microorganism to suppress disease effectively in the field.

Bacillus subtilis is available as a bacterial seed treatment to control fusarium, pythium, and rhizoctonia root rots. It reportedly reduces white mold of snap beans (*Sclerotinia sclerotiorum*).

***Bacillus subtilis* QST713** has been shown to reduce disease substantially in controlled environments (such as greenhouses or high tunnels) but shows variable efficacy in field trials. It has been used for clubroot (*Plasmodiophora brassicae*) on brassica crops and has reduced powdery mildew on greenhouse cucumbers.

Burkholderia cepacia is used as a seed treatment or as a soil drench after transplanting for fungal disease and nematodes.

Candida oleophila suppresses post-harvest fungal disease on fruit and vegetable crops.

Fusarium oxysporum provides *Fusarium* suppression for fruits and vegetable crops.

***Gliocladium* spp.** provide *Fusarium* and *Pythium* suppression.

Gliocladium virens is used for *Fusarium* and *Rhizoctonia* suppression. It should be incorporated into soil or soilless growing media at the rate of 1 to 1¼ ounces per cubic foot (1,001–1,251 g/m³) of media 3 days prior to planting. In planting beds, the rate is ¾ to 1 ounce per square foot (229–305 g/m²). It can also be applied to seed as a treatment coating. Best results are in well-drained soil with an acidic pH.

Pseudomonas cepacia has been shown to suppress pythium root rot on grasses, sclerotinia root rot on sunflower, and *Botrytis* on greenhouse tomatoes. Repeated applications at 10- to 14-day intervals appear to be necessary.

Pseudomonas chloraphis is good for soilborne and seed-borne fungal disease suppression.

Pseudomonas fluorescens helped to suppress fire blight, *Fusarium*, and *Rhizoctonia*.

Pseudomonas syringae suppresses fungal disease of stored fruits and vegetables.

Reynoutria sachalinensis is an extract of giant knotweed. It boosts plant vigor and growth while stimulating the plant's ability to develop resistance to plant pathogens (induced resistance). This developed resistance builds plant health as opposed to attacking the disease directly. It suppresses plant diseases including *Fusarium*, *Pythium*, *Rhizoctonia*, and powdery mildew.

Streptomyces griseoviridis suppresses *Fusarium*, *Alternaria*, *Pythium*, and *Phomopsis*.

Streptomycin provides suppression of fire blight, bacterial blight, and bacterial wilt.

Trichoderma harzianum* plus *T. viride provides root disease suppression. In vegetable crops, it is used against pythium root rot, dollar spot, brown patch, and *Fusarium* (tomato). In strawberries, it is used against *Botrytis*. On tree fruits, it works as a wound sealant and as a preventive against *Armillaria*.

Trichoderma atroviride* plus *T. viride can be used as a seed treatment, root drench, or foliar spray within a temperature range of 39 to 91°F (4–33°C). It is effective on wheat for snow mold fungi; on verticillium wilt of potatoes, tomatoes, strawberries, and fruit trees; on pythium damping off of peas; on black scurf of potatoes; on late blight of potatoes; on sclerotinia rots on flowers, potatoes, beans, carrots, and plants in the cabbage family; and on fusarium stem rot on tomatoes and flowers.

Antitranspirants

Antitranspirants, which are commonly applied to reduce transpiration and inhibit plant water loss, have been shown to suppress several foliar diseases, including powdery mildew and phytophthora blight. However reports about their disease suppression effectiveness are mixed. They are thought to provide a mechanical barrier on leaf surfaces that inhibits fungal germination on leaf tissue.

Potassium Bicarbonate

There are several commercial formulations of potassium bicarbonate. They have been used with varying degrees of effectiveness for powdery mildew on vegetable and fruit crops. Best results are obtained by using potassium bicarbonate dissolved in water at the recommended rate for the commercial formulation you have purchased and then adding ¼ to ½ teaspoon of insecticidal soap or vegetable oil per gallon (4 L) of the water–potassium bicarbonate mixture. Sometimes potassium bicarbonate can cause leaf burn, especially in sensitive young plants. Test first! Sodium bicarbonate, commonly known as baking soda, has also been used for powdery mildew, but it is not generally as effective as potassium bicarbonate.

Compost Tea

I have not had great results using compost tea for disease management, but other farmers I work with have had good results. The efficacy of compost teas depends on the materials used in the compost and the method of tea preparation (aerated or not aerated) and application (addition of spreader-stickers and adjuvants). It is worth experimenting with all of these variables for specific diseases. There are many sources of information on making and using compost tea; one good one is the website of the National Sustainable Agriculture Information Service (ATTRA) at https://attra.ncat.org. Researchers in Germany found that compost tea used as a preventive foliar spray inhibited *Phytophthora* on tomatoes and potatoes and downy mildew and powdery mildew on grapes. Other researchers have reported minor to major suppression of several plant diseases using compost extracts, including apple collar rot, apples cab, *Botrytis*, *Fusarium*, and *Phytophthora*. But some researchers also report that compost teas have no effect on foliar diseases.

MODERATE-IMPACT INTERVENTIONS

These are methods and materials with more chance of injuring beneficial microorganisms and natural enemies, but if they are used carefully, sparingly, and discriminately, they will not permanently disrupt the agroecosystem pest suppression you have designed.

Horticultural Oils

Horticultural oils include petroleum-based oils and vegetable oils (such as canola oil). There are also essential oils (such as mint oil), but they are usually not classified as horticultural oils and are used in small quantities. Petroleum-based oils can be toxic to soft-bodied predators, such as syrphid fly larvae and predatory spider mites.

HEAVIEST-IMPACT INTERVENTIONS

New research shows that some of the following high-impact intervention materials have problematic ecological impacts. For example, lime sulfur, Bordeaux mix, and copper kill nontarget beneficial natural enemies and microorganisms living and growing on leaf surfaces, creating an ecological vacuum effect that facilitates the multiplication of disease microorganisms.[34] If you decide to use these materials, you can find application instructions and information on the product labels and seek advice from your state's extension services.

Neem

Neem is especially effective against powdery mildew and downy mildew. It functions in two ways. As a preventive, it coats leaf and stem tissues so that spores are unable to adhere. As an inhibitor of fungi already present on leaves, it seals fungal patches so that spores may not be released. See also the "Neem Oil" section on page 133.

Sulfur

Sulfur is a naturally occurring mineral that plants require for optimum nutrition and it is an effective fungicide used against powdery mildew, rust, apple scab, brown rot, and other fungal diseases. It is best used as a preventive before heavy disease infection occurs. Don't use sulfur fungicides when temperatures are above 90°F (32°C) or within a month of horticultural oil applications, because burning of plant tissue could occur. An added advantage of sulfur is that it can suppress spider mites. It is toxic also to some predators and parasites (including predatory mites).

Bordeaux Mix

Bordeaux mix contains copper sulfate and lime and is used as a dormant disease control against black spot, rust, anthracnose, and bacterial diseases such as fire blight and bacterial wilts. Bordeaux mix must be used carefully to avoid damage to plants. It is highly toxic to fish and aquatic life. See copper precautions below.

Copper and Copper Hydroxide

Copper is a metal useful against many plant diseases such as bacterial blight, leaf spot, fire blight, and anthracnose. It is best used as a preventive management tool before heavy disease infection occurs. However, copper is very persistent and can protect foliage for longer periods than both sulfur and lime sulfur.

Copper materials for sprays differ in their chemical form, the amount of metallic copper (active ingredient) present, and whether the copper material is a liquid or a dry formulation. "Fixed" copper, such as copper hydroxide or copper oxychloride, has a longer residual time than soluble copper formulations and is generally used only for delayed dormant sprays. Soluble copper formulations, such as copper octanoate, are designed to reduce copper toxicity to leaves and fruits and can be sprayed when foliage is present. Do not spray copper during bloom! Apply no more than two times in a single season. If using copper, you need to know what form and amount you are using so you know when and how to spray it! Check with your supplier or local extension agent if you are unsure. Copper solutions, when diluted, are only moderately toxic to humans but are skin and eye irritants.

The use of copper is discouraged, especially over extended periods of time, because it is a heavy metal that remains in the soil indefinitely and can leach into ground- and surface water. Because of this persistence, it is also toxic to soil microorganisms. Copper sulfate is highly toxic when concentrated in the environment, especially to fish and aquatic life. Finally, copper fungicides can harm populations of beneficial leaf-inhabiting microbes that naturally suppress disease and, in some cases, contribute to greater perennial crop decline as a result.[35] Personally, I cannot justify the use of copper on my farm. Carl and I stopped using copper by the early 1990s. When it comes to managing plant diseases, I recognize that I am fortunate to farm in a dry climate that is less encouraging to disease.

Lime Sulfur

As the name indicates, lime sulfur is a combination of lime and sulfur. Adding lime can make the sulfur more effective as a fungicide, but the combination can also cause plant damage more easily and is more toxic to mammals, including humans. Lime sulfur is used to control anthracnose, brown rot and leaf curl on peaches, mildew, apple scab, and certain scale insects. It is an effective dormant-season fungal disease preventive. Lime sulfur has the ability to burn out and kill disease organisms that have already begun to infect leaves. When used during the season, it has provided the best organically acceptable option for control of diseases such as apple scab in high rainfall areas in the eastern US, but such use can cause significant amounts of foliage burn.

Plant damage caused by lime sulfur is worse during dry weather when temperatures reach 80 to 90°F (27–32°C). However, this foliage burn effect also seems to allow lime sulfur to burn out germinating fungal spores. Thus it can be sprayed preventively, but can also target already-germinating fungal spores for 24 to 36 hours after application. I stopped using lime sulfur as a dormant spray for peach leaf curl in my orchard because its persistence in the environment led me to worry about the negative effects on beneficial foliar and even soil microorganisms.

CHAPTER 8

Working with Plant Competition

Nature insists on keeping the ground covered to protect soil health and ecosystem functions. Bare soil is unnatural, whether in a farm field or a backyard garden. Weeds clearly want to be a part of the ground cover diversity and can play a vital role in helping us keep a growing root in the soil year-round. The key is to know which weeds fit into your garden or farm design and soil fertility system and which don't. From there, you can experiment to discover the benefits of specific weeds. For example, as I described in chapter 5, annual spring-blooming cool-season brassica weeds are a part of pest suppression strategies on my farm. I also eat some of my "weeds" and use them medicinally. Because my soil organic matter system provides slow, steady, optimal nutrient release, I can design most weeds and other non-crop vegetation into my farm system without causing too much competition with crop plants. I do not weed much and when I do, I weed very selectively (later in this chapter, I explain precisely what I mean by weeding selectively). This approach might not work in a low fertility or early-developmental-stage agroecosystem, but it usually does once a system reaches dynamic equilibrium.

In this chapter I explain how to develop a weed-tolerant system that will work for an individual farm or garden based on climate, soil, and crop production needs. This includes consideration of which weeds are more or less competitive with specific crops, and deciding which ones are most likely to peacefully coexist with the food crops you want to grow. Instead of getting rid of all weeds, all the time, focus on suppressing the ones that compete most with your crops and provide the fewest system benefits.

Figure 8.1. Blooming weeds such as this shepherd's purse provide habitat and easily accessible nectar for early-spring beneficial insects within crops, including this syrphid fly.

When we decide to adopt a systems approach to weed suppression, rather than a zero-weed-tolerance policy, it takes planning, understanding weed ecology, and some monitoring (just as do insect and disease pest suppression). Here are the general steps for working with plant competition as you develop your weed-tolerant agroecosystem:

1. Characterize the places where weeds are a problem. Is the area dry, wet, sunny, shady?
2. Decide what vegetation and ground cover you want for the area and make sure the species are well adapted to your soil and climate.
3. Identify your weeds. For each problematic weed, you'll also want to learn some key aspects of its ecology:
    - What is its life cycle—annual, biennial, perennial?
    - How does it reproduce—by seed or vegetatively?
    - What growing conditions does it prefer?
    - When is it most vulnerable to suppression strategies?

Tables 8.1, 8.2, and 8.3 provide information on common weeds, with details taken from my own experience and that of longtime weed researchers. These tables should be helpful to you as you do your research. Once you've assembled information about the weeds you're dealing with, you can apply what you've learned to limit the optimal conditions for the weeds that compete most strongly with crops. In general, this includes:

- Minimizing soil disturbances.
- Avoiding bare soil (cover soil with non-weedy, vigorous vegetation that fits your garden or farm plan).
- Developing a plan to suppress weeds that reproduce from underground stems or root pieces, and figuring out how to avoid their spread at times when you cultivate or till an area.
- Choosing and combining several management tools that target vulnerable stages in the weeds' life cycles.

In this chapter I first explain how to assess your weed situation and decide which weeds pose the greatest risk of outcompeting crops. This includes a discussion of the ecology of some common dominant weeds of farms and gardens. With that knowledge in mind, we move on to a discussion of how to design weeds into your farm or garden system. And to conclude, I cover ways to intervene when a weed is outcompeting crops, starting with minimal-impact methods and concluding with the heaviest-impact methods that should be used only in a crisis situation.

Assessing Weed Competition

One farmer's weed can be another gardener's friend. In order to decide whether a particular weed is too competitive or has benefits in our farm and garden systems, we fall back on ecological function. The first step is to understand the ecological niches specific weeds fill. A weed's ecological niche is based on the weed's characteristics and its interactions within the garden and farm system. These include life cycle, method of seed production, and seed longevity, as well as bloom time and pollen/nectar availability for insects. A weed's *economic niche* is calculated based on the time and costs to keep that weed below levels that interfere with your desired crop growth or yield. Specific weeds have different height, meaning shade potential, and root competitiveness with specific crops. Weed species also vary in quantity of seed produced, and the seeds differ in their longevity and length of dormancy. These are the tools we start with in designing a weed tolerance/suppression plan in our gardens and farms.

START WITH ID AND LIFE CYCLE

Like all seed-bearing plants, weeds can be *broadleaf* plants or *grasses*. Dandelion, redroot pigweed, and

Figure 8.2. It is important to be able to identify weeds at their vulnerable seedling stage. The two larger seedlings in the center are pigweed and lambsquarters. In the lower right are two prickly-leaved Canada thistle seedlings. The majority of the smaller weed seedlings with rounded, slightly serrated leaves are common field speedwell.

chickweed are typical broadleaf weeds. Annual bluegrass, quack grass, and downy brome are typical grass species. And like all plants, weed species have a specific life cycle: annual, biennial, or perennial. (Table 8.1 lists common weeds and their life cycles.)

Annual weeds germinate from seed, grow, mature, and die in one season. They survive between growing seasons as seeds in the soil waiting for disturbance (tillage) to signal that it is time to germinate.

Summer annuals germinate in the spring, then seedlings mature rapidly, flower, and seed. Common lambsquarters and purslane are examples.

Winter annuals germinate in the fall, overwinter as seedlings, then mature, flower, and seed the following spring. In warm climates, they may flower by mid- to late winter. Pennycress and purple deadnettle are examples of winter annuals.

Biennial weeds have a life span of 2 years. The first year's growth is vegetative. In the second year, the plant sends up a flowering stalk from a basal rosette of leaves and the plant flowers, goes to seed, and then dies. Burdock and bull thistle (*Cirsium vulgare*) are biennials.

Perennials live longer than 2 years and in some cases for many years. Perennials reproduce from seeds and also vegetatively from root buds, stolons, and rhizomes (horizontal

Table 8.1. Ecology of Common Weeds

Common Name	Botanical Name	Life Cycle	Seed Longevity	Seed Production	Soil Needs
Wild mustard	*Brassica kaber, Sinapis arvensis*	Winter and summer annual	Low–medium	High	Adaptable, cool and moist for germination
Shepherd's purse	*Capsella bursa-pastoris*	Winter annual	Long	High	Cool and moist
Henbit	*Lamium amplexicaule*	Winter annual	Medium	Medium	Cool and moist
Chickweed	*Stellaria media*	Winter annual	Long	Medium	Cool and moist; sensitive to drought
Lambsquarters	*Chenopodium album*	Summer annual	Long	Very high	Warm and moist
Pigweed	*Amaranthus* species	Summer annual	Very long	Very high	Warm and moist
Knotweeds, smartweeds	*Polygonum* species	Summer annual	Long, unless disturbed	High	Warm and moist
Prickly lettuce	*Lactuca serriola*	Summer annual	Low–medium	Medium	Warm and moist, but very drought-tolerant
Purslane	*Portulaca oleracea*	Summer annual	Medium–long	High	Warm and moist
Bindweed	*Convolvulus arvensis*	Perennial, creeping	Long	Variable: medium to high	Well drained, irrigated or dry; can tolerate drought
Yellow foxtail	*Setaria pumila*	Summer annual grass	Low	High	Warm and moist
Quack grass	*Agropyron repens*	Perennial, creeping	Low	Medium	Adaptable; prefers cool, moist soil

underground stems). Dandelion, Canada thistle, and quack grass are examples of perennials. Perennials that reproduce vegetatively from rhizomes are often called creeping perennials. Quack grass and field bindweed are creeping perennials.

Weeds whose life cycles are most closely synchronized with specific crops compete most problematically with those crops.[1] Winter annual weeds such as chickweed and henbit become dominant in areas where fall/winter greens such as kale and winter crops like wheat are grown regularly. Some perennials can dominate in these crops, too. Summer annuals (such as pigweed) often end up dominant among short-season annual crops (such as lettuce). Summer annual and some perennial weeds are commonly dominant in crops that have a long summer growing season (think winter squash, tomatoes, and corn).

Table 8.2. Weed Competition Potential

Common Name	Reduced Tillage Decreases	Tillage and/or Cultivation Timing to Suppress	Potential for Crop Shading	Root Type	Increases with Nitrogen Fertilizer
Wild mustard	Sometimes	Early spring or before seed set	Tall	Slight taproot	Yes; big response to nitrogen
Henbit	Yes	Early spring or before seed set	Short	Fibrous	Yes
Chickweed	Yes	Early spring and all summer before seed set	Medium	Fibrous	Yes
Lambsquarters	Sometimes	Early summer before seed set	Tall	Slight taproot	Yes
Pigweed	Yes; intolerant of shade	Early to late summer before seed set	Tall	Taproot	Yes; also stimulates seed germination
Knotweeds, smartweed	Yes and no; tillage decreases	Early summer before seed set	Tall	Fibrous	Yes
Prickly lettuce	Yes, in most cases	Spring and early summer before seed set	Tall	Slight taproot	No (increases with phosphorus)
Purslane	Yes; tillage cues germination	Early spring and summer before seed set	Short	Fibrous	Yes
Bindweed	No; likes reduced tillage	Early spring, but survives tillage; suppressed by shady crops and cover crops	Medium height, but climbs on crop foliage	Taproot, creeping stolons	No (increases with phosphorus or potassium)
Dandelion	No	Early summer after seeds germinate; older plants resist tillage	Short	Taproot	Yes (also increases with phosphorus)
Yellow foxtail	Yes	Late spring to early summer after seeds germinate	Tall	Fibrous	Yes
Quack grass	No	Early spring as leaves begin to grow; summer, when dry and hot	Tall	Creeping rhizome	Yes; big response

ECOLOGY OF COMMON DOMINANT WEEDS

Here are the biological details of some common weeds that I have developed weed suppression plans for on my own and on other farms. Some, such as pigweed and lambsquarters, are dominant on farms in most parts of North America (and the world). This is also an example of the kinds of details you might gather to develop suppression plans for your own dominant weed species wherever you garden or farm. I describe some of the important ecology for summer annuals first, then winter annuals, then perennials.

Pigweed

Pigweed (*Amaranthus retroflexus*) is a summer annual weed with a pinkish red taproot. Pigweed spreads both from germination of newly released seed and from germination of seed carried over in the soil seed bank from previous years. (For more

Figure 8.3. Low-growing speedwell is blooming below tall, space-filling, taprooted wild mustard. I choose common field speedwell as a less competitive annual ground cover in annual vegetables, but I selectively weed out the taller wild mustard.

information about the seed bank concept, see "Weed Seed Bank Management" on page 164.) Seeds can germinate anytime soil moisture is adequate during the growing season, and their germination rate is very high—pigweed seeds harvested in a Mississippi study showed 94 percent viability at time of harvest. However, after burial in soil for 30 months, the seeds were only 7 percent viable.

Pigweed thrives with extensive tillage and is diminished in reduced-tillage situations. It has the capacity to extract more nitrogen from and grow faster on nitrogen-poor soils, such as recently abandoned fields. In a Michigan study, seedlings of pigweed and several other annuals emerged within 2 months after an agricultural field was rototilled, and pigweed was one of the most dominant species in that field the first growing season. But in similar adjacent fields that had been abandoned for 5 and 15 years, pigweed was not present at all.[2] In my minimal-soil-disturbance vegetable production beds pigweed is uncommon, and it is almost nonexistent in the no-till orchard.

Lambsquarters

Lambsquarters (*Chenopodium album*) is a summer annual weed with a slight taproot. It is a *short-day plant*, meaning that it increases in vegetative growth when days are long and flowers when days are short. Researchers have found a close correlation between rates of nitrogen applied and increased lambsquarters seed germination.[3] Germination was 34 percent in plots with previous nitrogen fertilization, but only 3 percent in unfertilized plots. This nitrogen stimulation of seed germination could also be related to soil temperature, because available soil nitrogen levels increase as soil temperatures increase in spring. A high density of seedlings is normally found in crop fields early in the spring, and lambsquarters can continue germinating all summer. As soon as the crop or weed canopy starts to cover or shade the soil, lambsquarters germination is inhibited.[4] Lambsquarters normally decreases with undisturbed ground cover and reduced tillage management, but seed viability does not. In one study, after 6 years buried under undisturbed alfalfa, seed viability of lambsquarters was still 53 percent.[5] In my minimal-soil-disturbance vegetable production beds, lambsquarters is rare; it is nearly absent in the no-till orchard. Hence one way to diminish lambsquarters is to reduce tillage and fertilize selectively in the crop row only.

Chickweed

Chickweed (*Stellaria media*) is a winter annual that germinates in the fall or spring, when soil temperatures reach about 60°F (16°C). However, if the soil is very moist, seeds can sprout at higher temperatures. Germination flushes can occur after an irrigation or rain. Dry and hot weather and dry soil discourage chickweed germination. The seeds typically germinate close to the soil surface. The deeper the seeds are buried, the less likely it is that they will germinate or that the seedlings will survive. Mature seeds can germinate immediately upon being released without a dormancy period. Thus, chickweed can reseed and persist as long as the weather is humid and soil is moist. It loves my high tunnel in fall and early spring for that reason.

Chickweed completes its life cycle in 5 to 6 weeks. Cultivation can suppress chickweed, but is most effective if the soil is dry and plants are small. Cultivation when plants are large and soil is moist can encourage chickweed spread through re-rooting. Burying the seed with tillage can reduce seed germination; however, it may also bring weed seeds closer to the surface. Chickweed can become competitive in my high tunnels with winter/spring greens. But it also provides early bloom for beneficial insects, so I weed it selectively. It is rarely competitive in my minimal-soil-disturbance vegetable beds; in fact, it is present only under row covers among crops sown in early spring. The chickweed disappears when the weather gets dry and hot.

Wild Mustard

Wild mustard (*Brassica kaber*) is usually a winter annual, but can be a summer annual in cooler climates (see figure 8.3). It can emerge in late summer, early fall, or spring. Wild mustard plants produce 10 to 18 seeds per pod and 2,000 to 3,500 seeds per plant. Some seed is capable of germination as soon as it matures, but seeds may also remain viable in the soil for as long as 60 years, particularly if they are deeply buried. A Canadian study aligns with this information, because after 4 years under undisturbed alfalfa, more than 50 percent of wild mustard seeds were still viable. This study also showed that incidence of wild

Figure 8.4. I selectively weed chickweed in my high tunnel, but it is still present as a blooming ground cover beneath mature kale and other leafy crops. Since it is edible, I don't worry about whether or not it gets mixed in with my greens during harvest.

mustard was lower in undisturbed, no-till cropping systems compared with tillage cropping systems.[6] My experience is similar: Wild mustard is rare in my minimal-soil-disturbance vegetable production beds and nearly absent in the no-till orchard. I tolerate wild mustard on my field edges and in the orchard where, though it is uncommon, its flowers entice wild and honeybees.

Bindweed

Bindweed (*Convolvulus arvensis*) is a long-lived, hardy, and persistent creeping perennial that spreads from an extensive horizontal root system and also from seed. Bindweed roots and rhizomes produce buds that sprout to form new plants. Even shoot and root fragments as short as 2 inches (5 cm) can form new plants. A single bindweed plant can spread more than 10 feet (3 m) in a single growing season. Its extensive underground root/rhizome network allows for overwintering without foliage and then resprouting in the spring in new places. New shoot growth begins when day temperatures are near 57°F (14°C) and night temperatures are above 35°F (2°C).

Bindweed seeds have a hard seed coat and can persist in the soil for up to 60 years. Seed production varies depending on temperature and moisture conditions, ranging from 25 to 560 seeds per plant. Seeds can germinate throughout the growing season, but peak germination is in late spring. Bindweed prefers heavier, but well-drained, clay soils rather than sandy soils, but it is found in both irrigated and dryland crops and grasslands.

Suppressing bindweed can be a challenge, but it is doable with a long-term commitment Because it is a multistage process, it's worth describing in detail. Very young seedlings of field bindweed can be suppressed with cultivation, but only for about 3 to 4 weeks after germination. After that, perennial buds are formed, and cultivation is not as effective. Alfalfa, cereal grains, sorghum, and corn in a crop rotation or as cover crops partially reduced bindweed growth. Repeated cultivation or hoeing (every 2 to 3 weeks) has been partially effective in reducing established stands of field bindweed. Subsoiling provides some suppression of bindweed. In my minimal-disturbance vegetable beds, bindweed is present in some areas and can become a problem in dry beans because it hinders harvest. It does not appear to greatly affect crop growth and yield as long as irrigation is plentiful. But in dry, hot, poorly irrigated situations, bindweed can outcompete many of my annual crops, and it can reduce yield of some crops in very hot, dry years. I suppress bindweed in early spring by wheel-hoeing. After that, I mow strips of the living mulch closest to the crop beds very short and repeat the close mowing all summer so the bindweed does not spread into crops (I leave the center part of my row middles unmowed, as shown in figure 8.10). However, bindweed plants adapt to mowing by creeping along the ground and thus escaping cutting. In my experience, mowing alone will not suppress bindweed; it takes both repeated disturbance and avoidance of bare soil areas.

Canada Thistle

Canada thistle (*Cirsium arvense*) is a creeping perennial that reproduces both from seed and from vegetative buds in its root system. Generally thistle invasions start on disturbed ground, including ditch banks, overgrazed pastures, and tilled fields. One plant can colonize an area 3 to 6 feet (1–2 m) in diameter in 1 to 2 years. Canada thistle grows in a variety of soils, but is most competitive in deep, well-aerated, fertile, cool soils. It does best where soil moisture is adequate and is less common in light, dry soils.

Canada thistle can extend its roots 15 feet (5 m) horizontally, and vertical roots can grow 6 to 15 feet deep. Plants grow in circular patches, sometimes from just one clone. Canada thistle shoots sprout from its perennial root system in middle to late spring and form a rosette of first spring leaves. The greatest flush of new shoots occurs in spring, but another flush can happen in the fall. In fact, a flush can occur anytime during the growing season when the soil is moist, particularly if growth is disturbed by tillage.

Plants germinate from seed in mid- to late spring. Seedlings grow slowly and are sensitive to competition, especially if shaded. Canada thistle seedlings develop the ability to reproduce from their root systems at 7 to 8 weeks after germination. Seed can remain viable in soil for up to 22 years, and deep burial enables seed longevity. Canada thistle is mainly spread by roots, however, because it allocates most of its energy into root reproduction and spread.

Tillage breaks roots into smaller segments and stimulates the development of new plants. Shoots sprout from root pieces about 2 weeks after tillage disturbance. Even thin root pieces as small as ¼ inch (6 mm) long can form new plants. These small roots can survive buried in the soil for up to 100 days without light or nutrients.

The key to Canada thistle suppression is to stress the plant and force it to use stored root nutrients. This plant is a survivor! Grass and alfalfa cover crops can compete with Canada thistle. Some farmers have succeeded in suppressing this weed by using a summer fallow technique followed by a cover crop, as explained in the "Weed Seed Bank Management" section on page 164. In my minimal-disturbance vegetable production beds, Canada thistle is nonexistent, and it is nearly absent in the no-till orchard.

Quack Grass

Quack grass (*Elytrigia repens*) is a cool-season creeping perennial grass. Many researchers report that quack grass is *allelopathic* (producing a suppressive effect) to crops and other grasses. It has been shown to release biochemicals during decomposition that can inhibit the germination of alfalfa.[7] In one study, the allelopathic effect was highest 7 to 10 days after quack grass was cut, and the effects diminished after 3 weeks. (See also "Suppressing Weeds via Natural Phytotoxins" on page 167.)

Quack grass stores energy in roots and rhizomes in late summer and fall that allow it to survive over the winter. In early spring it still has a supply of root energy; however, it is most vulnerable to suppression just as the shoots are emerging in the spring. At that point the stored energy from the roots is being expended on new growth, but leaves are not yet creating new energy at full capacity. Hence early-spring tillage can suppress quack grass. Even with repeated summer tillage or cultivation, though, quack grass can return in the fall after most crops have been harvested or have died back. During the hot part of the summer, leaf and shoot growth is limited, but underground rhizomes flourish.

Quack grass thrives in undisturbed, no-till crops, including winter or yearlong cover crops, and perennial crops such as asparagus and strawberries. For quack grass suppression, some weed researchers recommend leaving field edges unplanted and cultivated to reduce movement of roots and rhizomes into fields, along with field-edge tillage and in-field cultivation repeated when regrowth reaches three to four leaves per plant. Tillage can be effective because it lifts roots and rhizomes to the soil surface, where they dry out and die. Quack grass can also be suppressed by tilling infested ground in midsummer, when the ground is very dry, followed by planting a cereal rye cover crop in the fall. In my minimal-disturbance vegetable production beds, quack grass becomes part of the living mulch row middles and increases over time; after 4 years it becomes competitive, and more spring tillage/cultivation of crop rows is required to suppress it enough for good annual crop yields. Quack grass is one of my biggest challenges in minimal-till vegetable production because it increases and spreads over time with no-tillage. It was suppressed after I planted a red clover and triticale cover crop for a year and then strip-tilled the cover crop for crop beds. I maintained the untilled red clover–triticale cover crop as no-till, perennial living mulch row middles on my Montana farm for 5 years. The quack grass took 4 to 5 years to rebound and become competitive with crops in the crop beds. Then I had to do more spring strip tillage in my crop beds, followed by close mowing of the living mulch row middles. I have found that when quack grass begins to dominate in my system, my crop beds need three passes with a tiller (2 to 3 inches [5–7.5 cm] deep) in the spring, spacing the passes 7 to 10 days apart, to suppress the quack grass.

Benefits of Weeds

Assessing the competitive potential of the weeds in your system is important, but your weed observations can also provide clues about the health of your soil, especially if one weed species is dominating an area. For example, dandelions might dominate in a compacted soil. Chamomile and goosefoot usually indicate a soil with a high pH. Thistles grow well in wetter, heavier soils. Sorrel and horsetail usually prefer acidic, low-pH soils. Barnyard grass indicates a high water table. Field horsetail indicates poorly drained, low-pH soils. Common mallow indicates high levels of soil nutrients, especially nitrate-nitrogen. Dominant redroot pigweed can indicate a soil high in potassium and manganese and low in phosphorus and calcium. Lots of lambsquarters can mean your soil is low in phosphorus and high in potassium.

Weeds and certain weed management strategies can also provide habitat for beneficial insects (as discussed in "Diversity Details" on page 97). Flowering weeds such as purple deadnettle and chickweed provide nectar and pollen for syrphid flies and parasitic wasps that are predators of aphids and cabbage worms according to my own on-farm research. Other researchers have found

Table 8.3. Potential Benefits of Weeds

Common Name	Bloom, Pollen, and Nectar for Beneficial Insects	Other System Benefits
Wild mustard	Spring–fall	Trap crop for flea beetle and cabbage worm
Henbit	Early spring and fall	Especially attractive to bumblebees
Chickweed	Early spring, summer, and fall	Especially attractive to syrphid flies
Lambsquarters	Summer	Edible and tasty spinach alternative in hot climates
Pigweed	Summer	Seeds support birds and other wildlife
Knotweeds, smartweed	Summer	Seeds support birds and other wildlife
Prickly lettuce	Summer	Seeds support birds and other wildlife
Purslane	Spring and fall	Edible and rich in omega-3 fatty acids
Bindweed	June until fall	Erosion control
Dandelion	Summer	When tilled in, it suppresses tomato root and crown rot (*Fusarium*)
Yellow foxtail	Midsummer–fall	Seeds support birds and wildlife
Quack grass	Summer	Good winter soil cover

similar weed benefits for beneficial insects, including University of California entomologist Robert Bugg and the Xerces Society. Weeds also provide winter cover for resident beneficial insects. One pest insect-eating carabid beetle, *Pterostichus melanarius*, preys upon weed seeds and also uses weeds in the fields as a sheltering spot for the winter, which facilitates its rapid movement into crops in the spring. Weed seed predation by vertebrates and invertebrates (such as carabid beetles) may contribute to high rates of seed mortality in cropping systems. However, seed burial during tillage may limit the overall efficacy of predation on weed seeds.[8]

Further, weeds can provide grow-your-own fertilizer. They can be composted, mowed/mulched, or tilled into the soil as part of a soil organic matter system nutrient-cycling plan. They can also maintain a growing root in the soil to feed our soil microbial community over the winter in late-cropping situations when it is hard to overseed a winter cover crop. One of my favorite winter ground covers is chickweed (*Stellaria media*). Per ton (907 kg) of dry weight, chickweed contributes 77 pounds (35 kg) of nitrogen, 34 pounds (15 kg) of phosphorus, a whopping 220 pounds (100 kg) of potassium, and 28 pounds (13 kg) of calcium to the soil. As a comparison, legume cover crops provide 34 to 73 pounds (15–33 kg) of nitrogen to the soil per ton of dry weight. Of course, it is easier to grow a ton of red clover cover crop than to grow a ton of chickweed, but still, the comparison helps us put weeds into a more complete ecological perspective and to understand their potential to add valuable nutrients to farm soils. There are other nutritious weeds, too. Lambsquarters, redroot pigweed, and mallow (*Malva rotundifolia*) each add 80 pounds (36 kg) of nitrogen to the soil per ton (907 kg) of dry weight. Lady's thumb (*Polygonum persicaria*) and sow thistle (*Sonchus oleraccus*) both add more than 70 pounds (32 kg) of calcium to the soil per ton (907 kg) of dry weight.

Table 8.4. Weed Competition Ability of Specific Vegetable Crops

High	Average	Low
Asparagus	Beans, *snap*	Carrots
Beans, *dry*	Beets	Celery
Broccoli	Cantaloupe and melons	Cucumbers
Brussels sprouts	Cauliflower	Onions
Cabbage	Collards	Parsley
Cauliflower	Eggplant	Peppers
Chard	Lettuce	Potatoes, *white*
Kale	Okra	Potatoes, *sweet*
Mustard greens	Peas	
Parsnips	Pumpkins	
	Radishes	
	Rutabagas	
	Spinach	
	Squash	
	Sweet corn	
	Tomatoes	
	Turnips	

Suppressing Potentially Competitive Plants

Managing the complicated plant competition relationships that occur on most farms and in gardens requires integrating multiple weed suppression strategies based on weed ecology principles. We begin with one of the 10 basic ecological principles (from chapter 1), and then dive deep into a consideration of the role of rotation and tillage in suppressing weeds, how to manage the weed seed bank, the use of cover crops and living mulches for weed suppression, and bed preparation tips to discourage weeds.

SELECTIVE WEEDING

Here's a principle you can apply at any time of year and at any scale, from the garden to the farm field. It's Carl's and my ecological principle 9:

> **Weed selectively.** Choose a ground cover mix; then identify and "chop and drop" the plants or "weeds" that are competitive with specific crops in specific growing situations and climates.

Selective weeding starts with figuring out which weeds may have benefits in your farm or garden system and are not too competitive with the crops you are growing. Once you've identified those weeds, you choose to *not* remove them. Weeds that you do choose to cut or pull can be left in place on the soil surface as a mulch and as another organic residue addition (as long as they are not yet starting to form seeds).

Choosing which weeds fit into your system will be based on the ecological information you have collected on your weeds, similar to the biological details I presented in the first part of this chapter. For example, on my Montana farm I allowed selected weeds to grow in the no-till living mulch and within crop rows to provide winter cover, early-spring bloom, and summer bloom and ground cover for beneficial insects, birds, bacteria, and fungi. One of these was fanweed, as I described in chapter 5. Fanweed is a brassica weed and supports cabbage aphids (especially during flowering and seeding). These aphids feed, maintain, and build cabbage aphid predator and parasite populations in my system so that they are present when I need them to protect my brassica crops later in the season. However, I only allowed more competitive weeds in living mulch row middles, selectively removing them from my crop rows. One of these was black nightshade (*Solanum nigrum*), which I left in the row middles as a trap crop (also described in chapter 5). Which weeds you decide to selectively remove and leave in place will depend on the crops you are growing, your climate, and your soil's ability to support both crops and weeds without diminishing your crop yield significantly.

ROTATION

Weed scientists consider crop rotation to be the foundation of weed management in organic vegetable production.[9] In chapter 7 I discussed creating rotations for disease suppression based on diversification of plant family and harvested crop part (that is, root, leaf, or fruit). Rotations for weed suppression focus on diversifying:

- Crop and cover crop planting date.
- Soil nutrient levels needed by crops and available (or not) in certain garden beds or farm fields.
- Water availability and specific crop water needs.
- Timing and type of tillage.
- Timing of harvest date.
- Crop type, both by harvestable part and root type (shallow- versus deep-rooted).

As in insect suppression, increasing crop species diversity alone is not sufficient. You'll be more successful if you diversify all of the ecological relationships involved in weed suppression. For example, in a 2019 meta-analysis of rotation and weed management experiments, researchers found that diversifying crop planting dates was more effective in suppressing weeds than was increasing crop species richness in rotations. Combining the two techniques was even more effective. And tillage also plays a part. These researchers further discovered that increasing rotational diversity suppressed weeds more under reduced tillage conditions than tilled conditions.[10] Hence it is important to simultaneously diversify several components within your rotation at once. For example, rotate crop families in specific beds or parts of fields, but also rotate between summer and fall crops and vary your crop (and cover crop) planting date. For example, spring-germinating weeds are suppressed well by fall-sown grains because the grain cover crop is already well established when spring weeds germinate. On the other hand, spring-planted competitive crops suppress species such as purslane that don't germinate until midsummer. Add to this summer/fall crop rotation and rotating in stronger weed competitors. For example, rotate in competitive crops (such as broccoli, corn, and mulched winter squash) with slow-germinating, less weed-competitive crops (such as carrots, seeded lettuce, and onions).

A weakness in many garden and farm rotations is a focus on "favorite crops." For example, if each year you grow a large amount of winter greens and brassicas, winter weeds can become dominant as the perfect growing conditions are repeatedly provided for them. This might not be a bad thing. Winter annual weeds become spring-blooming flowers that provide early pollen and nectar habitat for beneficial insects. But it can also mean a buildup of winter weeds as they repeatedly flower and produce seed. My high tunnel has become overrun with chickweed, which I used to leave undisturbed but now selectively weed out, removing approximately 80 percent of what germinates in my winter greens. It is vital to change things up and sometimes rotate perennials with annual crops. This temporary shift to perennials breaks weed cycles and decreases weed seed viability. A good example is a rotation that includes a year or two of grass-legume- or grass-wildflower-mix cover crops in a field or in garden beds. Planting such diverse strips for a year or two in some of your garden beds or some parts of your farm fields can also provide hay mulch (grow-your-own fertilizer) and a food source for beneficial insects as well as a dense soil cover that prevents annual weed seed germination. If annual weed seed is prevented from germinating, it will decrease in seed viability over time, though the speed of decrease varies with weed species (see table 8.1 for weed seed longevity).

Here are some of the best weed suppression rotation hints I use:

- Grow legume crops or cover crops before heavy-feeding crops (tomatoes, corn, potatoes, and broccoli). The legume crop will help feed the vegetable crops and also break weed cycles.
- Grow a perennial and/or a sod-forming crop or cover crop on 25 to 50 percent of your garden beds or farm production area each year to break annual weed cycles.
- Transplant summer crops into beds immediately after harvesting and severely mowing early-spring-seeded crops to minimize soil disturbance and suppress weeds that may have invaded seeded crops.
- Grow crops such as corn, wheat, clovers, and medics that form mycorrhizal associations before non-mycorrhizal crops like broccoli and other cabbage family (Brassicaceae) plants. This helps keep the microbial community balanced to maintain our soil organic matter system.
- Grow peas before fall brassicas: These two crops are a good succession cropping sequence in most climates. Fall brassicas benefit from the nitrogen fixed by the peas, and you can thwart the preferences of weeds that grow best in spring versus those that grow best in summer.
- Grow cucurbits (cucumbers, melons, and summer squash) after a spring cover crop or early greens (spinach, lettuce, or brassica greens). This allows you to rotate both different plant families and also spring/summer-planted crops in the same bed or field. Also, mowed or tilled-in residues from the early greens act as a green manure or mulch.
- Grow a cover crop or deep-rooted, long-season crop after root crops. Soil disturbance when root crops are harvested reduces soil structure, which can give annual weeds an advantage over the next crop, since soil disturbance is necessary for the germination of many weed

species' seeds. Growing soil-building legume crops or cover crops before a root crop is another option to help diminish the soil disturbance that occurs after root crop harvest.

- Grow a cover crop of cold-hardy clovers, hairy vetch, and/or perennial ryegrass by overseeding into summer crops (such as winter squash) in late summer to provide a winter cover crop after harvest and avoid bare soil that weeds would otherwise fill.

WEED SEED BANK MANAGEMENT

The weed seed bank is the reservoir of weed seeds in the soil. Without new input to the soil weed seed bank in your farm or garden's soil, it will generally decrease in amount and diversity over time, due to losses from germination, predation, microbial decay, and age-related degeneration.[11] Cover crops, mulches, mowing, cultivation, and black plastic tarps can be used to prevent weeds from flowering and seeding, thus limiting input of new weed seeds into the soil. Many traditional organic farms rely on *summer fallow* to deplete weed seed banks and also diminish the energy stored in perennial weed roots. Summer fallow is a process of purposely keeping cropland out of production during the growing season, allowing weed seeds to germinate, and repeatedly cultivating those germinating weeds. We can predict the peak emergence of problem weed species in a field based on each species' preferred soil temperatures and moisture for germination. When summer fallow is timed to coincide with peak emergence, large declines in the weed seed bank can occur.

The downside of summer fallow is that leaving soil bare decreases levels of organic carbon in the soil and disrupts the soil organic matter system. These negative effects on soil health can sometimes be mediated by planting a cover crop after the fallow period. Summer fallowing might be worth the soil health cost if you have a serious infestation of problem weeds. For example, one successful organic farm in western Oregon manages Canada thistle by plowing infested ground in midsummer, when the soil is very dry. In the fall they plant cereal rye. This has been an effective treatment providing close to 90 percent control.

TILLAGE

Healthy soils that are high in organic matter host a greater diversity of soil microorganisms and invertebrates, which can enhance both weed seed decay and predation.[12] Thus, although tillage is a widely used tool for suppressing weeds, it is not without its downside. Tillage has been shown to decrease weed-seed-eating ground beetles, including carabid beetles, as well as soil microorganisms and invertebrates that may help to suppress weed seeds.

On the negative side, tillage can encourage weed seed germination. It provides most of the cues that signal many kinds of weeds to germinate: light, high soil temperatures, fluctuation in day/night temperatures, and available nitrate-nitrogen in the soil. Tillage also alters the vertical distribution of weed seeds buried in the soil and creates microsites that can encourage the germination of certain weed species.[13]

On the positive side, tillage can decrease weeds, particularly weed seeds with long viability in buried soil. For example, in a 6-year study with a naturally occurring population of viable annual weed seeds, the numbers in the top 9 inches (23 cm) of soil decreased with various rates and timing of tillage: There was a 22 percent decrease per year in undisturbed soil, a 30 percent decrease per year in plots tilled twice a year (March and September), and a 36 percent decrease in plots tilled four times a year (March, June, September, December). This is a lot of data, so let me summarize. Tillage decreased the seed viability of certain weeds over time compared with no-tillage. But fewer weed seeds were able to germinate in untilled soil because they did not receive the light, higher temperature, and nitrogen release cues. Hence, if you use tillage to manage weeds, you will encourage weed seeds to germinate and eventually you can use this tillage strategy to diminish the weed seed bank. However, repeated tillage can make you dependent on tillage to manage weeds, by continually providing the conditions weed seeds like for germination.

Cultivation and hand-pulling are the best approaches for suppressing small annual weeds and taprooted perennial weeds, but not very effective on most creeping weeds that spread by rhizomes and stolons. See table 8.2 for details about when to till for maximum efficacy against specific annual, perennial, and stoloniferous perennial weeds.

PLANTING COVER CROPS AND LIVING MULCHES

As you are probably beginning to understand after reading about cover crops and living mulches to build soil and habitat and to help suppress disease, different species affect ecological relationships in various ways. Let's now examine how the same is true when it comes to the most effective species for weed suppression. (Refer back to "Living Mulch and Cover Crop Options" on page 68 for detailed information about grass, legume, and mustard family species for cover crops and living mulches.)

Cover Crops

There is a lot to consider when matching your cover crop choice to the goals of your particular weed suppression plan. Learning as much as you can about the ecology and life cycles of the weeds you want to suppress is essential. (Table 8.5 is a summary of the information presented in this section on interactions between cover crops and weeds.)

Cover crops suppress weeds via several mechanisms that often interact and affect more than one ecological function at a time. Common weed suppression mechanisms include resource and light competition, interference with weed life cycles, physical suppression by cover crop residues, release of phytotoxic biochemicals, and encouraging soil microorganisms to suppress weed growth and reproduction.

Shading and smothering. Cover crops can shade out weed species that require light for germination. They also compete directly with weeds for nutrients and water as they develop extensive canopies and deep root systems, so that weeds don't have the resources to invade. Combinations of grasses and legumes provide thick cover crop canopies that decrease the light cues needed for weed seed germination. Sowing at an especially heavy rate can also have a smothering effect.

Various cover crops have been shown to suppress specific weeds when managed in specific ways. For example, in one study the greater the amount of red clover biomass incorporated into the soil, the more early-season suppression of wild mustard occurred and the higher the sweet corn

Figure 8.5. Blooming red clover "weeds" moving from living mulch row middles into maturing onions and red cabbage in late summer at Woodleaf Farm, Oregon.

Table 8.5. Suppressing Weeds with Cover Crops

Cover Crop or Crop Combination	Weeds Suppressed*	Comments
MUSTARD FAMILY COVER CROPS		
Black mustard (*Brassica nigra*) and Indian mustard (*B. juncea*)	Lettuce and barnyard grass seed and seedlings were suppressed.	Plants in the mustard family can also reduce crop plant seed germination.
White mustard (*Brassica hirta*)	Fall-planted and incorporated in the spring suppressed early weed emergence for 30 days.	Incorporate into soil at least 2 weeks prior to planting crops.
Fall planted rutabaga and rapeseed (*Brassica rapas*)	General annual weeds suppressed.	Wait longer to plant if planting small-seeded crops such as lettuce.
Winter rapeseed (*Brassica rapas*)	Lambsquarters and redroot pigweed were reduced by 96% and 50% after fall-planted and spring-incorporated winter rapeseed.	
CEREAL GRAINS AND GRASSES		
Rye (*Secale cereale*) summer and/or winter cover crops	Reduced weed shoot biomass by 90%, 82%, and 60% during a 3-year study.	Rye decreased lettuce seed germination by 91%.
Sorghum (*Sorghum bicolor*) and barley (*Hordeum vulgare*) as winter cover crops	Reduced annual weeds by 50%, especially winter annuals: chickweed, shepherd's purse, and quack grass.	Barley decreased lettuce seed germination by 99%.
GRASS-LEGUME MIXES		
Rye (*Secale cereale*) and subterranean clover (*Trifolium subterraneum*)	A winter cover crop mix suppressed 65 to 85% of broadleaf weeds and 65 to 80% of grass weeds.	Root rot fungi thrive on fresh, green residues; allowing cover crops to rot for 1– 2 weeks before planting allows for them to be replaced by beneficial fungi.
LEGUMES		
Red clover (*Trifolium pratense*) and white clover (*T. repens*)	Reduced velvetleaf, pigweed, purslane, foxtails, crabgrass, and lambsquarters biomass by 65%.	Red clover attracts more beneficial insects than white clover.
Alfalfa (*Medicago sativa*) winter and/or summer cover crops	Reduced both annual and perennial weeds, including field bindweed.	There are fast-growing annual and slower-growing perennial alfalfa varieties.

* This summary represents information gleaned from my own on-farm research and from a review of the scientific literature.

yield.[14] An incorporated crimson clover cover crop decreased lambsquarters biomass by 65 percent and increased sweet corn yield by 131 percent.[15] Using another legume, alfalfa, as a cover crop or in a crop rotation has reduced populations of both annual and perennial weeds. Clover and alfalfa are great for suppressing weeds when planted as a cover crop and tilled before planting the cash crop. Red clover is my personal favorite for suppressing weeds. However, incorporating grass cover crops can also suppress weeds. Studies at Michigan State University indicate that weed suppression was greater after tilling in cover crops of oats, barley, rye, or Sudan grass than after mulching with wood chips.

Breaking weed life cycles. The duration and the percentage of total land taken out of annual crop production and put into cover crops is key to the level of weed suppression achieved (as well as soil and habitat building). Some growers remove up to half of their land from crop production to put into cover crops. In my farming I give up 30 to 50 percent of my cropland to plant short-term (1 year) and long-term (5 years) cover crops or perennial living mulches. Increased time in perennial grass/clover or no-till cover crops can break weed cycles and suppress certain weeds. Deep-rooted cover crops can accomplish the same thing. In Maryland suppression of winter annual weeds was greater following forage radish winter cover crops than it was following any other cover crop species.[16] In Michigan a cover crop of oilseed radish reduced both early-spring weeds in vegetable crops and weed seeds in the soil seed bank.

Timing is also critical because so many ecological relationships are involved between the cover crops and the weeds. Our goal is to manage the timing of cover crop planting and tillage, type of cover crop, and amount of residue left by a cover crop. These variables can all be managed to suppress the establishment, growth, and seed production of specific weeds.[17] Properly timed mowing or tillage of cover crops can disrupt specific weed growth and weed seed production. Some farms regularly mow their cover crops to keep any weeds from seeding.

Suppressing weeds via natural phytotoxins. Cover crop mixes also exude a diversity of chemicals that inhibit germination and growth of weeds.[18] Decomposing organic residues from some cover crops suppress weeds by releasing temporarily plant-toxic biochemicals (such as phenolic compounds) called *allelochemicals.* These biochemicals usually have temporary negative effects on other plants or other organisms. Allelochemicals are released from plants by leaching, root exudation, volatilization, or residue decomposition. For example, sunflowers are *allelopathic.* They release terpenes and various phenolic compounds from all their roots, leaves, stems, flowers, and seeds that inhibit the growth of other plants.

Allelopathic potential has been well documented for several cover crops, including cereal rye, hairy vetch, and red clover. A study of inhibition of wild mustard seedling growth by tilled-in red clover concluded that the inhibition was due, in part, to the biochemical effects of decomposing red clover residues.[19] Brassica cover crops, such as rapeseed and black mustard, also contain biochemicals that suppress germinating weeds. Although biochemicals in cover crops can be used to help suppress weeds, the biochemicals can also have adverse effects on vegetable crops. Fortunately, the effects of most types of biochemicals released by decomposing cover crops dissipate in time.[20] For example, the biochemical in brassicas that is most responsible for its phytotoxic effect loses 80 percent of its punch within 2 weeks. Hence, for any potentially allelopathic cover crops, it is wise to wait 2 to 4 weeks after mowing or turning the cover crop before planting a vegetable crop in the same area. Cover crops exhibit greater plant-toxic effect if they decompose anaerobically (without oxygen). Some growers take advantage of this to control weeds; they purposely bury the cover crop deeply by using a chisel plow and/or plowing deeply when they incorporate a cover crop to encourage anaerobic decomposition.

However, as with most of the systems discussed in other chapters, managing all the ecological relationships can be complex. For example, one study examined the simultaneous interaction of several weed suppression effects of a cover crop. After a red clover cover crop was tilled under, allelopathic biochemicals from the decomposing red clover provided weed suppression for up to 2 weeks. And increased microbial activity suppressed weed germination and growth for 30 days after the red clover cover crop was tilled in. However, the

Figure 8.6. Mowed wheat (*Triticum aestivum*) living mulch in between onion crop rows provides a mulch mat on the soil surface to suppress winter annual weeds, such as the blooming purple deadnettle weeds in foreground.

microorganisms examined in this study also acted to degrade the biochemicals released from the cover crop, thus indirectly helping germinating weeds survive. So weed suppression was a result of two different ecological relationships. As the red clover first began to decompose, weed suppression was due to released biochemicals. Later on, though, microbial suppression was most effective. Hence the result of these ecological interactions was that the red clover cover crop suppressed weeds in different ways at different times, but also may have helped out the weeds a little too.

Living Mulches

The presence of living mulches in row middles also helps to suppress many weeds. In one study, when hairy vetch and rye living mulch residues covered at least 90 percent of the soil, weed density was decreased by 78 percent.[21] Examples of species that can be used as both weed-suppressive cover crops or carefully managed living mulches include:

Red clover (*Trifolium pratense*) is adapted to cooler areas; plant in fall, spring, or early summer. It needs to be kept mowed as a living mulch to set it back so that it does not compete with crops. Red clover can be competitive as a living mulch if it's not stressed a bit by mowing.

Mustards (*Brassica* spp.) release biochemicals that suppress germinating weeds temporarily when mowed or tilled in. (They can also inhibit small-seeded vegetable crops, like lettuce, from germinating.)

Annual ryegrass (*Lolium multiflorum*) germinates and grows quickly. If planted as a living mulch, don't allow it to go to seed; mow or till it before it seeds.

Perennial ryegrass (*Lolium perenne*) is adapted to cooler areas and can be planted in either fall or spring. It can be challenging to manage because it is so vigorous and can grow tall as a living mulch. In small garden spaces, mow it regularly to keep it from taking over.

Annual wheat (*Triticum aestivum*) and **winter cereal rye** (*Secale cereale*) are both good weed-suppressing cover crops and living mulches. Wheat does not produce as much biomass as cereal rye, but it can be a good cover crop for smaller garden spaces. Rye may also have an allelopathic effect. However, managing rye as a living mulch can be challenging because it is so vigorous.

SELECTIVE FERTILIZING

Weeds usually also have very high rates of nutrient uptake and are adapted to take advantage of the quick nutrient release that occurs when soil is disturbed. In fact, weeds are so good at grabbing soil nutrients, they can have concentrations of nitrogen, phosphorus, potassium, calcium, and magnesium in their tissues 1.5 to 3 times higher than the crops with which they compete. Weeds are especially great at competing for easily available, more soluble soil nutrients (like nitrate-nitrogen). In fact, most annual weeds are more competitive with crops when soils have high nutrient levels, especially high nitrogen levels.[22] When nitrogen fertilizer was applied shortly after planting in a fava bean field, the ratio of crop biomass to total biomass (crop + weed) ended up in favor of the weeds. But, lower levels of early-season nitrogen resulted in some weed suppression.[23] This is because most annual weeds have evolved to germinate in the presence of and take advantage of quickly available soil nitrogen. Thus this ecological principle makes good sense for weed suppression:

> **Fertilize selectively.** Fertilize the crop selectively; avoid fertilizing the whole garden or field.

In fact, fertilizing only in the crop row as opposed to applying fertilizer on the whole field has been shown to reduce weed competitiveness while increasing crop competitiveness and overall yield.[24]

BED PREPARATION AND PLANTING STRATEGIES

Planting strategies that set back weed seed germination and weed seedling growth can discourage weeds right from the start. There are also strategies that speed up crop establishment. Here are some that I use:

- Lightly cultivate rows or beds just prior to planting to dry out newly germinated weed seedlings. Then pre-irrigate beds or rows only. By pre-irrigation, I mean watering the prepared crop bed or row before planting such that water is added only in the spots where you will plant crops.
- Limit cultivation depth to avoid bringing up new weed seeds from lower soil horizons.
- Pre-soak all seeds before planting. With large-seeded crops (such as corn, beans, and squash), plant after pre-irrigating beds and don't irrigate crop beds again until necessary.
- Pre-germinate seeds that are naturally slow to germinate. To pre-germinate seeds, soak them overnight in hot water from the tap (it does not need to stay hot all night), then lay them between moist paper towels or in a plastic container with holes in the lid and/or sides for aeration.
- Begin with transplants rather than seeds when possible to give your crops a head start over weeds.
- Cultivate when crops are small and when first weeds germinate; then delay irrigation long enough to allow the disturbed weeds to dry out.

Now that I've presented an overview of methods for suppressing weeds, I want to describe some of the on-farm experiments I carried out to understand how weed suppression methods interact with soil and habitat building within a farm or garden system.

My On-Farm Weed Suppression Experiments

I was interested not only in examining the effects of tillage on soil and microorganism health, but also in evaluating the effect of tillage on weed suppression and crop yields.

EFFECTS OF TILLAGE AND APPLYING COMPOST

In my first experiment, my goal was to evaluate which weed species would be encouraged over two growing seasons by different tillage and fertilizer treatments, in minimally tilled crop rows, in no-till crop rows, and in my farm-made compost. An area of original 50-year-old, untilled pasture on my farm served as the basis for comparison (the control).

The Compost Results

The compost I made at Biodesign Farm (described in "Making Compost" on page 75) was a mix of manure, straw, and red clover, which was heated to USDA National Organic Program standards during the composting process. To evaluate the weed seed reservoir in my finished compost, I filled 1-gallon (4 L) pots with the compost, set up the pots in my greenhouse, watered them well, and then observed what sprouted and grew. I discovered that the greatest diversity and abundance of annual weed species was present in the pots when compared with numbers and species of weeds counted in my crop fields or the untilled pasture area. In fact, there

Table 8.6. Weed Species Found in Biodesign Farm Weed Ecology Study

Common Name	Botanical Name	Weed Life Cycle*	Untilled Pasture (Control)	Minimal-Till Plots	No-Till Plots	Compost (in Pots)
Pigweed	*Amaranthus retroflexus*	A		X		X
Chickweed	*Stellaria media*	A		X		X
Black nightshade	*Solanum nigrum*	A		X		X
Pennycress	*Thlaspi arvense*	A		X	X	X
Lambsquarters	*Chenopodium berlandieri*	A		X	X	X
Prickly lettuce	*Lactuca serriola*	A		X		X
Purslane	*Portulaca oleracea*	A		X		X
Tumble mustard	*Sisymbrium altissimum*	A				X
Barnyard grass	*Echinochloa crus-galli*	A				X
Common mallow	*Malva neglecta*	B	X	X	X	X
Black henbane	*Hyoscyamus niger*	B				X
Lanceleaf plantain	*Plantago lanceolata*	P	X			
Common plantain	*Plantago major*	P		X		
Dandelion	*Taraxacum officinale*	P	X	X		
White campion	*Silene alba*	P	X	X	X	
Yarrow	*Achillea millefolium*	P	X			
Buttercup	*Ranunculus acris*	P	X			
Total perennials (P)			6	3	1	0
Total annuals (A)			0	7	2	9
Total biennials (B)			1	1	1	2

* A = Annual, B = Biennial, P = Perennial

were some species that I did not observe at all in my crop fields or in the pasture area. This suggested to me that the one compost ingredient I brought in from off the farm—the sheep manure from my neighbor's farm—was the source of the unique weeds. When I stopped using manure, annual weeds also decreased in my vegetable fields.

The Crop Row Results

In 2006 I tilled some experiment rows shallowly in the spring and then did no further weed cultivation after that. I also left other rows untilled altogether to measure the effects of no-tillage. I identified the weeds that sprouted and grew in the rows, and I found higher numbers of annual weed species in the minimally tilled plots compared with the no-till plots and the control area. The untilled pasture control had the most perennial weeds and no annual weeds at all (see table 8.6).

The lesson to be learned from this experiment is that tillage results in more annual weeds and that not tilling encourages perennial weeds over the long term. Another lesson is that applying compost can bring in new weed problems—even compost heated to 149°F (65°C) can be a source of weed seeds!

Figure 8.7. The Brussels sprouts plants from my 2007 weed ecology experiment look quite happy in this plot covered with EcoCover paper mulch.

EXPLORING COMPLEX INTERACTIONS

In 2007, I set up another experiment as part of my quest to understand the interactions among soil health and nutrient cycling, crop yield, and weed competition. I evaluated how these interactions were affected by several soil/weed management treatments.

The five treatments were: no-tillage, minimal-tillage, conventional tillage, tillage plus vinegar spray, and tillage plus paper mulch (EcoCover). The experimental area consisted of one 600-foot-long, 4-foot-wide (183 × 1.2 m) crop row split into replicated plots, three per each of the five treatments. At the start of the experiment, a 2-year-old red clover cover crop was growing in the row. I strip-tilled the treatment plots, except for the no-till plots, making three passes with the tiller in the tillage plots, but

Figure 8.8. This is the infrared radiant heat flamer I used to prepare no-till plots in my 2007 weed ecology study.

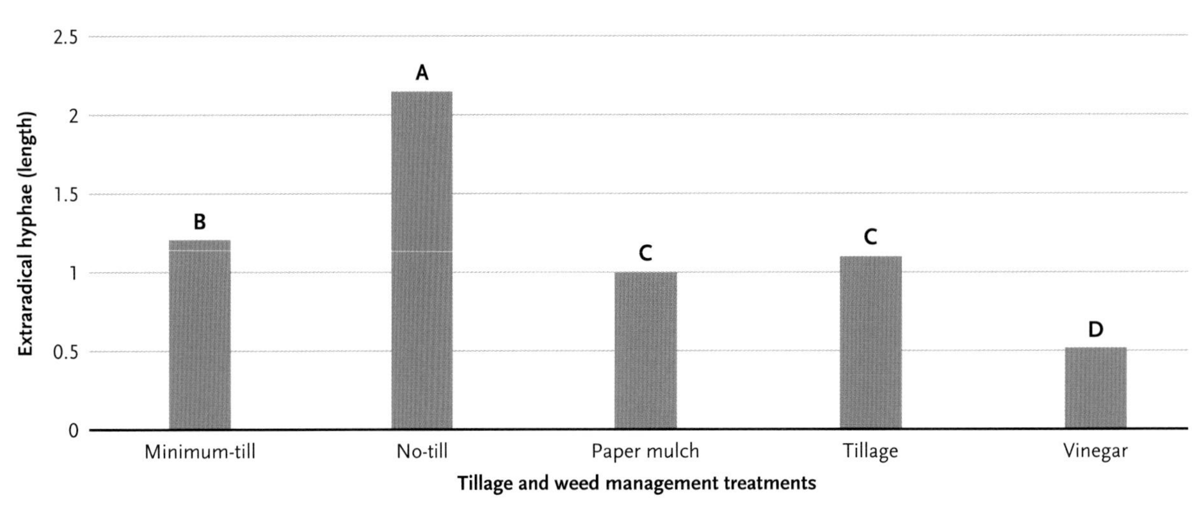

Figure 8.9. The effects of weed management and tillage treatments on mycorrhizal density in experimental plots at Biodesign Farm in 2007. Letters A through D signify statistically significant differences among treatments.

Figure 8.10. At Woodleaf Farm in 2021, living mulch residue blown into crop rows suppresses weeds in a bed of young pepper transplants and cycles nutrients. By late summer, creeping perennial grass, clover, and broadleaf weeds will infiltrate the beds but also provide winter cover (without needing to overseed a cover crop).

only one shallow pass in the minimal-tillage plots. Tillage plots were kept weed-free with cultivation to keep the soil surface bare all season. I applied a weed mat paper mulch over another set of tillage treatment plots and sprayed a vinegar herbicide on weeds twice during the growing season in still another set. In the minimal-tillage plots, I tilled only once and did not cultivate weeds that grew back during the season. In the no-till treatment plots, I closely mowed the red clover in the spring and then burned the remaining growth with an infrared flamer. I left the perennial red clover intact as a living mulch in the row middles on either side of all treatment plots. I planted all of the prepared plots with Brussels sprouts.

Here are the highlights of the results of this experiment:

Yield. Brussels sprouts plants in tilled plots covered with paper mulch had more biomass and the highest average yield per plant. Next best were the plants in the conventional tillage rows and the rows that were tilled and sprayed with vinegar before planting. I concluded that tillage increased yields by minimizing crop competition with other plants.

The minimal-tillage rows took fourth place, and the no-till plots had the lowest yield and smallest plants. Using the flamer was not my best vegetation suppression idea. Although these plots had the highest level of mycorrhizae, the red clover grew back from its strong roots within several weeks and competed heavily with young Brussels sprouts.

Soil temperatures. Paper mulch and conventional-tillage plots had the highest soil temperatures from April through the end of August. Thus tillage resulted in increased soil temperatures, especially early in the season. Soil temperatures were lowest in the no-tillage and minimal-tillage plots all season.

Mycorrhizae presence. The greatest mycorrhizal density was observed in no-tillage plots; the second greatest density occurred in minimal-tillage plots. The tillage/vinegar treatment resulted in the lowest density, followed by conventional-tillage and tillage/paper mulch plots. It was clear that tillage led to a decreased presence of mycorrhizae.

Soil aggregate stability. Tillage plots tended toward lower soil particle aggregate stability, but the lower levels were not statistically significant in this 1-year experiment.

Soil organic matter. SOM levels in the plots in April, one week after initial tillage, were not statistically different. But, by September levels were slightly higher in minimal-tillage and no-tillage plots and lower in tillage plots.

In summary, my experiment showed there is a trade-off between tillage and no-tillage. The weed suppression due to tillage resulted in better yields, but the tillage led to decreased soil/microbial health. There is a possible sweet spot with a shallow, minimal-tillage approach. Yields were not quite as high as in plots that are tilled more frequently and to greater depths, but soil and mycorrhizae health were much better.

The results of this on-farm experiment, as well as my yield and pest records from 1993 through 2010, were a great help to me in designing the farm system that I use today employing permanent living mulch in row middles and very limited strip-tilling to prepare crop beds in the spring. I like the balance this provides for an economically sustainable yield without sacrificing long-term soil health.

I continue to try to understand how much tillage and/or reduction in weed/living mulch crop competition is necessary for economical yields with different crops in different climate conditions and soil fertility levels. It is an ongoing process: I learn from my dynamic agroecosystem even as I add and subtract management methods and strategies! Figure 8.10 shows my current methods and strategies.

Intervention When Weeds Impact Crops

Although I now have particular strategies and methods that generally work well for suppressing weeds in my farm system, I want to stress that it's not possible to establish such a system overnight! You'll also need to experiment and observe to figure out how best to suppress weeds in your system. And even after strategies are in place, a system naturally shifts out of a more balanced dynamic equilibrium over time, and then weed management interventions are necessary. Fortunately, there are choices, and in this section of the chapter, I review the methods and tools available. As with my approach to insect and disease management, I first present interventions that have minimal ecological impact, then those with moderate impact. There are no appropriate heaviest-impact strategies acceptable for use on organic farms, so none are presented.

MINIMAL-IMPACT INTERVENTIONS

The methods and materials that have the least chance of negatively affecting beneficial soil microorganisms and soil health range from an integrated, systems approach using many small tools to solarization and hand cultivation.

Integrated Soil and Crop Management Method
This approach combines various methods described in this chapter, including selective fertilizer, water management, mulches, smother cover crops, insect biological control (weed seed predation), and vegetation competition to suppress weeds.

Solarization
It's possible to kill weed seeds by heating the soil in a limited area by tightly covering the surface of premoistened soil with clear or black plastic 1.25 to 2 mils thick for 4 to 6 weeks in summer. Soil temperature must reach 100 to 140°F (38–60°C) to kill most seeds. Black plastic is better for cool climates. Solarization is most effective in very hot-summer areas. Solarization suppresses annual weeds best and is especially effective in decreasing winter annuals. Most perennials are usually not well suppressed for long by solarization.

Cultivation
It is possible to bring a weed problem under control by hand-weeding or cultivating using hand or tractor-drawn tools. The goal is to work shallowly, causing the least soil disturbance. Repeated cultivation is most effective for weed suppression, and early timing is the key. The smaller the weed seedlings, the better. Disturbing soil may *increase* the density of certain weeds such as pigweed. And of course our goal for soil building is minimal soil disturbance.

Organic Mulches
Mulch choice depends on the crop receiving the mulch and the predominant weeds. Small-seeded annual weeds are the easiest type to suppress with mulch. Many types of organic mulch keep soils cool and moist; these may not be good choices for warm-season vegetable crops in cool climates.

Bark chips, wood chips, sawdust, straw, lawn clippings, and leaves are popular mulch choices; these are discussed in detail in "Mulch Options" on page 65.

As noted earlier in the chapter, some mulches can be allelopathic, but this toxic effect is usually short-lived (days to several weeks) as chemical substances break down. Do not mulch seedlings with shredded leaves or wheat and rye straw mulch, because there are numerous reports on the phytotoxic effects of wheat/rye straw and leaves of certain tree species as they decompose. There are also reports that coffee grounds and apple pomace can suppress seedling growth. With advanced planning, these materials might be used in preparing a weed-free space for planting later.

You might also consider paper mulch. Recycled or pressed heavy paper and newspaper mulches are possible alternatives to plastic mulches. Recycled paper mulches manage early-season weeds but break down in several weeks, particularly in wet weather. Thus they may not provide reliable season-long weed control.

Reusable Landscape Fabrics
Landscape fabrics are woven and nonwoven polypropylene and polyester and laminated plastic fabrics. They allow air and water penetration, except the laminated plastic types, and usually last as long as 10 years if handled carefully. Brands differ in their weed control effectiveness and longevity. In general, only the laminated plastic weed fabrics are effective for controlling rhizomatous weeds such as quack grass and nutsedge.

In an effort to balance yield with soil and microbial community health, I occasionally will spread reusable landscape fabric after shallowly tilling crop rows where I will plant warm-season crops such as tomatoes and sweet peppers (see figure 8.11). I have reused some of the fabric for 10 years. Be sure to purchase the landscape fabric that has a smooth underside—it is easier to lift off the beds in the fall.

MODERATE-IMPACT INTERVENTIONS
Lower-impact herbicides such as herbicidal soap, vinegar, corn gluten meal, synthetic plastic mulch,

Figure 8.11. Landscape fabric spread over minimal-tilled, strip-till crop rows and held in place with metal pins.

and use of a flamer present a greater chance of injuring beneficial soil microorganisms, soil-dwelling insects, and soil health. But when used carefully and sparingly, they will not severely disrupt agro-ecosystem pest suppression.

Herbicidal Soap

Potassium salts or an ammoniated soap of fatty acids sprayed on foliage causes plants to dehydrate. This material is not effective on deep-rooted species, however, and weeds must be young and succulent at time of application. It works best for young annuals or when applied to the first springtime leaves of perennials. The ideal time to apply is in hot, dry weather. Repeat applications are usually necessary. It is a contact herbicide that can damage crop plants, so it important to target only the plants you want to kill.

Soap with Vinegar

Vinegar (8 to 20 percent acetic acid content) mixed with household soap can suppress small annual weed seedlings. It is best on young, succulent annuals or the first spring leaves of perennials; repeat applications are usually required. Studies in California report suppression of field bindweed for 6 to 10 weeks with one application when first true leaves emerge. My own studies at Biodesign Farm showed a decrease in redroot pigweed, chickweed, and common mallow weed seedlings compared with untreated controls. I also documented a decrease in mycorrhizal fungi in plots sprayed twice in one season with vinegar. I use vinegar and soap once a year in my asparagus beds just before they begin to sprout in the spring to temporarily suppress perennial grassy weeds and give the asparagus a head start.

Citric Acid and Clove Oil

A spray of citric acid mixed with clove oil has the same "burn-down" effect as a vinegar spray. Injury symptoms often appear within a few hours of application. This is a contact herbicide that effects any green parts of the plants it contacts. Like vinegar it is most effective on young, succulent, annual weeds. There are commercial products available that contain a mix of citric acid and clove oil and products that contain clove oil alone. I have not tried using these materials in my garden or on my farm.

Corn Gluten Materials

Unlike herbicidal soap and citric acid, corn gluten is not a contact herbicide. It can actually stimulate crop growth, but can also encourage weeds if they are present. Corn gluten materials are a by-product of corn milling that has a high (10 percent) nitrogen content. When mixed into the soil before weed seeds germinate, it reportedly inhibits root development by reducing water uptake by seeds. Mix into the soil (rather than surface-apply) at rates of 20 to 40 pounds per 1,000 square feet (98–195 kg/m^2) before weeds emerge. Corn gluten will not suppress weeds post-emergence.

Various research studies show different levels of effectiveness—some show no reduction in weeds. When I tested corn gluten meal at Biodesign Farm, it caused decreased growth of certain species compared with the control. However, it still took just as long to weed treated plots as untreated plots. Thus I recommend you test this product on a small scale for your own situation before you rely on it to help with suppressing weeds.

Plastic Mulches

Using plastics in our world is becoming more problematic, and single-use plastics don't make any ecological sense. But if you can reuse plastic tarps (as is done with plastic used for solarization), it might make sense. Certainly plastic is well used in commercial horticulture.

Plastic mulch is available in many thicknesses and colors. It is impermeable to water and can restrict air movement into soil. Black plastic gives nearly 100 percent control of broadleaf annual weeds when left in place for several weeks, and 60 to 75 percent control of rhizomatous perennial grass weeds when left in place for several months. Clear plastic encourages, rather than controls, weed seed germination and growth. Green and brown or infrared transmitting (IRT) mulches occupy a niche between black and clear plastic mulch. Unless daytime temperatures are greater than 55°F (13°C) during the first 4 to 6 weeks of the growing season, weeds may become well enough established under IRT plastic to compete with crop plants.

Flaming

A flamer is best suited for suppressing germinating or very young broadleaf weeds; flaming is less successful on grasses. A flame is used to cause plant cells to burst, creating a "burned" look to the tissue. The goal is not to set plants on fire! Flaming is most effective used with slow-germinating crops such as carrots and parsley. Established weed seedlings are most susceptible to flaming at the one-to-five-first-true-leaf stages; bigger weeds are not well suppressed. Hot, dry weather enhances the efficacy of flaming.

Flaming can be effective at the pre-plant, pre-emergent, and post-emergent stages. For the pre-plant approach, pre-irrigate unplanted beds to germinate weeds and then use flaming to set back the weeds prior to planting crops. Flaming is usually done after irrigation and before crops emerge. Unfortunately, flaming can kill beneficial ground-dwelling predators, such as spiders and carabid beetles. For this reason, if you use this technique, it is best to do so as pre-emergent flaming after you have disturbed the soil for preparing beds and seeding crops, but before vegetation (which will encourage spiders and carabid beetles) has begun to cover the soil.

PART TWO

Crop Management and Troubleshooting

An Ecological Approach

In part 1, I focused on how to manage systems and ecological relationships. At times, however, farmers and gardeners also need detailed cultural and pest management information to avoid and troubleshoot problems with individual crops. Thus this part of the book is a slight departure from the minimalist approach presented in part 1. Here I present a way of thinking about problem solving that seems more crop-specific—but in fact understanding the biological details of each crop you grow is an important part of managing all of the overlapping ecological principles in play on your farm or in your garden.

I include systems approach techniques in the problem-solving sections of the chapters that follow, along with frequent reminders of the potential ecological impacts inherent in the suggested interventions for specific crop problems. Sometimes these interventions, such as using row covers or spraying neem, are needed to avoid losing a crop, but the goal is always to first do no ecological harm.

CHAPTER 9

Vegetable Crop Growing and Troubleshooting Guide

In this chapter I present crop-specific recommendations, troubleshooting tips, and intervention details to solve problems for the most widely grown vegetables. Crops are listed in alphabetical order, and for each I first describe good management practices and then provide a troubleshooting list of the symptoms of common pests and problems. Thus this chapter is designed to be used as a reference for planning and for troubleshooting, not for reading through from Asparagus to Tomatoes! Check the entries below for each crop's preferences and for the whole-plant, foliar, and/or fruit symptoms you are seeing to identify the pest or problem. Once you have identified the cause of a problem, you can refer to the discussions of possible interventions for specific insects in chapter 10, and for specific diseases in chapter 11.

A Quick Guide to Growing Vegetables

Growing vegetables successfully requires paying attention to the temperature, light, soil and fertility, water, and special ecological requirements for each crop. Here, I begin with general details about how to plan and design a garden/farm ecosystem that includes all the vegetable crops you want to grow. This includes understanding the value of crop rotation and the general characteristics of vegetable crops that are important to their growth: whether they are cool- versus warm-season crops, their temperature and soil pH needs, and their nutrient requirements. For most farmers and gardeners, extending the harvest is an important aspect of managing vegetable crops. A guide to season extension could turn into a book of its own, but in this part of this chapter, I provide tips on succession planting, soil mixes for indoor seed starting, and season extension tools.

THE BASICS

Crop rotation is the practice of growing different types of crops in an area over time. For example, in each row or area of the farm I grow crops from

Table 9.1. Harvested Plant Part / Plant Family Rotation

Year	Plant Part Harvested	Plant Family
Year 1	Fruit	Tomatoes, peppers, eggplants, corn, broccoli, squash, melons, cucumbers, pumpkins
Year 2	Leaf	Spinach, chard, cabbage, kale, lettuce
Year 3	Root	Carrots, beets, onions, garlic

Table 9.2. Temperature and Soil pH Tolerances of Specific Vegetable Crops

Crop	Temperature Tolerances	pH Tolerances
Asparagus	Very cold-hardy	Tolerates > 7.0
Beans, *snap*	Susceptible to freeze injury below 32°F (0°C)	Tolerates < 6.0
Beets	Half cold-hardy	Tolerates > 7.0
Broccoli	Very cold-hardy	Tolerates > 7.0
Brussels sprouts	Very cold-hardy	Tolerates > 7.0
Cabbage	Very cold-hardy	Tolerates > 7.0
Cantaloupe	Very susceptible to freeze injury below 32 to 34°F	Tolerates > 7.0
Carrots	Half cold-hardy	Tolerates < 7.0
Cauliflower	Half cold-hardy	Tolerates > 7.0
Celery	Half cold-hardy	6.0–7.0
Collards	Very cold-hardy	Tolerates > 7.0
Corn, *sweet*	Susceptible to freeze injury below 32°F	Tolerates > 7.0 and < 6.0
Cucumbers	Very susceptible to freeze injury below 32 to 34°F	Tolerates < 6.0
Garlic	Half cold-hardy	Tolerates < 6.0
Kale	Very cold-hardy	6.0–7.0
Lettuce	Half cold-hardy	6.0–7.0
Mustard greens	Very cold-hardy	6.0–7.0
Okra	Very susceptible to freeze injury below 32 to 34°F	6.0–7.0
Onions	Half cold-hardy	6.0–7.0
Parsnips	Half cold-hardy	Tolerates < 6.0
Peas	Half cold-hardy	Tolerates > 7.0
Peppers	Very susceptible to freeze injury below 32 to 34°F	Tolerates > 7.0 and < 6.0
Potatoes, *white*	Half cold-hardy	6.0–7.0
Potatoes, *sweet*	Very susceptible to freeze injury below 32 to 34°F	Tolerates < 6.0
Pumpkins	Very susceptible to freeze injury below 32 to 34°F	Tolerates > 7.0 and < 6.0
Rutabagas	Very cold-hardy	6.0–7.0
Spinach	Half cold-hardy	Tolerates >7 .0
Squash	Very susceptible to freeze injury below 32 to 34°F	Tolerates > 7.0
Tomatoes	Susceptible to freeze injury below 32°F	Tolerates > 7.0 and < 6.0
Turnips	Very cold-hardy	Tolerates < 6.0

different plant families that produce different harvestable plant parts each cropping season or year. There are many kinds of rotations, including special rotations designed to suppress weeds (as described in chapter 8) and long rotations designed to suppress specific diseases (as described in chapter 7). Generally, I follow a FRUIT–LEAF–ROOT rotation, and I also plan rotations by plant family. Table 9.1 summarizes my rotation scheme.

Farmers classify vegetable crops as warm-season crops or cool-season crops. Understanding the general needs of these two types of plants is fundamental to growing them well. Table 9.2 provides general temperature requirements for individual crops.

Warm-season crops germinate when soil temperatures are in the range of 65 to 75°F (18–24°C). These crops, which include tomatoes and sweet corn, prefer air temperatures between 60 and 80°F (16–27°C) for optimal growth. Plant growth is slowed and plants can be injured when air temperatures drop below 40°F (4°C). Air temperatures below 32°F (0°C) can kill warm-season plants.

Cool-season crops, such as the Brussels sprouts in figure 9.1, germinate when soil temperatures are in the range of 45 to 55°F (7–13°C). Plants can tolerate a light freeze (30°F [–1°C]), and grow poorly when air temperatures are greater than 80°F (27°C).

Most vegetable crops grow best at a soil pH of 6 to 7, but some can tolerate higher pH or lower pH. These are good choices if your soil pH is above 7.0 or below 6.0. Crops that can tolerate pH greater than 7.0 include asparagus, beets, cabbage, muskmelon, sweet corn, pumpkins, tomatoes, peas, summer squash, and spinach. Crops that can tolerate pH below 6.0 include sweet corn, pumpkins, tomatoes, snap beans, carrots, cucumbers, parsnips, peppers, eggplant, and watermelon. Table 9.2 lists the soil pH tolerances of individual crops.

Figure 9.1. Harvesting mature Brussels sprouts under the snow at temperatures of 5 to 15°F (–15 to –10°C). Once the sprouts of this very hardy crop thaw out, they taste great.

As discussed in chapter 2, some crops are heavy feeders that need a steady supply of nutrients including nitrogen, potassium, phosphorus, and calcium in an easy-to-use plant-available form. But others aren't so picky. The crop entries in this chapter provide information on specific crop nutrient requirements, and you can also refer to table 4.5 for an overview of which crops are heavy feeders, which crops have average requirements, and which crops are light feeders.

SUCCESSION PLANTING

Part of growing your own diet is to plant crops at intervals in order to maintain a consistent supply of produce throughout the season (succession planting). Depending on a crop's speed of maturity, you can replant and interplant into standing crops throughout the season. The beauty of succession planting is that it maintains previous crop roots in the ground for soil microbe habitat while new crop roots are establishing. Succession planting maintains the quality of crops like lettuce that reach maturity quickly and then get old and bitter. Seeding new lettuce plantings provides continuous fresh, new leaves. Succession planting helps crops escape certain pests because it allows you to start several plantings of the same crop in different areas of your field or garden. It's also an excellent technique to push the limits of short growing seasons with early and late plantings of frost-tolerant crops.

Figure 9.2. Shade cloth protects this row of cool-season kale and cabbage from a hot week that reached 100°F (38°C).

The key is to keep making new plantings at regular intervals. Baby lettuce, radishes, and spinach can be planted on 7-day intervals. Peas, sweet corn, and bush beans can be planted on 10-day intervals. Beets, turnips, and head lettuce can be planted on 14-day intervals. Summer squash, carrots, and Swiss chard can be planted on 30-day intervals.

Here are some of my most reliable succession-planting tips:

- Interplant cool-season crops and crops that germinate quickly and are good plant competitors, like Swiss chard, into late-summer crops when their harvest is almost complete.
- Interplant summer crops with fall crops like spinach, mustard, or Swiss chard.
- Plant several varieties of one crop that have shorter and longer days-to-harvest periods, sowing them at the same time.
- Begin the growing season in late winter/early spring with varieties that germinate and grow in cooler temperatures.
- Use heat-tolerant varieties for summer planting and cool them with shade cloth.
- Time successive plantings more closely together as cooler fall weather approaches, since they will germinate and grow more slowly at cooler temperatures.

STARTING SEEDS IN MICROBIALLY AMENDED GROWING MIX

Starting seeds indoors is one of the most common ways to get a head start on the growing season. When I first started gardening and farming, I used *sterile media*, which is soilless growing mix for transplants that contains peat plus a coarse material for aeration, such as perlite. In sterile media, weed seeds, disease microorganisms, and insects are absent. These sterile media provide few nutrients to support plant growth, however, and few of the beneficial microorganisms that help to make nutrients from organic residues available to plant roots. Also, though sphagnum peat moss has been shown to contain disease-suppressive fungi (*Trichoderma* spp.), sterile media do not suppress root disease as well as microbially amended media. Finally, if we are building a microbially dependent soil organic matter system in our garden and field agroecosystem soils, we want our transplants to be ready for that environment. We really want a transplant root system that has already been introduced to a microbe-rich environment and established root-soil-microbes associations.

Compost-amended mixes also provide beneficial soil microorganisms to help promote plant growth and suppress disease.[1] Many researchers report that compost-amended potting mixes suppress the root rots that seedlings are susceptible to.[2] Compost-amended media reportedly create habitat for beneficial soil microorganisms both during seedling growth and when seedlings are transplanted into the field. Other researchers report that germinating seeds are colonized by amendment-derived microbes and that this microbial community changes as seedlings develop in the field with their own unique rhizosphere.[3] In fact, the microbial community in compost-amended soil is different from that in non-amended soil; it usually comprises both soil- and compost-derived microbial species.[4]

The compost and/or organic residue fertilizers we put into our growing media and mixes help transplants establish quickly in the field. In one study, compost and organic amendments influenced the root bacterial community structure in both the transplant media and the rhizosphere zone of tomato plants 4 weeks after transplanting in the field. These researchers compared a sterile, peat-and-vermiculite-based mix with the same mix plus other organic amendments: sesame meal, alfalfa meal, and composted dairy manure made by

Helen's Growing Mix

Here is my recipe for a microbially active growing mix:

- 2 gallons (8 L) coconut coir or peat moss
- 1 gallon (4 L) composted local pine and fir wood bark
- 2 cups (0.5 L) alfalfa meal
- 2 gallons (8 L) finished compost
- ½ gallon (2 L) microbially rich soil

I sift the wood bark to remove large chunks, and I gather the loamy or sandy loam soil from my own farm fields. I prepare the mix in a 30-gallon garbage can by adding a little of each ingredient at a time and turning with a shovel to mix thoroughly. I use my own hay mulch compost (I describe my composting method in chapter 4). I add a small amount of water as I mix just to moisten the mix slightly and keep the microorganisms happy. It should not be wet and soggy, just moist to the feel. My compost does have weed seeds, so I pre-germinate the weed seeds and then let them dry out before sowing crop seeds.

Figure 9.3. Hooped and nonhooped row covers protecting early lettuce transplants and seeded spinach. Using row cover enables me to plant when nighttime temperatures are still in the low 20s F (down to –7°C).

Figure 9.4. Winter cabbage and lettuce (left) and spinach, spring-seeded peas, and new lettuce (right) grow well in a high tunnel in early spring. Stumps of harvested radicchio plants are sending up flower stalks, which I leave to bloom as an early pollen and nectar source for beneficial insects.

either vermicomposting (earthworm-driven) or thermogenic (self-heating) composting techniques. Transplant biomass and early yield were highest for plants in mixes with all four amendments. But there was a trade-off between high germination percent with vermicompost and increased longer-term plant growth with alfalfa or sesame meal amendments.[5] Other researchers have noted reduced germination rates with higher rates of alfalfa meal.[6] In my experience, however, higher rates of alfalfa meal are not necessary when you add both compost and alfalfa meal to your growing mix.

SEASON EXTENSION TOOLS

Some season extension techniques are very simple, such as mulching to keep the roots of fall/winter crops like kale, cabbage, and Brussels sprouts warm. Observe the microclimates around your garden—find warm areas for fall planting and cool areas for summer planting of cool-season crops. Set up a few raised beds, where the soil will warm up faster in the spring.

Some other methods require the use of row cover or setting up a cold frame or other structure. It takes some time, but it helps crops flourish and resist pests.

Hooped row covers or low tunnels covered with plastic or row cover. Plant spinach in low tunnels in fall to overwinter; when plants reach the two-true-leaf stage, they will resume growth in late winter/early spring to produce the first crop of the new season. In my dry, cold climate, I mow my living mulch and use it to mulch my last planting of outside spinach in the fall. Then in early spring I pull off what's left of the mulch and enjoy a super-early spinach crop.

High tunnels or hoop houses. Plant cold-loving crops inside a high tunnel in late summer and fall, then cover the plants with row cover alone or supported on hoops. This allows me to pick fresh lettuce even after a week of temperatures in the single digits Fahrenheit (down to –17°C) later in the winter.

Cold frames. Seed leeks in a cold frame in late winter, then transplant them into a low tunnel as soon as the soil can be worked. Cold frames are also great for early lettuce, spinach, and other cool-season crops. Movable cold frames are nice for beds that you want to plant early and then remove once temperatures warm.

Cold-tolerant fall and spring crops. Lay reusable shade cloth, clear plastic, or black plastic on the soil for a week before planting early-spring or fall crops to warm the soil and improve seed germination.

Figure 9.5. Fall kale and broccoli crop mulched with mowed living mulch to keep roots warm as temperatures drop.

Asparagus

Choose a plant-competition-free spot for asparagus because it will persist in one spot for 15 to 20 years. Although its cousins grow wild, cultivated asparagus is not happy about competing with perennial weeds, particularly creeping grasses.

CROP MANAGEMENT

Temperature

Asparagus is a very cold-hardy, cool-season perennial that requires a winter dormant rest period. Asparagus grows best at day temperatures of 75 to 85°F (24–29°C) and night temperatures of 60 to 70°F (16–21°C).

Light

Asparagus does best in full sun, but will grow reasonably well with light shade for several hours during the day.

Soil and Fertility

A light, deep, well-drained soil with a pH of 6.0 to 6.7 is preferred, but asparagus will tolerate soil pH over 7.0. Asparagus requires average to low levels of phosphorus and potassium. Mulching with organic residues such as composted deciduous leaves helps maintain soil organic matter content and often helps to prevent disease. Remember to add a nitrogen source once a year in early spring as well. I add nitrogen by mowing my fresh green living mulch in early spring and blowing it into the crop row just before asparagus begins to sprout.

Water

Asparagus has deep roots, is relatively drought-tolerant, and requires little supplemental water except when the new crowns are becoming established after planting.

Special Ecological Preferences

You can either plant 1-year-old crowns or start asparagus from seed. I have well-established beds from seedlings I grew in containers in my greenhouse for 3 months before transplanting to a prepared bed in the spring. Soak both seeds and crowns in compost tea for 5 minutes just prior to planting. If your soil is heavy, plant asparagus on slightly raised beds. Do not harvest from new plantings until they are at least 2 years old. Harvest of young plantings can stress plants, making them more susceptible to pest problems.

Once established, asparagus does well with annual clover living mulch sown in the crop rows. I suppress the clover with added mulch from the mowed row middles around the asparagus plants every spring and fall. Mow the spent asparagus foliage and ground cover in late winter to early spring. This will suppress the reseeded annual clover and allow for new asparagus spears to emerge. If perennial grasses begin to move into the ground cover, I set them back by applying a vinegar spray to the bed surface in the spring before asparagus spears emerge.

TROUBLESHOOTING PESTS AND PROBLEMS

Yellow leaves, slow growth.

Cause: Nitrogen deficiency.

Comments: Heavy, wet soil can result in similar symptoms.

Yellow leaves, wilting, and dieback of crown.

Other symptoms: Small spears. Lesions at the base of spears or below the soil line.

Cause: Fusarium wilt disease.

Comments: Resistant cultivars include Jersey Giant, Jersey King, Jersey Knight, and Viking KB3.

Foliage turns reddish brown and drops.

Other symptoms: Small reddish spots on main stems and branches.

Cause: Rust disease.

Comments: Use resistant cultivars such as Martha Washington, Jersey King, Jersey Knight, and Viking KB3.

Defoliated plants and deformed spears.
Cause: Several insect pests, including asparagus beetles, cause this kind of injury. Frost damage can also cause deformed spears.
Comments: Asparagus beetles are most active in cool weather. Handpicking is effective, especially in small gardens. Lady beetle larvae and other predators eat both eggs and larvae of asparagus beetles.

Beans

Snap beans come in many shapes and colors and are eaten in an immature stage. Dry or field beans are available in even more shapes and colors and are harvested for storage in a dry, mature form. Lima beans are a different species from snap and dry beans and are more sensitive to cold, especially cold soils.

There are both pole types and bush types of snap and lima beans and dry beans. Pole types produce later harvests over longer periods, but must be supported. Bush types set pods more quickly and have a shorter, more concentrated harvest period.

CROP MANAGEMENT

Temperature
Beans require a warm, frost-free growing period, but they do not like extreme heat. Optimal growing temperatures are 65 to 80°F (18–27°C). Most bean cultivars grow less well at temperatures above 85°F (29°C), but will tolerate temperatures as low as 50°F (10°C). Soil temperatures at planting are the key to good bean growth. Beans do not germinate well and are more susceptible to pests at soil temperatures below 60°F (16°C). Optimal germination occurs when daytime temperatures reach 75 to 80°F (24–27°C).

Light
Beans do best in full sun.

Soil and Fertility
Choose lighter soils for earliest plantings. If you have cool, heavy soils, cover seeded beans with clear plastic until they emerge. If you have not grown beans or other legumes in your soil recently, purchase the proper bacterial inoculant (to promote nitrogen fixation) and apply it before planting. Beans prefer a soil pH of 5.5 to 7.0. At a pH above 7.0, iron and zinc deficiencies may occur. Beans are susceptible to certain micronutrient deficiencies. Foliar seaweed applications when plants are young may help. Beans are one of the vegetable crops most sensitive to salt injury; they grow poorly in soils with high sodium content. Bean roots can also be injured by the high soluble salt levels in raw or poorly decomposed manures, which leads to poor plant growth.

Water
Maintain moist, but not wet, soil.

Special Ecological Preferences
Most disease problems in beans can be avoided by avoiding cool, wet soils. Compost addition and raised beds are reportedly the two most important cultural tools to prevent disease in beans. Raised beds are especially helpful in wet climates. Rapid germination is essential to avoid root rot problems. Soaking seed and pre-germinating it between moist layers of paper towels helps to prevent root rot and many pest problems. Protect early and late crops with row cover in cold climates.

TROUBLESHOOTING PESTS AND PROBLEMS

Yellow, curling leaf margins; wilted plants.
Other symptoms: Flowers and pods drop.
Cause: Leafhoppers (potato).
Comments: Not usually a serious problem.

Yellow, distorted new growth.
Cause: Tarnished plant bug or mosaic virus.

Comments: Many mosaic-resistant or -tolerant cultivars are available. For curly top virus, Great Northern and Accelerate are resistant.

Yellow leaves; slow growth.
Cause: Nitrogen deficiency.

Curled, yellow, withered leaves.
Cause: Aphids.

White, stippled leaves, which later become bronzed.
Other symptoms: Stunted growth.
Cause: Mites.

Small reddish brown specks, especially on leaf undersides.
Other symptoms: Leaves yellow and drop.
Cause: Rust.
Comments: Kentucky Wonder, Bountiful, and Opus are less susceptible than other varieties.

Golden yellow, crinkled, mottled, curled leaves.
Cause: Mexican bean beetle.

Blossoms drop.
Cause: Excessive heat or rain.
Comments: Wait for new blossoms to form. If blossoms form but pods fail to develop, zinc may be deficient.

Pitting and russeting of pods.
Cause: Chilling injury.
Comments: Occurs in most cultivars at temperatures below 45°F (7°C).

Water-soaked or brown patches on pods.
Other symptoms: Yellow, blotched seeds.
Cause: Bacterial blight.
Comments: Great Northern is blight-tolerant.

Holes chewed in pods.
Cause: Caterpillars, including cabbage looper, armyworm, corn earworm, and European corn borer.

Beets

Beets prefer cooler temperatures for best growth, but can grow all summer if mulched in hot climates.

CROP MANAGEMENT

Temperature

Although beets prefer cooler temperatures for best growth, they can grow in hot temperatures if seed is pre-germinated. Seed germinates at temperatures of 45 to 85°F (7–29°C). Optimal growing temperatures are 50 to 65°F (10–18°C). Beets grow poorly at temperatures above 80°F (27°C), but will tolerate temperatures as low as 40°F (4°C). However, seedstalks form if beets are exposed to 50°F (10°C) or cooler temperatures for 2 to 3 weeks after the first true leaves have formed. On the other hand, roots can tolerate colder temperatures than beet tops. I often harvest fall beets after night temperatures are dropping into the low 20s F (down to –7°C) but before the ground freezes.

Light

Full sun is necessary to avoid leaf spot diseases and grow sizable roots.

Soil and Fertility

Beets require a deep, well-drained soil with a pH of 6.5 to 7.5 and a moderate amount of nitrogen. They are susceptible to nitrogen deficiency. They may suffer boron deficiency in light and/or sandy soils.

Water

Uniform moisture is best. Water deeply and regularly during dry periods.

Special Ecological Preferences

Pre-germinating seed or soaking in warm water for 24 hours before planting speeds germination and helps prevent root rots and seed maggot problems in cool, spring soils. Beets compete poorly with other plants when small, so keep them weed-free and thin them well. Once they have begun to form roots that are 1 inch (2.5 cm) or larger, they compete relatively well with shallow or fibrous-rooted weeds.

TROUBLESHOOTING PESTS AND PROBLEMS

Leaves turn yellow, older ones first.
Other symptoms: Stunting.
Cause: Nitrogen deficiency.

Purplish patches on leaves.
Cause: Phosphorus deficiency.
Comments: Common in cool spring soils. Plants usually outgrow the problem when soils warm.

Tan spots with dark borders on leaves.
Cause: Cercospora leaf spot.
Comments: Scarlet Supreme and Red Ace are less susceptible to this disease.

Meandering white or translucent tunnels in leaves.
Cause: Leafminers.

Internal black, dead, hard spots in roots.
Cause: Boron deficiency.
Comments: For future crops, prevent boron deficiency by adding borax to the soil in liquid form (1 tablespoon per 100 square feet [0.1 m^2] in 1 gallon [4 L] water). Or add kelp (10 pounds per 100 square feet [49 kg/100 m^2]). Foliar seaweed sprays can be started as soon as first true leaves appear.

Misshapen roots.
Cause: Calcium deficiency or hard, heavy soil.
Comments: To prevent future problems, add gypsum or calcitic lime if soil pH is below 6.2.

Light-colored roots with wide zoning of rings.
Cause: High temperatures and fluctuating soil moisture.

Broccoli

Broccoli is the pickiest of the brassica crops (especially to get to market in perfect condition) because it does not like it too cold when plants are young or too hot when heads are developing. But the great nutrition and taste make broccoli worth the effort.

CROP MANAGEMENT

Temperature
Broccoli is a cool-season crop and tolerates lows of 40°F (4°C) once it is beyond the five-leaf transplant stage. However, broccoli is more susceptible to cold and freezing injury than Brussels sprouts or cabbage. Optimal growing temperatures are 60 to 65°F (16–18°C); poor growth occurs at temperatures above 75°F (24°C). Applying shade cloth during hot periods can improve growth. Broccoli is a good candidate for hooped row cover to mitigate cold temperatures, especially for spring-planted transplants. Prolonged exposure to temperatures of 35 to 45°F (2–7°C) when transplants are at or below a four- to five-leaf stage may initiate flowering, or bolting, that results in tiny, open broccoli heads. High temperatures of 85 to 90°F (29–32°C) cause broccoli plants to bolt (form seedstalks prematurely) and/or yellow as well.

Light
Broccoli can handle (and actually prefers) a little shade at transplanting and in hot, sunny climates. It also responds well to row covers for 3 to 5 weeks after transplanting to mitigate temperature extremes and enhance root growth (as well as to discourage flea beetles).

Soil and Fertility
Broccoli prefers fertile well-drained soils with a pH of 6.0 to 6.8. Like other brassicas, it's a heavy feeder and is susceptible to several nutrient deficiencies including potassium, nitrogen, and boron deficiency. Broccoli is more likely to show deficiency

symptoms overall when grown in sandy soils with low organic matter that have either acidic or alkaline soil pH.

Water

Broccoli has very shallow roots and thus needs regular irrigation. Mulching helps to conserve moisture, control weeds, and discourage flea beetles. However, waterlogged soil can encourage clubroot disease.

Special Ecological Preferences

Broccoli is very weed-tolerant once well established as five- to seven-leaf transplants. In fact, in hot, sunny climates, broccoli may benefit from leaving certain shallow-rooted annual weeds in place to provide shade and diversity. Broccoli also does very well growing with living mulches and/or undersown with cover crops, particularly legume cover crops. Remember, however, to avoid planting cover crops in the Brassicaceae family, such as radishes or mustard, with or before their broccoli cousins. This is especially important if black rot disease has been a problem in the past, because transfer of the bacteria can occur from a brassica cover crop to the broccoli crop. Also, all plants in the cabbage family are non-mycorrhizal, so a brassica cover crop followed by a brassica crop does not make good sense for the balance of the soil microbial community.

Most diseases of the cabbage family are spread by water, so it is a good preventive practice to keep irrigation water off leaves in wet, humid climates and especially when broccoli heads are developing. However, flea beetles are discouraged by sprinkler irrigation.

TROUBLESHOOTING PESTS AND PROBLEMS

Yellow, stunted plants.

Other symptoms: Plants wilt during bright, hot days and recover at night.

Causes: Usually a sign that roots are injured, from various causes.

Comments: Pull up a plant and check the roots for root knot nematodes (irregularly shaped galls); cabbage root maggots (roots riddled with slimy, winding tunnels); or clubroot disease (thick, irregular, club-shaped roots). Emerald Jewel and Monclano broccoli varieties are resistant to clubroot disease.

Yellow V-shaped spots on leaf margins.

Other symptoms: Blackened leaf veins stems show internal black streaks.

Cause: Black rot disease.

Comments: Many black-rot-tolerant varieties are available. Avoid planting cover crops in the Brassicaceae family before broccoli due to transfer of black rot bacteria.

Yellowish leaves; lower leaves drop.

Other symptoms: Stunting; twisted stems.

Cause: Fusarium yellows virus disease.

Comments: Usually occurs soon after transplanting. Several fusarium-resistant cultivars are available.

Leaves turn yellow, older ones first.

Other symptoms: Slow growth.

Cause: Nitrogen deficiency.

Older leaves whitish and may droop.

Cause: Cold injury and/or wind injury.

Comments: Row cover is the best preventive for both. Cold injury can make transplants more susceptible to diseases and insects.

Yellow, curled leaves; stunted plants.

Cause: Aphids.

Light-colored spots on leaves; spots turn papery.

Other symptoms: White mildew on leaf undersides.

Cause: Downy mildew disease.

Figure 9.6. These broccoli plants are wilting due to root damage caused by clubroot disease. The end of the row is affected probably due to water puddling in this part of the row.

Comments: Many varieties are tolerant of downy mildew.

Small pitted spots in leaf; shot-hole injury.

Cause: Flea beetles.

Comments: Varieties with dark green, glossy leaves are usually more susceptible to flea beetles.

Large irregular holes.

Cause: Cabbage worms, or possibly grasshoppers.

Comments: Several types of caterpillars feed on brassicas, including imported cabbage worm, cabbage looper, and diamondback moth. Varieties with dark green, glossy leaves are less susceptible to cabbage looper and imported cabbage worm feeding (but unfortunately more susceptible to flea beetles).

Leaves with meandering tunnels.

Cause: Leafminer (pea and vegetable).

Hollow stems; small, uneven heads form.

Cause: Excess nitrogen; excess irrigation; potassium deficiency.

Hollow stems, distorted heads, and discolored stem tissue.

Cause: Boron deficiency.

Black areas in center of head.

Cause: Rot due to water collecting inside the head.

Comments: Cold weather and/or rapid growth can also cause black to brown areas in the center of the head.

Small heads that form prematurely (bolt).

Cause: Exposure to cold; high-temperature stress; poorly adapted variety for your climate; exposure of plants to fluctuating temperatures.

Comments: Many heat-tolerant cultivars are available.

Brussels Sprouts

The Brussels sprout is one of the hardiest members of the cabbage family. This crop is susceptible to many of the same pests and problems as broccoli.

CROP MANAGEMENT

Temperature

Brussels sprouts can tolerate temperatures below 40°F (4°C) during active growth and below 20°F (–7°C) once mature and acclimated.

Light

Full sun to part shade. Does better in part shade in hot climates.

Soil and Fertility

Well-drained soil with a pH of 5.5 to 6.8.

Water

Keep well watered during hot weather.

Special Ecological Preferences

I harvest Brussels sprouts all winter even after the sprouts freeze; they thaw well and taste great. Growing and troubleshooting problems of Brussels sprouts are much like those of broccoli. Brussels sprouts are more susceptible to heat injury than broccoli—they develop tan to dark brown leaf margins when heat-damaged. In warmer climates, grow Brussels sprouts in heavier, moister soil and use shade cloth or natural shade.

TROUBLESHOOTING PESTS AND PROBLEMS

Refer to the "Troubleshooting Pests and Problems" section for broccoli, beginning on page 190.

Cabbage

Green, red and savoy types of cabbage can all be planted for both spring and fall harvests or grown in a high tunnel for winter harvest. Cabbage is susceptible to many of the same pests and problems as broccoli.

CROP MANAGEMENT

Temperature

Like broccoli, cabbage is a hardy, cool-season crop. Cabbage likes cool, moist conditions. Optimal growing temperatures are 60 to 65°F (16–18°C). Cabbage grows poorly at temperatures above 75°F (24°C), but with shading, mulch, and moisture, it tolerates higher temperatures (producing smaller heads than usual, though). It will tolerate temperatures as low as 40°F (4°C) with no problem, and if hardened will tolerate significantly lower temperatures. However, do not plant cabbage outdoors too early. Cabbage is a biennial, and prolonged exposure to cool temperatures when transplants are at a four- to five-leaf stage may initiate flowering (bolting). Ten or more continuous days of 35 to 45°F (2–7°C) weather will cause bolting in most cabbage cultivars, or contribute to smaller heads later on.

Light

Plant cabbage in full sun in cool climates to part shade in hot climates.

Soil and Fertility

Cabbage does best in well-drained soils with a pH of 6.0 to 6.8. It's a moderate to heavy feeder and is susceptible to several nutrient deficiencies including boron, molybdenum, calcium, potassium, and copper. Nitrogen and phosphorus are required at average to high levels. Spring plantings do best in lighter, sandier soils that warm up more quickly.

Fall plantings have fewer problems if grown in heavier soils.

Water

Cabbage has shallow roots and thus needs an abundance of water, especially in hot climates. However, uneven or excessive moisture after heads are formed can cause splitting.

Special Ecological Preferences

Rotation and good sanitation are the best pest preventives. Do not grow cabbage where any other brassicas have grown for the past 3 years. Once your seeds have germinated, try to keep air temperatures at 60°F (16°C) in your greenhouse or seed-starting area. Hot temperatures will cause transplants to be lanky.

TROUBLESHOOTING PESTS AND PROBLEMS

Refer to the "Troubleshooting Pests and Problems" section for broccoli, beginning on page 190.

Disease- and Insect-Tolerant Cabbage Cultivars

Black rot: Many tolerant cultivars are available.
Fusarium yellows: Bravo and Grandslam.
Downy mildew: Loughton shows some tolerance.
Clubroot: Resistant green cabbage varieties include Kilaton, Kilagreg, and Tekila. Resistant Chinese cabbage varieties include Pacifiko and Emiko.
Cabbage worms: Ruby Perfection, Red Ace, Early Jersey Wakefield, Mammoth Red Rock, Chieftain Savoy, Savoy Perfection Drumhead, and Danish Ballhead are less susceptible to cabbage looper and imported cabbage worm feeding. Cabbage butterflies avoid laying eggs on red cabbage varieties.
Flea beetles: Early Jersey Wakefield tolerates these beetles better than most cabbage cultivars.
Root maggots: Early Jersey Wakefield and Red Ace are somewhat resistant.

Figure 9.7. Young red- and green-leaved cabbage plants. Look closely to see that the green cabbage leaves have holes from cabbage worm feeding, but red cabbage leaves have almost no feeding damage.

Carrots

Carrots are a cool-season crop, hardy enough to be undisturbed by light frosts in the spring and heavy frosts in the fall. A multitude of carrot cultivars exists, from short, stumpy cultivars adapted to shallow soils to long, thin cultivars that require more exacting carrot growing conditions.

CROP MANAGEMENT

Temperature

Optimal carrot growing temperatures are 60 to 70°F (16–21°C). As temperatures increase, most cultivars develop shortened roots, while lower temperatures encourage longer, more pointed carrots. Carrots grow poorly at temperatures above 75°F (24°C), but will tolerate temperatures as low as 45°F (7°C) and as high as 90°F (32°C) if kept well irrigated and mulched. I helped carrots maintain good quality through a heat wave with 3 weeks above 100°F (38°C) with shade cloth, heavy mulching, and irrigating just before the hottest part of the day.

Light

Full sun to part shade. Shade is important in hot, very sunny climates.

Soil and Fertility

Carrots perform best in well-aerated, deep, loose sandy loam soils with a pH of 5.5 to 6.8. No other crop is as sensitive to poor soil structure. Misshapen carrots are more often a result of hard, compact soil than any pest problem. Carrots require moderate to high levels of potassium and phosphorus, but only moderate levels of nitrogen. They are very sensitive to salt injury and are not a good crop to choose in areas where high soil sodium levels exist.

Water

An abundant and even water supply is necessary for good root development. Water is especially important to give emerging seedlings an edge in the plant competition struggle.

Special Ecological Preferences

Do not work carrot soil when it is wet, add a generous amount of well-decomposed organic residues or compost (never use fresh compost or manure!) before planting, and try planting in raised beds. Do not let soil become saturated.

Carrots are not good at the plant competition struggle and require earlier weed intervention and thinning from gardeners and farmers. Carrots are a good candidate for pre-germination seed treatments to speed germination. I never plant carrot seed without pre-germination seed treatments. Carrots do well with flaming for weed suppression if you get the timing right. Probe in your carrot row to check germinating seeds and flame once the first seeds have germinated. Protect seedlings with row cover if temperatures are predicted to fall below 30°F (–1°C). Carrots can be left in the ground and harvested late into the fall/winter if mulched.

TROUBLESHOOTING PESTS AND PROBLEMS

Yellow, dwarfed young leaves.

Other symptoms: Bushy growth, purpled leaves.
Cause: Aster yellows virus disease.

Yellow leaves, older first.

Other symptoms: Stunted plants.
Cause: Nitrogen deficiency.
Comments: Waterlogged soil will produce the same symptoms.

Dark spots with yellow borders.

Cause: Cercospora leaf blight, if younger leaves are affected; possibly alternaria leaf blight if older leaves are affected.
Comments: Danvers 126 is resistant to Cercospora. Danvers 126 and Bolero are less susceptible to alternaria fungus disease.

Poor seedling emergence.

Cause: Crusted, hard soil and/or high temperatures.

Yellow, stunted plants.
Other symptoms: Plants wilt during bright, hot days and recover at night.
Cause: Usually a sign that roots are injured. Pull up the plant and check roots for wireworms, root knot nematodes, rust fly, weevil, or disease.

Bumps on roots; deformed roots.
Cause: Root knot nematodes.

Small, irregular holes scattered over surface of root.
Cause: Root knot nematodes; wireworms.
Comment: Damage usually occurs later in the season and is worse in dry years.

Roots chewed.
Other symptoms: Meandering, rusty-red scars along the root.
Cause: Carrot rust fly.

Dark tunnels on upper and outer parts of roots.
Cause: Carrot weevil.
Comments: Tunnels often show a zigzag pattern.

Dark cankers on roots; root deterioration.
Cause: Parsnip canker.

Small, woody, hairy, pale roots.
Cause: Aster yellows virus disease.

Jagged cracks in roots, water-soaked appearance.
Cause: Freezing injury or uneven irrigation (which leads to uneven growth).
Cause: Magnesium deficiency, phosphorus deficiency, low temperatures, and/or excessive heat.

Forked roots.
Cause: Heavy, compacted, or stony soils; injury to root tips; overcrowding of roots.

Poor root color and taste, green shoulders on roots.
Cause: Exposure to sunlight or high temperatures.
Comments: To prevent this problem, hill up rows with 2 inches of soil or mulch to cover root shoulders.

Figure 9.8. Forked carrot root due to overcrowding and poor thinning of carrot crop.

Cauliflower

Cauliflower is susceptible to many of the same pests and problems as broccoli. It also has similar light, temperature, soil, and nutrient needs. Refer to the "Broccoli" entry on page 190 for growing and troubleshooting information.

Cauliflower does have its own unique ecological preferences. For one thing, cauliflower is fussier about temperatures than the other brassicas. Though a cool-season crop like broccoli, cauliflower is not as cold-tolerant and will not hold up to temperatures below 45°F (7°C). Curds turn brown in response to cold and freezing injury. If temperatures are predicted to go below 45°F, protect cauliflower heads with row cover. Exposure of flower curds to bright sun and high temperatures will cause purplish brown patches and irregular, loose texture. To shade the developing heads from sunlight, tie the outer leaves over the heads, or use self-blanching types. Heat-resistant cultivars include Amazing and Flame Star. There are many cold-tolerant cultivars, including Snow Crown. Cashmere and Majestic are black-rot-resistant. Majestic and White Corona show some tolerance to downy mildew. Super Snowball A tolerates flea beetles better than most cultivars.

Celery

Celery has a reputation as a fickle plant. Its native ancestors grew in marshes, so it does not like to dry out.

CROP MANAGEMENT

Temperature

A cool-season crop, celery withstands a light frost, but temperatures below 55°F (13°C) for more than 10 days when plants are young may cause premature, skinny stalk formation.

Light

Celery grows best in full sun, but it tolerates partial shade.

Soil and Fertility

Celery is a heavy feeder. It does best with generous amounts of well-decomposed organic residues or compost and prefers a soil pH of 6.0 to 6.8. Celery roots are fragile; they don't grow well in heavy or very light soil. Raised beds with amended soil improve celery growth.

Water

Celery is a moisture-loving crop; supply water evenly and regularly.

Special Ecological Preferences

Seeds need direct sunlight in order to germinate and are slow to germinate.

TROUBLESHOOTING PESTS AND PROBLEMS

Yellow, dwarfed young leaves.
Other symptoms: Bushy growth, purpled leaves.
Cause: Aster yellows virus disease.

Mottled, yellow leaves.
Other symptoms: Twisted stems, dwarfing.
Cause: Mosaic virus.
Comments: Tall Utah 52-70R Improved is resistant.

Leaves turn yellow, older ones first.
Other symptoms: Stunted plants.
Cause: Nitrogen deficiency.

Yellow leaves, reddish stalks.
Cause: Fusarium yellows virus disease.
Comments: There are many tolerant cultivars.

Irregular yellow-brown leaf spots.
Other symptoms: Sunken, elongated tan spots on stalks.
Cause: Cercospora leaf blight.

Brown mottling of leaves.
Other symptoms: Crosswise cracks on stalks.
Cause: Boron deficiency.
Comments: Utah 52-70R Improved is tolerant of low boron.

Blackening at center with death of growing points.
Cause: Calcium deficiency.
Comments: Tall Utah 52-70R Improved is tolerant of low calcium.

Dwarfed, rotted plants.
Cause: Carrot rust fly.

Chard

Chard competes well with other plants and weeds, germinates easily, and is easy to grow.

CROP MANAGEMENT

Temperature

Swiss chard is a cool-season crop, a type of beet developed for its green leaves and rainbow-colored stems, which grow vigorously from late spring through to fall frost. Optimal growing temperatures are 60 to 65° F (16–18°C). Chard is not too picky about heat and cold but it does not grow as well when temperatures are over 80°F (27°C). It will withstand light frost when young and very heavy frosts when mature. A row growing in partial shade continued to produce tasty foliage for me during a July and August in which temperature were in the high 90s (35°C or more) for 8 weeks, while both spinach and lettuce did poorly.

Light

Full sun to partial shade. During the summer, plant chard in a partially shaded location or provide shade.

Soil and Fertility

Chard prefers light, well-drained soil with a pH of 6.5 to 7.5. It is not demanding of nutrients.

Water

To keep leaves tender, provide chard with plenty of water.

Special Ecological Preferences

Because it germinates and fills in quickly, chard is one of the fall crops I use to overseed summer crops. I often let spring plantings or overwintered chard plantings go to seed to reseed themselves for a second crop. Note, though, that this is not a sound crop rotation practice and should not be done if you have had disease problems in chard.

TROUBLESHOOTING PESTS AND PROBLEMS

Purplish patches on leaves.
Cause: Phosphorus deficiency.
Comments: Most likely to occur when soils are cool in spring. Plants usually outgrow the problem when soils warm.

Brown leaf tips.
Cause: Reaction to bright sun and heat; temperatures above 80°F (27°C).

Tan spots with dark borders.
Cause: Cercospora leaf spot.

Yellow, rotted leaves.
Other symptoms: Light-colored spots on upper leaf surfaces; white mildew on leaf undersides.
Cause: Downy mildew.

Stunted, crinkled leaves.
Cause: Curly top disease.

White or translucent, irregular tunnels in leaves.
Cause: Leafminers (vegetable).

Chewed and leaves.
Cause: Armyworm.

Figure 9.9. White irregular tunnel in a leaf surface made by a vegetable leafminer.

Corn

Corn is the perfect succession crop. Cultivars are classified as early, medium, and late with a days-to-harvest period ranging from 65 to 100 days. Cultivars are further classified into three groups: "normal sugary" (traditional corn flavor), "sugar enhanced" (increased tenderness and sweetness), and "supersweet."

CROP MANAGEMENT

Temperature

Optimal growing temperatures are 60 to 85°F (16–29°C), and growth is minimal under 50°F (10°C). Growth rates increase as temperatures increase, up to 90°F (32°C). Corn is very sensitive to cool soils and will not germinate when the soil temperature is below 50°F. To avoid many pest problems, wait to plant until soil temperatures reach at least 60°F (16°C). To increase germination success and avoid pest problems, cover soil with clear, 1.5-mil plastic 2 weeks prior to planting. Pre-germinate seed and after planting, then re-cover with plastic until seedlings emerge.

Light

Sweet corn demands full sun.

Soil and Fertility

Corn needs rich, light soil with a pH of 5.5 to 6.8. Corn is a relatively heavy feeder; provide added organic residues before planting and side dressings of organic residues when plants reach 1 foot (30 cm) tall and again when they begin to silk. I mow and blow living mulch into corn rows at these times.

Water

Irrigate corn consistently and well, especially when seedlings are emerging and while plants are young and growing rapidly.

Special Ecological Preferences

Unfortunately, most pests seem to prefer the supersweet cultivars. They also tend to germinate less well, and they must be isolated by at least 25 feet (8 m) from all other sweet corn or cross-pollination may result in reduced kernel sweetness. I recommend watching wind patterns, too. I grew Floriani red flint corn 300 feet (91 m) upwind from my sweet corn, and just one unusual wind event enabled cross-pollination of the flint corn with the sweet corn. It was not a happy mistake!

Rotate your crops so that at least 2 years pass before corn is grown in the same area. Corn is a good crop to add to vegetable rotations because it is one of the rare grasses grown as a vegetable crop. After harvest, cut or mow stalks and let them decompose on the surface. If poor stands have been a problem, plant later in the season and/or pre-germinate seed to avoid cool, wet soils that will slow germination. Plant rows of corn parallel to the prevailing winds, so that air moves freely up and down the rows.

TROUBLESHOOTING PESTS AND PROBLEMS

Wilted, stunted plants and/or spotty, poor stands.

Cause: Soil-dwelling pests including wireworm, root maggots, corn root worms, or white grubs.

Ragged leaf margins on young plants.

Other symptoms: Plants cut at the base of the stem.
Cause: Cutworms.

Holes in leaves.

Cause: Many kinds of caterpillars feed on corn leaves and ears, including corn earworm, European corn borer, and armyworms.

Round, raised, brown spots on leaves.
Cause: Rust.
Comments: Many rust-tolerant cultivars are available.

Mottled, stunted new leaves.
Other symptoms: Poor kernel formation at the base of the ear.
Cause: Mosaic virus.
Comments: Many mosaic-tolerant cultivars are available.

White-black, puffed-out growths on ears, tassels, and stem.
Cause: Smut.
Comments: Resistant cultivars are Silver Dollar and Silver King.

Older leaves purple on young plants.
Cause: Phosphorus deficiency.

Leaves turn yellow, older ones first.
Other symptoms: Stunted plants.
Cause: Nitrogen deficiency.

Large chewed areas at top of ears.
Other symptoms: Dry, brown, exposed kernels at ear tops.
Cause: Birds.

Damage to developing kernels at ear tips.
Cause: Corn earworm.
Comments: Corn varieties with long, tight husks are less susceptible to earworms. Resistant varieties include Country Gentlemen, Staygold, Golden Security, and Silvergent.

Damage to ear, but not at tip.
Other symptoms: Damage to stalks.
Cause: European corn borer and/or armyworms.
Comments: Corn earworms feed only at the tips of ears.

Cucumbers

Cucumbers are easy to grow as long as they receive consistent watering, lots of sun, and warmth. There are three main types of cucumber—slicing, pickling, and seedless—and many varieties of each.

CROP MANAGEMENT

Temperature

Soil temperatures of 60°F (16°C) are necessary for germination, and transplants do best if daily mean temperatures reach 60°F before planting out. Optimal growing temperatures are 70 to 75°F (21–24°C).

Light

Cucumbers insist upon full sun.

Soil and Fertility

Deep, light, loamy soil with a pH of 6.5 to 7.5 is ideal for cucumbers, especially for early crops.

Water

Because the plants' relatively short roots must supply long, vining stems and large leaves, cucumbers need regular irrigation.

Special Ecological Preferences

Cucumber flowers must be well pollinated or fruit will not develop properly. The crop is susceptible to many pests when plants are young, so use floating row cover to provide early plantings with protection from cool temperatures and pest insects. To increase soil temperatures, conserve soil moisture, and get a jump on weeds before planting, lay a sheet of clear 1.5-mil plastic over the cucumber-patch-to-be. Remove the plastic sheet just prior to planting and lightly scuffle-hoe the germinated weed seeds.

TROUBLESHOOTING PESTS AND PROBLEMS

Wilting plants; pale green areas on leaves.

Other symptoms: Foliage turns brown and crinkly.

Cause: Squash bug.

Sudden wilting of all or part of a vine.

Cause: Squash vine borer.

Leaves wilt, turn yellow, and curl.

Cause: Aphids.

Comment: Usually not serious unless populations are high.

Vines wilt and die gradually.

Other symptoms: Wilting begins with young leaves; no leaf yellowing present.

Cause: Bacterial wilt.

Comments: H-19 Little Leaf, County Fair 83, and Saladin are more tolerant to bacterial wilt.

Small holes in leaves on young plants.

Cause: Striped cucumber beetles; spotted cucumber beetles.

Comments: Varieties more tolerant of cucumber beetle feeding usually have lower levels of the attractant compound cucurbitacin. They include Gemini, H-19 Little Leaf, Saladin, Liberty, and Wisconsin SMR58.

Distorted, yellow areas on leaves.

Other symptoms: Mottling and wrinkling of older leaves.

Cause: Mosaic viruses.

Comments: There are many varieties resistant to mosaic diseases.

Older leaves mottled yellow between veins.

Cause: Downy mildew.

Comments: There are many varieties resistant to downy mildew.

Powdery white spots mostly on upper surface of leaves.

Cause: Powdery mildew.

Water-soaked spots on leaves.

Other symptoms: Spots turn gray, tissue dies and drops out, leaving foliage with a shot-hole appearance.

Cause: Angular leaf spot.

Comments: Fanfare, Raider, and Sweet Slice are resistant.

Yellow, water-soaked spots that turn brown.

Cause: Anthracnose.

Comments: Resistant cultivars are Raider, Sweet Slice, and Regal.

Dark brown spots with concentric rings.

Other symptoms: Spots appear on older leaves first; vines defoliate eventually.

Cause: Alternaria leaf blight.

Fruit with brown and sunken spots.

Other symptoms: Edges of spots show zones of lighter-colored rings.

Cause: Alternaria leaf blight.

Small, brown, angular spots on fruit.

Cause: Angular leaf spot.

Circular, black, sunken cankers on fruit.

Cause: Anthracnose.

Fruit wilts and shrivels.

Cause: Bacterial wilt.

Eggplant

Eggplant is one of the more fussy vegetables to grow. The key to growing eggplant successfully is to understand its need for warmth and sunlight.

CROP MANAGEMENT

Temperature

This warm-season vegetable is more susceptible to cold injury than either tomatoes or peppers. Poor growth and fruit set occur below 65°F (18°C), whereas optimal growing temperatures are 70 to 85°F (21–29°C). Do not plant out before average daily temperatures exceed about 65°F. Like tomatoes and peppers, eggplants do not like extreme heat; growth diminishes above 95°F (35°C).

Light

Eggplants are sun lovers that want at least 6 hours of direct light each day.

Soil and Fertility

Eggplants do best at a pH of 6.0 to 6.8, but will tolerate a pH as low as 5.5. They prefer well-drained, fertile sandy loams, high in organic matter. Eggplants require higher nitrogen levels and moderate amounts of phosphorus and potassium. Like tomatoes, eggplants are susceptible to magnesium deficiencies. Foliar- or root-applied Epsom salts will correct this problem.

Water

Eggplants have a high moisture requirement; do not let them dry out. They are good candidates for raised beds and drip irrigation.

Special Ecological Preferences

Eggplants perform well with mulches. In cooler climates, clear or black plastic mulch has reportedly increased yields by almost 300 percent. During cool springs, I roll out reusable black weed mat alongside newly transplanted eggplants before I cover the plants with row cover in order to heat the soil. Sometimes I even set up the weed mat a few weeks in advance of planting to warm the soil. The row cover also protects young plants that are susceptible to wind injury.

TROUBLESHOOTING PESTS AND PROBLEMS

Large holes in leaves; skeletonized leaves.

Other symptoms: Whole plants and leaflets stripped.

Cause: Colorado potato beetle.

Large holes in leaves.

Cause: Grasshoppers.

Tiny pits in leaf surface.

Other symptoms: Leaves may have a lace-like appearance and turn dull, dry green.

Cause: Flea beetle.

Leaves turn yellow.

Other symptoms: Gradual wilting and browning of areas between leaf veins.

Cause: Verticillium wilt.

Yellow mottling of leaves.

Other symptoms: Curling and malformation of leaves.

Cause: Tobacco mosaic virus or cucumber mosaic virus.

Comments: Tobacco-mosaic-tolerant varieties include Black Bell, Epic, Megal, and Dusky Hybrid. Megal is also tolerant of cucumber mosaic.

Kale

Kale is a member of the cabbage family, and it can be grown like broccoli, though its culture is easier because it is not as preferred by pests or as picky. It has similar light, soil, and nutrient needs as broccoli, but it is hardier. Kale can tolerate lower temperatures—below 40°F (4°C) when young and into the low 20s (down to –7°C) when mature—and lower soil pH (5.5 to 6.8) than can other brassicas. In fact, frost improves the taste. Late plantings of kale are susceptible to cabbage aphid, and cabbage worms do find kale plantings eventually. Refer to the "Broccoli" entry on page 189 for growing and troubleshooting information.

Kohlrabi

Kohlrabi is a member of the cabbage family that resembles an aboveground turnip, with a swollen area of stem just above ground level. Kohlrabi can tolerate lower temperatures—below 40°F (4°C)—and lower soil pH (5.5 to 6.8) than can other brassicas. The trick to producing good-quality kohlrabi is to encourage rapid growth. Keep young plants well watered. Kohlrabi can be grown like cabbage, but it is not as susceptible to pests, which seem to prefer some of the other brassicas. Refer to the "Broccoli" entry on page 189 for growing and troubleshooting information.

Lettuce

Lettuce will happily grow just about anywhere with ample sunlight, cool nights (50 to 60°F [10–16°C]), loose soil, and plenty of moisture. It is one of the least picky crops to grow.

CROP MANAGEMENT

Temperature

Optimal lettuce growing temperatures are 60 to 65°F (16–18°C). Most lettuce cultivars grow poorly at temperatures above 80°F (27°C), but will grow well at temperatures as low as 45°F (7°C) and tolerate temperatures to 25°F (–4°C) once mature. Temperatures above 90°F (32°C) inhibit seed germination and cause early seedstalk formation (bolting). Choose bolt-resistant cultivars if you plan to grow lettuce through the summer.

Light

Full sun to partial shade. Shade cloth helps plantings stay cool in mid- to late summer.

Soil and Fertility

Keep the pH at 6.0 to 6.8. To maintain rapid growth, provide an average amount of nitrogen in forms that release slowly as well as forms that release more quickly.

Water

Moisture control is the key to healthy, pest-free lettuce crops. Water well during and for 2 weeks after germination, but do not let the soil become saturated. Raised beds help to provide good drainage. Keep the moisture supply consistent and even to avoid growth spurts.

Special Ecological Preferences

For best quality, lettuce needs to grow rapidly and without interruption. Sprinkler irrigation during the heat of the day also helps in dry, hot climates.

TROUBLESHOOTING PESTS AND PROBLEMS

Outer leaves on mature plants wilt or suddenly collapse.

Cause: Sclerotinia disease.

Lower leaves and midribs rot.

Other symptoms: Leaves near the base of the plant are affected first.

Cause: Rhizoctonia disease (bottom rot).

Slimy, rotted head.
Cause: Gray mold.

Bleached center leaves are dwarfed, curled, or twisted.
Cause: Aster yellows virus disease.

Yellow or pale leaves; stunted plants.
Cause: Leafhoppers.

Yellow, distorted new growth.
Cause: Tarnished plant bug.

Yellow leaves; slow growth; bitter taste.
Cause: Nitrogen deficiency, waterlogged soil, and/or temperatures above 80°F (27°C).

Yellow, curled, or stunted leaves.
Cause: Aphids.

Light green or yellow, crinkled leaves with enlarged veins.
Cause: Big vein.
Comments: Big vein is caused by a virus-like organism. Avoid cold, wet soils. It's most pronounced at air temperatures between 40 and 60°F (4–16°C). Infected plants show no symptoms when air temperatures are well above 60°F.

Mottled, ruffled leaves; stunted plants.
Cause: Mosaic viruses.
Comments: Many resistant/tolerant cultivars are available.

Midrib discolored.
Cause: Temperatures above 85°F (29°C).
Comments: Ithaca is less susceptible to heat-induced midrib discoloration.

Leaf margins turn brown and die.
Cause: Tipburn.
Comments: Tipburn is prevalent in hot weather. Keep plants watered and growing rapidly. Many varieties are less susceptible to tipburn.

Browning and dwarfing of plants.
Other symptoms: Velvety white growth on leaves.
Cause: Downy mildew.
Comments: Many varieties are tolerant of downy mildew.

Powdery white dust on upper leaf surfaces.
Cause: Powdery mildew.

Outer leaf margins tan; blistering of leaves.
Cause: Freezing injury.

Large hole in leaves.
Other symptoms: Excrement found at the base of the head.
Cause: Feeding by caterpillars, including cabbage looper, beet armyworm, and corn earworm.

Large holes in leaves with slimy trails.
Cause: Slugs and snails.

White or translucent, irregular tunnels in leaves.
Cause: Leafminers (vegetable).

Onions

There are many shapes, colors, and tastes of onions, including tender spring onions; red, yellow, and white storage types; bunching onions; and gourmet shallots.

CROP MANAGEMENT

Temperature

Onion is a hardy, cool-season plant that grows well over a wide range of temperatures, from 45°F (7°C) to a high of 85°F (29°C). Best growth occurs at 55 to 75° F (13–24°C). Onions prefer cool temperatures for early development and warm temperatures near maturity. Hot summers can result in smaller bulbs and flowering (bolting). Onions can also be induced to flower after a cold snap (temperatures below freezing). They do best with constant temperatures and do not like fluctuations.

Light

Onions are very light-sensitive. They need full sun and at least 12 hours and up to 16 hours of light daily during bulb formation.

Soil and Fertility

Onions require fertile, deep, well-drained soil high in organic matter with a pH of 5.5 to 7.0. They need high levels of nitrogen and potassium and moderate to high levels of phosphorus.

Water

Even moisture is required during growth. Cut back on irrigation as onions begin to mature, especially near harvest.

Special Ecological Preferences

Onion bulb formation is controlled by day length and varies with the cultivar; 12 to 16 hours of daylight are needed. Growers in northern areas should plant cultivars that form bulbs under longer day conditions. You can plant seeds, sets, or field-grown transplants. Soak plants in compost tea for 15 minutes before planting, and/or dust roots of sets and plants with alfalfa meal after soaking and before planting. Onions do well on raised beds or ridges, especially if the soil is heavy. Onions do not compete well with other plants. Keep them weed-free until bulbs are forming. Summers are increasingly hot in my area, and so I mulch onions with mowed living mulch several times throughout the season to help suppress weeds and to keep the bulbs cooler as they form. Be careful not to injure bulbs while weeding or harvesting. Dry onions at 90 to 120°F (32–49°C) for 2 to 3 days immediately after harvest, then store them at 32°F (0°C) to prevent diseases in storage.

TROUBLESHOOTING PESTS AND PROBLEMS

Wilted, yellow plants.

Cause: Wireworm or fusarium yellows virus disease.

Comments: Check roots for wireworm feeding or disease. Several onion varieties are less susceptible to fusarium, including Copra and Valiant.

Yellow, dying plants.

Cause: Root maggots (onion or seed corn maggot).

Comments: In general, white cultivars are more susceptible to root maggot attack than are yellow and red cultivars.

Yellow leaves, slow growth.

Cause: Nitrogen deficiency or waterlogged soil.

Brown leaf tips; white blotches on leaves.

Cause: Thrips (onion).

Comments: Hot, dry weather intensifies injury; heavy rains tend to decrease thrips populations. Thrips populations are usually higher near alfalfa or small grains.

Tips of leaves die.
Other symptoms: Pale green to brown leaf spots; leaves later turn black with a purple, furry mold.
Cause: Downy mildew.

Leaf tips turn brown and die.
Cause: Hot, dry weather and/or excessive heat.

Papery, white leaf spots with a lengthwise split.
Other symptoms: Brown leaf tips.
Cause: Botrytis leaf blight.
Comments: This fungal disease is favored by humid weather and temperatures between 50 and 70°F (10–21°C). Tokyo Long White, Norstar, and Frontier are tolerant of leaf blight.

Small, sunken, light-colored spots on leaves.
Other symptoms: Spots have a ringed appearance, with purple centers; leaves turn yellow and collapse.
Cause: Alternaria (purple blotch disease).
Comments: Worsened by cool soils.

Gray, water-soaked appearance of outer bulb scales.
Cause: High temperatures (above 85°F [29°C]).

Bleached, soft bulbs.
Cause: Sunscald.
Comments: Occurs when onions are harvested on bright, sunny days and left to cure in direct sun. White onions are especially sensitive to sunscald.

Water-soaked, soft bulbs.
Cause: Freezing injury.

Dark green or black spots with concentric rings on bulb and neck.
Cause: Onion smudge.
Comments: Early Yellow Glove, Downing's Yellow Globe, and Southport Red Globe are more tolerant of onion smudge. Wet soil and temperatures above 75°F (24°C) promote this disease.

Rotted bulbs.
Causes: Purple blotch (bulbs turn dark reddish purple, then brown/black) or white rot (bulbs are black and totally rotten).

Parsnips

Parsnips have very simple needs. They are tolerant of plant competition once established and can even germinate and grow within some competing vegetation. Parsnips are one of the hardiest root crops. Growing temperatures are 60 to 65°F (16–18°C). Parsnip roots grow poorly at temperatures above 75°F (24°C), but will tolerate temperatures as low as 35°F (2°C). They will grow in full sun to part shade. Culture is much like that of carrots, except that parsnips require less nutrients and a higher pH (6.0–6.8).

TROUBLESHOOTING PESTS AND PROBLEMS

Yellow, dwarfed young leaves, bushy growth.
Cause: Aster yellows virus disease.

Dark spots with yellow borders.
Cause: Cercospora leaf blight (if younger leaves are affected) or alternaria leaf blight (if older leaves are affected).

Yellow, stunted plants.
Other symptoms: Plants wilt during bright, hot days and recover at night.
Cause: Usually a sign that the roots are injured. Pull up a plant and check the roots for wireworms, root knot nematodes, weevils, or disease.

Dark cankers on roots; root deterioration.
Cause: Parsnip canker.
Comments: Andover is resistant to canker.

Peas

Peas are easy to grow with one caveat: Rapid germination is essential to avoid root rot problems. So either pre-germinate seeds or wait until soil temperatures are optimal for germination (50 to 65°F [10–18°C]).

CROP MANAGEMENT

Temperature

Peas are a cool-season, moisture-loving crop. Optimal growing temperatures are 60 to 75°F (16–24°C). Most cultivars grow poorly at temperatures above 75°F (24°C) but will tolerate

Figure 9.10. Harvesting baskets of shelling peas from plants competing successfully with slowly encroaching living mulch row middles, weeds, and lettuce.

temperatures as low as 45°F (7°C). Pea foliage can withstand a light frost, but pods and flowers are injured by heavy frost. Sugar Snap is tolerant of colder conditions.

Light

Peas grow well in full sun to partial shade. Seed later plantings in a cool, shady area of the garden to lengthen their production season.

Soil and Fertility

Peas prefer a pH of 5.5 to 6.8, but are not too picky about soil fertility. Peas are susceptible to certain micronutrient deficiencies. Foliar seaweed applications when plants are young often improve plant health.

Water

Maintain moist but not wet soil; do not plant in wet soils.

Special Ecological Preferences

Rapid germination is essential to avoid root rot problems. Choose lighter soils for earliest plantings. Use the proper bacterial inoculant (to promote nitrogen fixation), applied before or at the time you plant. Soaking seed in compost tea pre-plant and/or pre-germination of seed may help to prevent disease. Studies indicate that an increase in soil organic matter is the factor most positively correlated with pea growth, yield, and nodule formation. Planting on raised beds is reported to be one of the most important cultural tools to prevent disease in peas grown commercially. Once well established, peas can tolerate a lot of plant competition and still yield well.

TROUBLESHOOTING PESTS AND PROBLEMS

Yellowed, withered, thickened, and curled leaves.
Other symptoms: Flowers may drop.
Cause: Pea aphids.

Yellowing, crinkled, mottled, and curled leaves.
Cause: Mosaic virus.
Comments: Use resistant/tolerant cultivars including Knight, Maestro, and Oregon Sugar Pod II.

Yellowing leaves, stems yellow inside.
Other symptoms: Dwarfing and wilting.
Cause: Fusarium wilt.
Comments: Many resistant cultivars are available.

Velvety, white growth on leaves.
Cause: Downy mildew.
Comments: Knight and Green Arrow are resistant to downy mildew.

Powdery, white spots and patches on leaves.
Cause: Powdery mildew.
Comments: Many resistant cultivars are available.

White or translucent, irregular tunnels in leaves.
Cause: Leafminer (pea).

Blossoms drop.
Cause: Excessive heat or rain.
Comments: Wait for new blossoms to form. Wando, Little Sweetie, and Sugar Snap tolerate heat.

Brown spot or cavity on surface of seeds.
Cause: Manganese deficiency.

White, powdery spots on pods.
Cause: Powdery mildew.

Holes in blossoms; tunnels in seeds.
Other symptoms: Holes and whitish wormlike grubs in seeds.
Cause: Pea weevil.

Holes chewed in pods.
Cause: Armyworms; possibly other chewing caterpillars.

Peppers

There are two main types of peppers: hot peppers and sweet (bell) peppers. Sweet peppers are picked green (immature) or colored (mature); mature color is usually red or yellow, but can also be orange, brown, white, lavender, striped, or dark purple. There are mini and long types of sweet peppers, too.

CROP MANAGEMENT

Temperature

Peppers are a warm-season crop that is very sensitive to cold injury. Optimal temperatures for growth are 70 to 85°F (21–29°C). Both extreme heat (above 95°F [35°C]) and cold (below 38°F [3°C]) damages pepper flowers.

Light

Peppers thrive in full sun; partial shade is good in hot, sunny, dry climates. Intense sunlight makes fruits more susceptible to sunscald.

Soil and Fertility

Peppers prefer well-drained, rich, light, loamy soil, with pH between 6.3 and 7.0 (they will tolerate up to 7.5). They are average to heavy feeders; they struggle when phosphorus is low.

Water

Peppers will use all the water you give them, but be careful not to let them get waterlogged.

Special Ecological Preferences

Peppers seem to appreciate shade when establishing. I cover transplants to increase temperatures and protect from frosts, but I am not afraid to leave row covers on for up to 7 weeks, because peppers grow so well under row cover in my sunny climate. In hot, very sunny climates, peppers might do better in a slightly shady microclimate of your garden or field. My pepper plants are always more robust with higher yields at the end of my field near the river where they get a bit of afternoon shade from tall cottonwood trees. They do very well with mulch once the weather warms up. As transplants, they want warm soils and benefit from pre-warming the soil with plastic (use the same sheet of plastic you used to warm the soil for early-spring plantings of peas). Peppers do not compete well with other plants and weeds. Keep them weed-free until they grow large enough to close their canopy. Peppers are susceptible to several foliage and fruit diseases that require wet foliage, so it is best to irrigate in a way that keeps water off foliage.

TROUBLESHOOTING PESTS AND PROBLEMS

Plants have blossoms but no peppers.

Cause: Extreme heat and cold.

Comments: Wait for new blossoms to form. If peppers still fail to form and plants are vigorous with dark green leaves, the problem may be too much nitrogen.

Fruit blossoms drop prematurely.

Other symptoms: Fruit is misshapen and discolored.

Cause: Low temperature during blossoming and/or uneven growing conditions (temperature and water extremes).

White or tan-colored, sunken spots near tips of fruit.

Other symptoms: Spots most often appear on the earliest fruit during dry periods; up to one-third of the earliest fruits may become dark and shriveled.

Cause: Blossom-end rot.

Sunken, white, papery-looking areas on side of fruit.

Cause: Sunscald.

Comments: Affected areas are on the side exposed to the sun.

Small, dark brown, wartlike spots on fruit.
Other symptoms: Spots are most prevalent during damp weather.
Cause: Bacterial spot.

Water-soaked spots on fruit, stem girdling, plant wilting and death.
Other symptoms: Spotting is followed by rotting; spots have black-specked centers.
Cause: Phytophthora blight.

Hole near stem end of pepper.
Other symptoms: Sawdust-like excrement usually visible at the stem end; larvae feeding at the seed core; peppers often color early and rot.
Cause: European corn borer.
Comments: Once you find large larvae in peppers later in the season, it is too late to treat for this pest.

Potatoes

Soil health is the key to good potato production. They need fertile, loose, well-drained soil; hard or compacted soil leads to misshapen tubers.

CROP MANAGEMENT

Temperature

Leaves and vines will grow at temperatures as low as 45°F (7°C), but optimal growth occurs at 60 to 77°F (16–25°C) and tuber formation requires 60 to 70°F (16–21°C).

Light

Full sun.

Soil and Fertility

Potatoes require a loose, well-drained soil with good aeration. They do poorly in heavy, compacted soils. Potatoes are heavy feeders and require moderate to high levels of nitrogen, phosphorus, potassium, calcium, and sulfur. Timing of nutrients is also important. Nitrogen and sulfur are needed early, while phosphorus, potassium, and calcium are needed later on. Use composted fertilizers, because fresh manure aggravates certain potato pest problems. Gypsum is a good source of calcium and sulfur and is used by commercial potato growers to get large, disease-free potatoes. It is a good way of adding calcium without raising soil pH. Potatoes are tolerant of acid soils and will produce on soils with a pH between 5.0 and 6.8.

Water

Regular irrigation. Potatoes are sensitive to fluctuating soil moisture levels; if irrigation is not well distributed throughout the season, growth cracks and knobby tubers will result. Near harvest, potatoes can be allowed to dry.

Special Ecological Preferences

Potatoes are sensitive to uneven moisture and drought. Kennebec is one variety that is resistant to drought during growth. Sanitation and crop rotations are important pest preventive measures. Potato growers in Virginia have found that killing vines just prior to harvest cuts down on Colorado potato beetles the following season. Do not plant potatoes where tomatoes, eggplants, peppers, or potatoes have grown within 4 to 5 years. Avoid following strawberries, brambles, and sod with potatoes. Use only certified seed. Warm seed potatoes for 2 weeks at 65 to 70°F (18–21°C) before planting to encourage rapid emergence. Plant seed pieces when soil temperatures reach 40°F (4°C). Soak seed pieces in well-composted compost tea solution for several hours before planting. Plant in raised beds 8 to 10 inches (20–25 cm) tall and keep covering with mulch or soil as plants emerge and grow. Correct storage prevents disease development after harvest. Store at 50 to 60°F (10–16°C) for 2 to 3 weeks to heal cuts and bruises, then drop temperatures to 40°F (4°C) and maintain humidity at 90 percent.

TROUBLESHOOTING PESTS AND PROBLEMS

Large gray-brown spots with concentric rings.
Other symptoms: Spots merge and cover the entire leaf.
Cause: Alternaria disease (early blight).
Comments: Resistant cultivars are Butte and Kennebec.

Water-soaked brown lesions on leaves.
Other symptoms: Lesions appear on lower leaves first.
Cause: Late blight.
Comments: There are several tolerant cultivars.

Light green to yellow leaves with lower leaflets rolled upward.
Cause: Leaf roll virus.
Comments: Katahdin and Yukon Gold are resistant.

Yellow, mottled leaves accompanied by crinkling.
Cause: Mosaic virus.

Yellowed, stippled, stunted leaves.
Cause: Leafhoppers.

Large holes in leaflets.
Other symptoms: Whole leaflets consumed.
Cause: Colorado potato beetle.

Small holes in leaflets.
Cause: Flea beetles.

Sprouts die before emergence.
Cause: Rhizoctonia disease.

Yellow, stunted plants.
Cause: Root knot nematodes.

Yellow leaves; gradual wilting, usually at flowering.
Cause: Verticillium wilt.
Comments: Russet Burbank is considered moderately resistant.

Plants wilt suddenly.
Cause: Wireworms.

Brown to black raised spots on tubers.
Cause: Rhizoctonia disease.

Rough, raised lesions on tubers.
Cause: Potato scab.
Comments: There are many resistant cultivars including Russet Burbank, Norland, Superior, Keuka Gold, Lehigh, Pike, and Marcy.

Enlarging brown to black blotches on tubers.
Other symptoms: Internal tissue is brown and dry just below the tuber surface.
Cause: Alternaria disease (early blight).

Tiny surface trails (or tunnels) just beneath tuber skin.
Cause: Flea beetle larvae (tuber flea beetle).

Green coloration of tubers.
Cause: Exposure to sunlight.

Radishes

Radishes are an easy, fast crop to use as a ground cover and keep a growing root in the soil. For information on troubleshooting pests and problems, see the "Broccoli" entry on page 190.

CROP MANAGEMENT

Temperature

Radishes like cool, moist conditions. They can germinate and grow at temperatures as low as 40°F (4°C). Optimal growth occurs when temperatures are 50 to 65°F (10–18°C); growth is poor above 75°F (24°C).

Light

Radishes grow best in full sun.

Soil and Fertility

Do not fertilize too heavily; radishes are light feeders. They prefer light soils with a pH of 5.5 to 6.8. Radishes are fussy only about high soil salt levels; avoid salty (high sodium) soils and raw, uncomposted manures.

Water

Irrigate heavily during germination and for the first 2 weeks after emergence. Radishes don't like drought. Adequate irrigation prevents root diseases such as radish scab.

Special Ecological Preferences

Radishes are susceptible to high-temperature injury. Plant early to avoid hot temperatures. Cherry Belle, Crimson Giant, and Summer Cross No. 3 are heat-resistant varieties.

Rhubarb

Rhubarb is a perennial vegetable that has few pests and is easy to grow.

CROP MANAGEMENT

Temperature

Rhubarb likes cool, moist summers and requires a winter cold enough to freeze the top few inches of soil and induce a dormancy period.

Light

Rhubarb grows well in full sun to part shade.

Soil and Fertility

Rhubarb prefers a soil pH of 6.0 to 6.8 and deep, sandy, well-drained soil high in organic matter. It begins to produce small stalks if underfertilized.

Water

Rhubarb requires moist conditions and an even water supply to get the best quality stalks.

Special Ecological Preferences

Choose a weed-free location, add composted organic residues, and plant rhubarb crowns in hills. Mulch with several inches of high-nitrogen compost each spring and water down the mulch well. Do not harvest new plantings for at least 2 years. When you do start to harvest, choose only thick stalks. Harvest for 5 to 7 weeks, and then stop so that plants do not become stressed.

TROUBLESHOOTING PESTS AND PROBLEMS

Small, round spots on leaves.

Other symptoms: Leaf spots may vary in color from yellow to brown to red; sunken spots on stalks.

Cause: Fungal diseases (*Ascochyta rhei* and *Ramularia rhei*).

Comments: Continue to remove and destroy diseased leaves and or stalks after harvest.

Spinach

There are two types of spinach, distinguished by leaf form: savoy (crinkled leaf) and flat-leaved.

CROP MANAGEMENT

Temperature

Spinach likes cool temperatures—average temperatures of 60 to 65°F (16–18°C) suit it fine, and it is unhappy when temperatures reach much above 75°F (24°C). Spinach germinates best at soil temperatures of 45 to 75°F (7–24°C), but will germinate even as low as 35°F (2°C). Spinach can survive air temperatures of 20°F (–7°C) if properly hardened. However, prolonged exposure of younger plants to temperatures below 40°F (4°C) initiates bolting (premature seedstalk formation). High temperatures and long day length also cause bolting.

Light

Grow spinach in full sun to partial shade. Later plantings or plantings in hot climates prefer more shade.

Soil and Fertility

Spinach prefers loam soils, but will produce in most soil types. Earlier crops do best in lighter soils, while later crops do better in heavier soils. Choose a soil with good drainage and high organic matter. Spinach is sensitive to low pH and will not produce well on soils with a pH below 5.5. Keep the pH between 6.0 and 7.0. Spinach requires moderate levels of potassium and phosphorus and good levels of nitrogen. Spinach is also sensitive to low levels of calcium and boron.

Water

Provide plentiful irrigation for growth and cooling when temperatures are warm.

Special Ecological Preferences

Spinach is a definite cool-season crop that can be grown in the spring and fall and overwintered. Provide shade if growing during warm summer months. Bloomsdale Longstanding, Olympia, and Tyee are resistant to bolting. Fordhook Giant and Unipack 151 resist heat injury. Increasing the exchangeable cation percentage of calcium to 85 percent of total cation capacity, compared with magnesium and potassium, has made growing spinach almost foolproof on my eastern Oregon farm (see "Cation Exchange Capacity" on page 85). It also tastes sweeter and grows exuberantly. Other farmers have also suggested this higher exchangeable cation percentage for spinach.

TROUBLESHOOTING PESTS AND PROBLEMS

Yellow areas on upper surface of leaves.

Other symptoms: Stunted growth; blue-gray patches on undersides of leaves.

Cause: Downy mildew (also called blue mold).

Comments: Many cultivars are resistant or tolerant to one or more of the many races (including some newly identified races) of downy mildew.

Yellow spots on upper leaves.

Other symptoms: White pustules on the undersides of lower leaves.

Cause: Rust (white rust).

Comments: Several cultivars are tolerant of white rust.

Young plants stunted and yellow.

Other symptoms: Gradual wilting of older plants.

Cause: Fusarium wilt.

Young central leaves turn yellow and curl.

Other symptoms: Mottling, browning, and death of larger leaves.

Cause: Mosaic virus (also called blight or yellows).

Comments: Resistant/tolerant cultivars include Winter Bloomsdale, Melody, and Indian Summer.

Yellow, deformed leaves that eventually die; stunted plants.

Cause: Curly top virus.

Yellow curled leaves and stunted growth.
Cause: Aphids, usually green peach aphids.

Light-colored winding tunnels in leaves.
Other symptoms: An enlarged blotch at the tips of leaves, which may turn white or brown.
Cause: Leafminer (spinach).

Ragged holes in outer leaves, usually between leaf veins.
Cause: Armyworms.

Squash

There are four species of squash: *Cucurbita pepo*, *C. maxima*, *C. moshata*, and *C. mixta*. Within these species there are summer and winter types. Summer squash fruits are harvested and eaten while still immature; winter squash are used when fruits are mature.

CROP MANAGEMENT

Temperature

Squash require warm temperatures and thrive in the heat. Seeds need 60°F (16°C) soil temperatures for germination. Transplants do best if daily temperatures reach 60°F before planting, and squash growth rates increase as temperatures increase to 90°F (32°C). Early summer squash plantings may take 8 to 9 weeks to reach harvest size, but later plantings growing in warmer soils can be ready to harvest in 6 weeks.

Light

Squash insist upon full sun.

Soil and Fertility

Squash prefer well-drained loam soils well supplied with organic matter. Composted and decomposing organic residues added generously to squash beds will go a long way toward preventing many pest problems. Squash will produce in soils with a pH of 5.5 to 6.8, but much prefer a pH above 6.0.

Water

Keep squash well watered, but do not let soil become saturated. Squash have a low root-to-shoot ratio. A shallow root system supports a very large mass of aboveground vegetation. Therefore, during dry, hot days, squash may begin to wilt, regaining turgidity as soon as evening falls and evapotranspiration slows.

Special Ecological Preferences

Squash is susceptible to many pests when plants are young, so protect early plantings from cool temperatures and pest insects by covering them with floating row covers. To increase soil temperatures, conserve soil moisture, and get a jump on weeds before planting, lay a sheet of clear 1.5-mil plastic over the squash-patch-to-be. Pull the plastic up and remove just prior to planting and cultivate very lightly.

One of the most important things to remember about squash plants is that fruit will not develop properly unless pollinating insects do their part. Each squash plant produces separate male and female flowers. Male flowers are produced first, then female flowers begin to be formed later. Bees are the main pollinizers that transfer pollen from male flowers to female flowers, so encourage these insects.

TROUBLESHOOTING PESTS AND PROBLEMS

Plants wilt; pale green areas form on leaves.
Other symptoms: Foliage looks brown and burned.
Cause: Squash bug.
Comments: Butternut-type cultivars, and Table Queen Acorn, Royal Acorn, Early Prolific Straightneck, and Early Summer Crookneck cultivars are more tolerant of squash bug than other varieties.

Figure 9.11. Squash growing on a trellis.

Vines or parts of a runner wilt suddenly.
Cause: Squash vine borer.
Comments: The Sweet Mama hybrid resists vine borers.

Leaves wilt, turn yellow, and curl.
Cause: Aphids.
Comments: Cocozelle (zucchini) is less preferred by aphids than other squash varieties.

Vines wilt and die gradually.
Other symptoms: Symptoms appear on newer leaves; no leaf yellowing is present.
Cause: Bacterial wilt.
Comments: Choose varieties that are less attractive to beetles (see below), because beetle feeding spreads the bacteria that cause wilt.

Small holes in leaves, causing a lacy appearance.
Cause: Striped cucumber beetles; spotted cucumber beetles.
Comments: Bennings Green Tint, Blue Hubbard, Early Butternut Hybrid, Waltham Butternut, Ashley, Chipper, Eversweet, Gemini, Improved Long Green, and Table King are more tolerant of cucumber beetles.

Yellow spots on leaves.
Other symptoms: Mottling, distortion, and wrinkling of older leaves.
Cause: Mosaic and curly top viruses.
Comments: Multipik and Superpik are tolerant to mosaic diseases. Curly-top-resistant squash varieties include two specialty heirloom varieties: Umatilla Marblehead and Yakima Marblehead.

Older leaves mottled yellow between veins.
Cause: Downy mildew.
Comments: Some squash and pumpkin cultivars tolerate specific strains of downy mildew, but not all strains.

Figure 9.12. Striped cucumber beetles chewing small holes in squash seedlings. If you spot them, take action to protect plants, because these beetles also spread bacterial wilt disease.

Water-soaked spots on leaves.
Other symptoms: Spots turn gray and tissue dies and drops out, leaving foliage with a shot-hole appearance.
Cause: Angular leaf spot.

Dark brown spots with concentric rings.
Other symptoms: Spots appear on older leaves first; vines defoliate eventually.
Cause: Alternaria leaf blight.

Small spots on fruits; spots turn white.
Cause: Angular leaf spot.

Brown and sunken spots on fruit.
Cause: Alternaria.

Tomatoes

There are two main types of tomatoes to choose from: determinate and indeterminate. Determinate types grow to a certain height and then stop producing new leaves and stems, putting all their energy into producing fruit heavily over a 4-to-6-week period. Indeterminate types continue to grow, producing new vines and fruiting clusters all season long.

CROP MANAGEMENT

Temperature

Optimal temperatures for vegetative growth of tomatoes are 79 to 90°F (26–32°C). Germination and growth are slow in soils below 57°F (14°C). Chilling injury occurs at 32 to 45°F (0–7°C) and frost damage occurs at 28 to 30°F (–1 to -2°C). However, tomatoes don't like it too hot, either. Blossoms will die at temperatures over 95°F (35°C). Best fruit development occurs at 55 to 75°F (13–24°C).

Light

Tomatoes require full sun all day.

Soil and Fertility

Tomatoes thrive on a deep, uniform soil with a pH of 6.0 to 6.8. Avoid soils with shallow hardpans, compacted layers, and high water tables. These soil conditions inhibit root development and stress plants. Tomatoes require higher levels of nitrogen and phosphorus than other vegetable crops. Too much nitrogen, however, can delay fruiting and result in soft fruit. Plants need moderate to high levels of potassium and calcium. If pH is low, add calcium in the form of calcitic lime. If pH is between 6.2 and 6.8, add gypsum to increase calcium levels. There is evidence that higher calcium levels may decrease tomato fruit susceptibility to some diseases; see table 7.2 on page 139.

Water

Water management is critical in tomato production. Tomatoes do best with a constantly moist soil with no fluctuations, neither drying out nor becoming saturated. Apply water if the soil is crumbly and you cannot form a soil ball that sticks together lightly. Do not add water if your soil looks wet, feels sticky, forms a tight soil ball, and/or will ribbon out between fingers when squeezed.

Special Ecological Preferences

Certain cultivars are capable of setting fruit at lower-than-usual temperatures for tomato, and other cultivars can set fruit even during high heat. Other cultivars are resistant to specific diseases, including fusarium and verticillium wilts, mosaic viruses, and root knot nematodes. In seed catalogs the abbreviation *VF* refers to verticillium and fusarium resistance. *VFN* means the cultivar is also resistant to nematodes. Some cultivars resist the cracking of fruits that occurs in conditions of heavy or fluctuating rainfall and temperature, while others resist sunscald, the result of high temperatures and exposure to sun.

Tomatoes do well on raised beds with drip irrigation and mulches. Raised beds provide better control of water and air movement. Use of drip irrigation keeps water off leaves, thus decreasing disease infection and spread. Mulches prevent fruit touching the ground where soilborne disease microorganisms may exist. They also prevent microbe-containing soil from splashing up onto tomatoes during rain events. Staking tomato plants to keep them off the ground also helps to prevent pest damage.

TROUBLESHOOTING PESTS AND PROBLEMS

Young leaves turn yellow between veins.
Cause: Micronutrient deficiencies.
Comments: Usually seen on soils with high pH. Check soil pH and apply sulfur if needed.

Purplish leaves on young plants.
Cause: Phosphorus deficiency.

Comments: This deficiency is often due to cool spring soils that make phosphorus unavailable to plant roots. Plants usually outgrow symptoms.

Older leaves pale, small.

Other symptoms: General yellowing of foliage.

Cause: Nitrogen deficiency.

Older leaves with small, pimple-like brown spots.

Other symptoms: Spots develop a light tan to white center as they age. Leaves turn yellow then brown, wither, and die.

Cause: Septoria leaf spot.

Comments: Brandywise and Stellar are tolerant cultivars.

Yellow, mottled leaves.

Other symptoms: Leaf curling and malformation; new growth develops a fernlike appearance.

Cause: Tomato mosaic virus; tobacco mosaic virus; curly top virus.

Comments: There are many mosaic-virus-resistant varieties. Curly-top-virus-resistant tomato cultivars include Roza, Columbia, and Saladmaster.

Leathery, black-brown spots on leaves.

Other symptoms: The spots show concentric rings. Lower leaves and stems are affected first. Fruits develop sunken, dry spots with concentric rings near the stem end.

Cause: Alternaria (early blight).

Comments: This disease occurs during humid/wet, warm conditions. Only 3 hours of leaf wetness are required for infection at temperatures between 70 and 77°F (21–25°C). There are many resistant varieties including Early Cascade, Mountain Magic, and Stellar.

Black-green water-soaked areas on older leaves.

Other symptoms: Damage also appears on stems. Fruits develop water-soaked spots, enlarging to large brown areas that remain firm.

Cause: Late blight.

Comments: This disease usually manifests during humid or wet, cool weather (60 to 70°F [16–21°C]). Several cultivars are resistant to late blight, including Brandywise, Mountain Magic, Mountain Merit, and Stellar.

Small round holes in leaves.

Other symptoms: Leaves stripped off plants.

Cause: Colorado potato beetle.

Figure 9.13. Spots on a tomato leaf due to early blight disease. Also a solanaceous flea beetle is feeding at top right.

Yellowing and "flagging" of shoots and leaves.
Other symptoms: Yellowing appears at the base of the plant first. A cross section of the infected stem has a reddish brown vascular system.
Cause: Fusarium wilt.
Comments: This disease occurs when temperatures exceed 80°F (27°C). There are many resistant varieties.

Yellowing and gradual wilting.
Other symptoms: Leaf margins curl upward; light brown discoloration of vascular system.
Cause: Verticillium wilt.
Comments: This disease occurs at temperatures below 75°F (24°C). Many resistant varieties are available, including Early Girl and Mountain Gem.

Whole leaves consumed.
Cause: Hornworms.
Comments: There are two types of hornworms; both are 3 to 4 inches (7.5–10 cm) long with white diagonal stripes. Tobacco hornworm has a red horn projecting from the head. Tomato hornworms have a green-black horn. Handpick these large worms; they are rarely a serious problem.

Large chunks of green fruit consumed.
Cause: Hornworms.
Comments: Colorado potato beetles can also consume large areas of green fruit, but their feeding is confined to the surface of fruits.

Small holes on fruit surface.
Other symptoms: A messy, watery, rotten internal cavity. Fruit appears to collapse like a popped balloon.
Cause: Corn earworm (also called tomato fruit worm).

Dark pinpricks surrounded by light-colored spots on fruits.
Other symptoms: A white, spongy area inside fruits below the spots.
Cause: Stink bugs.

Scarring and malformation of fruit, especially at blossom end.
Cause: Catfacing.
Comments: A physiological disorder caused by prolonged cool weather during blossoming. Using row cover when late-spring temperatures are cool (below 50°F [10°C]) can help reduce catfacing.

Yellow or white blister-like patch on fruit.
Other symptoms: The patch usually forms on the side of the fruit exposed to the sun.
Cause: Sunscald.

Yellow streaks or shoulders on fruit; uneven ripening.
Cause: Mosaic virus.
Comments: Many resistant varieties are available. If no distorted leaves are present, yellow shoulders may be due to weather extremes.

Fruit cracks at stem end.
Other symptoms: Cracks may be radial or in concentric circles around the shoulders.
Cause: Uneven irrigation and/or temperature extremes.

Black sunken area at blossom end of fruit.
Other symptoms: Often seen in the first-ripening fruit; hard, dark areas may form inside the fruit.
Cause: Blossom-end rot.

Gray to brown blotches develop on green fruit.
Other symptoms: Internal browning and uneven ripening.
Cause: Gray wall.
Comments: A physiological disorder aggravated by low light and low temperatures, high soil moisture, high soil nitrogen, low soil potassium, and excessive soil compaction.

CHAPTER 10

Vegetable Crop Insect Pests and Interventions

Most of the insect pests covered in this chapter are common throughout North America, and many are found worldwide, such as aphids. About half of the major North American vegetable pests are indigenous. Europe is the principal origin of the nonindigenous vegetable pests found in North America, followed by Asia and Central America. This is not surprising, because Europeans brought their crops and the crops' pests when they immigrated to North America.

I describe each insect and the damage it causes and then offer interventions, focusing first on those with minimal ecological impact, then moderate-impact options, and the heaviest-impact materials as a last resort. For some pests, I list only minimal- and moderate-impact interventions because heaviest-impact interventions are never really appropriate. For many of the pests, there are no effective moderate-impact interventions available, and thus I offer only minimal interventions and heaviest interventions.

Aphids

There are several species that feed on vegetable crops, including cabbage and green peach. Though different in appearance and crop host, these aphids can be managed similarly. The hosts of **cabbage aphid** (*Brevicoryne brassicae*) are Brassicaceae family plants: cabbage, cauliflower, Brussels sprouts, broccoli, kohlrabi, kale, turnip. They prefer the youngest leaves and flowering parts of brassica crops and are often found deep within the new leaves of kale or Brussels sprouts. They do not normally affect seedlings but build up as plants mature, especially later in the season when conditions are damp and cool.

The hosts of **green peach aphid** (*Myzus persicae*) include a very wide range of vegetable crops, especially when grown in greenhouses or high tunnels, including cabbage, endive, mustard greens, parsley, turnip, tomato, potato, spinach, pepper, beet, celery, lettuce, and chard. They are green to yellowish green, and are most common in late spring/early summer and fall. They prefer tender, new leaves and leaf undersides and seedlings rather than mature plants.

Figure 10.1. Grayish green cabbage aphids on cabbage leaves. The tan-colored, puffy ones have been parasitized by a wasp.

MINIMAL-IMPACT INTERVENTION

Avoid excessive rapid growth of succulent foliage by balancing mineral nutrient levels, specifically avoiding excessive rapid-release nitrogen fertilizer.

Encourage predators and parasites by providing habitat diversity. Specific predators and parasites, such as syrphids, spiders, lady beetles, lacewings, earwigs, and parasitoid wasps (*Aphelinus* and *Aphidius* spp., specifically *Diaeretiella rapae*) prey on cabbage aphid. Studies reinforce the benefit of greater vegetation complexity. For example, in one study, allowing weeds to grow between cabbage

Figure 10.2. Cabbage-aphid-free broccoli growing amid white clover in my forest garden experiment.

rows was associated with lower cabbage aphid abundance and enhanced populations of generalist natural enemies.[1] Both cabbage and green peach aphid populations were lower on broccoli and zucchini where crops were grown with clover living mulches, compared with clean-cultivated crops.[2] When I conducted a forest garden experiment at Woodleaf California from 2012 to 2015, one observation was that cabbage aphid populations were 68 percent lower and syrphid fly larvae were 50 percent higher on broccoli grown amid a clover-weed living mulch in the row middles between fruit trees compared with broccoli growing in a high tunnel without clover living mulch.

MODERATE-IMPACT INTERVENTION

Insecticidal soap sprayed in the evening or early morning can suppress aphids. Repeat applications are often required. Soap sprays may cause leaf burning under hot, high-sunlight conditions, especially in Brussels sprouts and cabbage and young pepper seedlings.

Figure 10.3. Armyworm larvae feeding damage on Swiss chard.

HEAVIEST-IMPACT INTERVENTION

Insecticidal soap plus neem oil sprayed in the evening or early morning kills aphids, but neem also kills other organisms. Repeat applications are often required.

Armyworms

Adults are 1½-inch (4 cm) moths with dark gray wings. Larvae are tan, green, or black, 1½ inches long, and have an inverted-Y marking on their heads. Caterpillars overwinter and emerge in early spring, when they do their heaviest damage. After several weeks, they enter the soil, pupate, and emerge as moths. Generally populations are not high enough to warrant more than minimal-impact interventions, which are usually effective.

HOSTS

Corn, celery, beet, pepper, lettuce, and other vegetables.

MINIMAL-IMPACT INTERVENTION

Two types of traps can be used. Blacklight traps catch male and female moths; pheromone traps catch only the males. Both traps are effective as indicators of population levels, but blacklight traps catch and kill many other nontarget moths.

Rotate crops, apply parasitic nematodes pre-planting, and plant early to avoid increasing worm populations as the season progresses. Use floating row covers. Be sure to remove covers when plants bloom so that pollination is not hindered.

Scout your plants every 2 to 3 days; apply *Bacillus thuringiensis* var. *kurstaki* (*Btk*) in granular or ES (emulsifiable suspension) form if 20 percent of leaves have holes. Be particularly attentive to spraying the undersides of leaves. *Btk* is most effective on small larvae, ¼ to ¾ inch (6–19 mm) long, in their first instar. Mix the ES in plenty of water to ensure good coverage.

Cabbage Worms

Each growing season there are several generations of the three main cabbage worm species in most parts of North America: **European cabbage worm**, **cabbage looper**, and **diamondback moth**. Some species are more prevalent or even absent in some areas. Generally they overwinter on brassica plants (including brassica weeds and cover crops), emerge in the spring, multiply on crops, and are most problematic later in the growing season.

HOSTS

Cabbage worms feed on all Brassicaceae family crops. Chinese cabbage, turnip, mustard, and kale are less usually preferred than cabbage, collards, Brussels sprouts, broccoli, and cauliflower.

MINIMAL-IMPACT INTERVENTION

Diversify brassica crop plantings with living mulches, cover crops, and specific weeds allowed to grow within crops to encourage predators and parasites of cabbage worms. A number of ground-dwelling carabid beetles eat imported cabbage worms and reduce worm pest populations.[3] Research confirms the benefits of diversity:

- In several studies, the density of all ground-dwelling predators, including carabid beetles, was higher in weedy cropping systems than in weed-cultivated systems.
- In one study more carabid beetles were caught in plots where Brussels sprouts were growing in white clover living mulch than on bare ground plots.

Figure 10.4. A green European cabbage worm is trying to hide in the inner leaves of a head of cabbage where it has been feeding, leaving behind large holes.

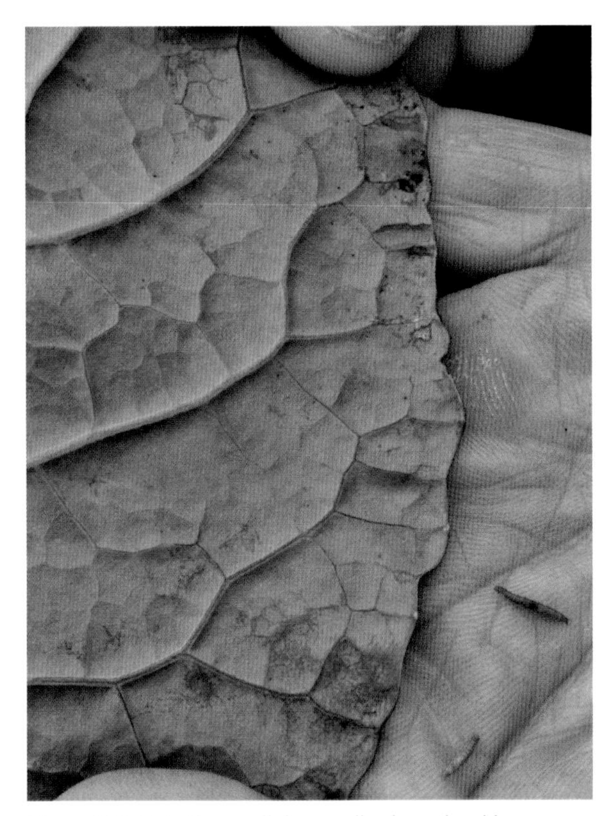

Figure 10.5. Note the small diamondback moth cabbage worms on the person's palm.

- Cabbage worm egg and worm densities and damage to broccoli at harvest were significantly lower in broccoli undersown with clover living mulches compared with broccoli grown without living mulches and instead cultivated for weeds. Spider counts were also higher on broccoli in living mulch habitats than in cultivated broccoli plots.
- Despite competition from living mulches, total broccoli yields were not lower in living mulch plots undersown with strawberry clover or white clover. However, yields were lower in yellow sweet clover living mulch plots when compared with weed-cultivated treatments, so this may not be as good a living mulch choice.[4]
- Allowing weeds to grow between cabbage rows was also associated with lower cabbage worm abundance and higher populations of predators and parasites.[5]
- My own on-farm research in Montana found higher predator populations and lower cabbage worm damage in my clover/weed living mulch system.

Bacillus thuringiensis (*Bt*) is very effective against cabbage worms if you begin spraying early, when you find one worm per two plants. Repeat applications are often required. In my on-farm experiments, *Bt*-sprayed Brussels sprouts resulted in 98 percent cabbage-worm-damage-free Brussels sprouts, but *Bt* was sprayed ten times to achieve that percentage. In plots where no cabbage worm sprays were applied and I relied only on predators in the living mulch providing biological control of cabbage worms, undamaged Brussels sprouts were only 88 percent, but the lower undamaged yield was balanced out by saving the expenditure to purchase and apply *Bt*.

HEAVIEST-IMPACT INTERVENTION

Spinosad is very effective, but there have been reports of cabbage worm resistance developing in British Columbia organic brassica fields where spinosad is used repeatedly.

Carrot Rust Flies

Adult carrot rust flies (*Psila rosae*) are housefly look-alikes with yellow heads. Larvae are yellow-white maggots. Pupae overwinter in the soil or sometimes as larvae inside carrot roots. Flies emerge in spring. Eggs are laid in moist soil near host plants; larvae feed on roots and on plant tissue near the soil line. The cycle occurs over 2 to 3 months, and several generations may occur in a season depending on weather.

HOSTS

Carrot, celery.

MINIMAL-IMPACT INTERVENTION

Since carrot rust flies are relatively weak fliers, traveling less than 1,000 yards (914 m) in search of egg-laying sites, isolating crops from previous years' fields can significantly reduce carrot fly problems, especially if the new crop is placed upwind of the previous crop. Plant late to avoid the time in spring when larvae are emerging. Avoid planting carrots immediately after plowing in a grassy area or pasture that has been in sod for several years. Cover carrot stands with row covers as soon as seedlings emerge. If carrot rust fly has been a past problem, don't delay harvest of fall carrots and don't overwinter carrots in the soil. Apply parasitic nematodes to soil when soil moisture levels are highest and temperatures are cooler.

Carrot Weevils

Carrot weevil (*Listronotus oregonensis*) adults are dark-brown, ¼-inch-long (6 mm) beetles with distinctive snouts. Larvae are creamy white grubs with reddish brown heads. Adults emerge early in the spring to lay eggs in stalks near the soil line. Larvae feed on roots, leaves, and stem tissue, sometimes causing stems to break off or fall over. They tunnel down through the crown toward the root, sometimes killing young plants and leaving feeding scars on carrot roots. Because larvae are sheltered inside carrot roots and stems, foliar insecticide sprays are not very effective in controlling this pest.

HOSTS

Carrot, celery, parsley.

MINIMAL-IMPACT INTERVENTION

Follow a 3-year rotation, use floating row covers, and clean up carrot debris at the end of the season. *Bacillus thuringiensis* var. *tenebrionis* shows some activity against adult carrot weevils. The parasitic nematode species *Steinernema carpocapsae*, *S. riobrave*, *S. feltiae*, *Heterorhabditis megidis*, and *H. bacteriophora* have been shown to be effective against carrot weevil larvae. *S. carpocapsae* was the most effective in sandy soil, reducing carrot weevil populations by 88 percent over an 8-day period, while *H. bacteriophora* was most effective in heavier soils, decreasing adult carrot beetle populations by 81 percent. Apply parasitic nematodes when planting and water in. Do not let them dry out.

Ground beetles appear to provide biological control for all carrot weevil life stages. In one laboratory study, smaller carabid beetles consumed the greatest amount of carrot weevil eggs. One large carabid species, *Pterostichus melanarius*, consumed newly emerged and overwintered carrot beetle adults, but many species and sizes of carabids consumed carrot weevil larvae. Reduced tillage encourages these weevil-consuming ground beetles.

Colorado Potato Beetles (CPB)

Adult Colorado potato beetles (*Leptinotarsa decemlineata*) are hard-shelled, with alternating black and yellow stripes. They are ⅜ inch (1 cm) long and are frequently found on young plants, especially in the spring. Larvae are soft-bodied, humpbacked, pinkish red grubs with two rows of black spots down each side of the body. CPB overwinter as adult beetles in plant debris. Adults surface in spring after rain and when the ground temperature has reached 57°F (14°C) at their hibernation depth. They then feed on young potato leaves or foliage of other host plants. After mating, the female immediately begins to lay bright orange eggs on the undersides of leaves of host crops and solanaceous weeds such as eastern black and deadly black nightshade. Larvae hatch, feed on leaf undersides, and then move to upper leaf surfaces. There are several generations per season.

HOSTS

All solanaceous crops (eggplant, pepper, potato, tomato), especially eggplant and potato.

MINIMAL-IMPACT INTERVENTION

Create habitat for beneficial insects: Lady beetles and stink bugs are the most effective natural enemies of CPB. The two-spotted stink bug, *Perillus bioculatus*, is also a CPB predator. Five consecutive releases of these predaceous stink bugs, starting when eggs were first present, maintained experimental plots in New Brunswick, Canada, virtually free from Colorado potato beetle. Deep straw mulch around susceptible solanaceous plants has been shown to suppress the first generation of CPB adults and larvae in late spring/summer. Experiments with living mulch row middles of rye/vetch mowed and blown into crop rows indicate that young plants must be covered enough that they are hidden from adults to receive the greatest benefit from mulching. Further, as the plants grow above the mulch level, they become susceptible to second-generation CPB damage.

During spring, cover plants with row cover to exclude the mobile adult beetles.

During the late summer or early autumn and in spring, apply beneficial nematodes to the soil to control the larvae.

For beetle larvae, a USDA study in Washington State showed that *Bacillus thuringiensis* San Diego (*Bt*SD) and *B.t. tenebrionis* (*Btt*) gave nearly results as good as the chemical insecticides applied as the control treatment, but without killing beneficial insects. *Btt* was most effective on newly hatched larvae. *Bt*SD and *Btt* affect a limited range of leaf-eating beetle species and are considered one of the most effective suppression materials for CPB.

MODERATE-IMPACT INTERVENTION

Beauveria bassiana suppresses both larvae and adults. Use either as a soil drench or a foliar application. In a Washington State study, suppression using *Beauveria* was inadequate until potato plants grew enough to cover bare soil in the row. *Beauveria* requires high humidity for germination and infection.

HEAVIEST-IMPACT INTERVENTION

Several insecticides approved for use on organic farms, such as pyrethrin and neem oil, provide some suppression of the adult beetles. Spinosad is effective for the larval through adult stages of CPB.

Corn Earworms and European Corn Borers

Larvae of the **corn earworm** (*Helicoverpa zea*), which is also called the tomato fruitworm, enter corn ears primarily through the silk channel. By contrast **European corn borers** (*Ostrinia nubilalis*) and fall armyworms enter through the husks or the cob. Adult corn earworms are light gray moths with dark, irregular lines on their wings. Larvae are yellow or white to green with brown heads and longitudinal white bands along the body. Adult moths usually do not overwinter in colder, northern climates; they migrate from the South and arrive in midsummer. Adult moths lay eggs on or near the silks of corn. (On other vegetable crops, eggs are laid on leaves near the developing fruits.) After 2 to 5 days, larvae hatch and tunnel into the corn ear. Four weeks later they move down the plant to pupate in the soil. Pupation takes 10 to 14 days. There are generally two generations in the North each year, with as many as six in the extreme South.

HOSTS

Corn, pepper, tomato.

MINIMAL-IMPACT INTERVENTION

An integrated approach is necessary to suppress ear worms and borers. Rotate crops, apply parasitic nematodes before planting, and plant early to avoid increasing worm populations as the season progresses, especially for European corn borer.

Maintain habitat for beneficial insects: big-eyed bugs, lacewings, spotted lady beetles, minute pirate bugs, and damsel bugs feed on corn earworm eggs and small larvae. Parasitic wasps such as *Trichogramma* species also attack earworm eggs.

On small plantings, try floating row covers applied before corn plants reach their full height; leave them in place until corn silks begin to emerge. This can help prevent moths laying eggs on the plants. Cover loosely with one large solid piece of row cover and secure it lightly, so that the corn will have plenty of room to grow. The trick to using row cover is to keep it secure enough so that the wind will not carry it away, especially as corn gets taller. More important, remember that corn needs to be pollinated, in general by wind. Remove row cover when 10 percent of plants begin to show their first silks.

Use pheromone traps or ask local extension agents to find out when earworm, corn borer, and armyworm adults are present in your area. Scout your plants every 2 to 3 days to catch the first signs of corn borer and armyworm feeding. These leaf feeders give whorl leaves a shot-hole appearance. Spray *Bacillus thuringiensis* var. *kurstaki* (*Btk*) if 20 percent of your corn leaves have shot-hole feeding injury. *Btk* is most effective on small larvae, ¼ to ¾ inch (6–19 mm) long, in their first instar. Rapid breakdown of *Btk* in sunlight and the necessity that insects ingest it make it less effective for corn earworm, because adult moths lay eggs on corn silks rather than on leaves, and silks are not as easy to spray. Hence heavier-impact interventions may be required.

Inspect corn silks for larvae. If worm populations are climbing higher, you can spray silks with *Btk* when about 10 percent of your plants are showing the first silks, then continue weekly until tassels turn brown.

HEAVIEST-IMPACT INTERVENTION

Spray silks with spinosad when the first silks begin showing on 10 percent of your plants and continue weekly until tassels turn brown.

Corn Rootworms

There are different corn rootworm (*Diabrotica*) species in different parts of the United States. Adults of all species are about the same size, 1/4 to 1/2 inch (6–13 mm) long, but northern corn rootworm adults are a uniform pale green, while western corn rootworm beetles are yellow or tan with three black stripes. Adult beetles are present from early summer through early fall. Larvae are white with brown heads, 1/2 inch long, and slender. In the North there is one generation per year. Larvae hatch in early summer from eggs laid at the base of host plants the previous fall. Adult corn rootworm beetles look a bit like cucumber beetles and sometimes can be found in cucurbits in the middle to late season when their favorite food, corn pollen, is no longer available. Unless populations are large, rootworm feeding doesn't seem to harm cucurbits.

HOSTS

Corn and sometimes cucurbits.

MINIMAL-IMPACT INTERVENTION

Additions of organic matter have been shown to decrease rootworm populations through increased egg predation. Do not plant corn in the same place for 2 years running. Try covering plants with floating row covers as described in the "Corn Earworm" entry on page 225. Remember to remove the row covers after corn begins to silk so that pollination may occur.

Cucumber Beetles

Both **striped cucumber beetles** (*Acalymma vitatum*) and **spotted cucumber beetles** (*Diabrotica undecimpunctata*) are 1/5 inch (5 mm) long and 1/10 inch (3 mm) wide with yellow wings and a black head and antennae. The striped beetle has 3 longitudinal black stripes on the wing covers while the spotted beetle has 12 black spots. Adult beetles overwinter in leaf litter and emerge in late spring to early summer. They lay eggs in the soil at the base of host plants. Striped cucumber beetles may carry the bacteria that cause bacterial wilt disease (see "Bacterial Wilt of Curcubits" on page 246). Spotted cucumber beetles feed on more plant species than striped cucumber beetles, including beans, corn, and potatoes. Both beetles feed mainly on leaves, pollen, and flowers, but they sometimes feed on immature cucurbit fruits, leaving scars on the skin. Field trials have not reported much success using organically acceptable insecticides to control cucumber beetles, so an integrated approach with only minimal-impact interventions is presented here.

HOSTS

Cucurbits (cucumber, squash). They also sometimes feed on other crops, such as beans and corn.

MINIMAL-IMPACT INTERVENTION

These options work best if employed together as an integrated approach.

Crop rotation and intercropping with other crops and/or ground covers. In research studies, intercropping cucumbers with corn and broccoli suppressed striped cucumber beetle damage

Figure 10.6. Striped cucumber beetles on squash leaves.

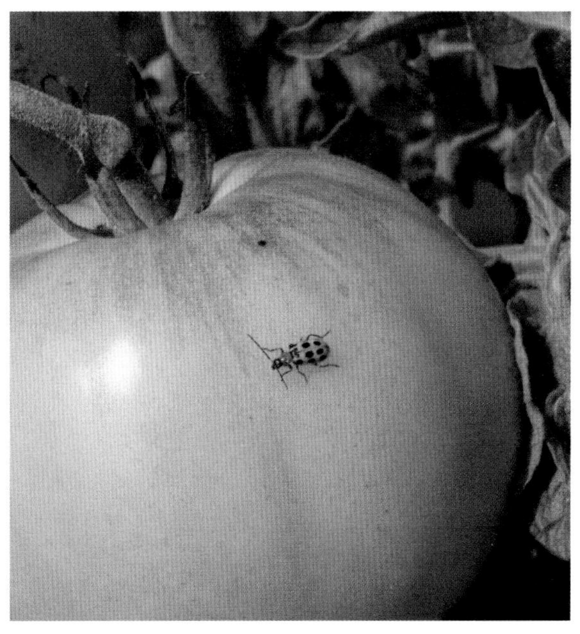

Figure 10.7. A spotted cucumber beetle on tomato.

compared with plots planted only in cucumbers. This increased crop diversity also reduced the incidence of bacterial wilt disease.[6] Another study found that intercropping watermelons and muskmelons with radish, nasturtium, tansy, buckwheat, cowpea, or sweet clover also reduced cucumber beetle feeding.[7]

Transplants rather than direct seeding. Seedlings are more susceptible than larger plants to cucumber beetle feeding damage and to bacterial wilt disease. Plant larger transplants to minimize risk and cover them with row covers until they are larger (row covers must eventually be removed to allow bees and other pollinators to visit the flowers). Young plants are vulnerable to stunting and bacterial wilt disease, but older plants can tolerate up to 25 percent defoliation from beetle feeding with no reduction in yield.

Traps. Unbaited yellow sticky traps and traps baited with volatile cucurbit floral extracts, known as TIC (trimethoxybenzene, indole, and trans-cinnamaldehyde), have been shown to trap cucumber beetles. You can make a similar product yourself: Save dried squash fruits. Powder them in the food processor. Bait your own yellow sticky traps with the powder. Note that yellow sticky traps also catch and kill some beneficial insects.

Perimeter trap crops. Main cucumber or squash crops are surrounded by plantings of a different, more attractive squash variety. One study reports that cucumber beetles generally aggregate at field edges.[8] Attractive trap crops may further encourage beetles to the edges of main plantings. Blue Hubbard and Buttercup winter squash varieties and dark green zucchini varieties are reported to be particularly attractive to cucumber beetles.[9]

Straw and hay mulch, or living mulches. Researchers report that cucumber plants grown in mulched soils had fewer cucumber beetles than those in unmulched soils. In a Maryland study, red clover living mulch plots had fewer striped cucumber beetles and aphid pests, and more big-eyed bugs and minute pirate bugs.[10] Colorado researchers suggest that mulch can slow beetle movement from one plant or field to another. Mulches also provide habitat for wolf spiders and other ground-dwelling predators that prey on cucumber beetles.[11] In fact, high spider populations increased cucumber yield by 25 percent in a study at the University of Kentucky. Ground beetles and spiders have been reported to reduce densities of striped cucumber beetle. Straw and hay mulch have been shown to provide food for springtails and other insects that eat decaying plant material; these decomposers are important alternative prey for spiders, helping to build spider numbers before cucumber beetle populations become high.[12] In addition, mulches enhance the soil organic matter system and the beneficial soil microorganisms that trigger plants' internal defenses.

Plastic mulches. Use of plastic mulch may reduce beetle feeding damage. In a Virginia study, metallic-colored plastic mulches repelled cucumber beetles. The University of Nebraska–Lincoln extension service suggests using reflective mulches below your plants to discourage the beetles from laying their eggs.

Cutworms

There are many species of cutworms (*Noctuidae* spp.). The larvae of most are 1¼ to 1¾ inches (3.2–4.4 cm) long; adults are gray/brown moths. Larvae feed at night and hide during the day. Moths are present in the spring and early summer.

HOSTS

Most young vegetable plants are attractive to cutworms.

MINIMAL-IMPACT INTERVENTION

Ground beetles, rove beetles, spiders, toads, and snakes feed on cutworms. Till gardens in early spring or late fall to expose cutworms. Use transplants and protect with cardboard or plastic collars.

Figure 10.8. This spider is searching for cucumber beetles to eat on squash leaves badly damaged by the beetles' feeding.

Apply parasitic nematodes to the soil 1 week before planting as long as soil temperatures are above 50°F (10°C).

Check for cutworms at night with flashlight and apply granular forms of *Bacillus thuringiensis* var. *kurstaki* (*Btk*) mixed with bran to the soil. *Btk* sprayed on the foliage of young plants is ineffective.

Flea Beetles

Though **brassica-feeding flea beetles** (*Phyllotreta* spp.) and **solanaceous-feeding flea beetles** (*Epitrix* spp.) have different host crops, they can be managed similarly. Both types of flea beetles are tiny black insects that do best during stable, warm spring weather; populations decrease during periods of fluctuating hot and cold temperatures with intermittent rains. Seedlings of crops are most vulnerable to flea beetle feeding when young or stressed, particularly by inadequate irrigation or moisture. Healthy adult plants can usually withstand a little flea beetle feeding. Flea beetles overwinter as adults in leaf litter, hedgerows, mulches, weeds, and wooded areas. Tiny white larvae hatch from eggs and then feed on roots of newly planted seedlings. Larvae later pupate in the soil.

HOSTS

The brassica-feeding flea beetles consume only cabbage family crops and weeds, including cabbage, cauliflower, Brussels sprouts, broccoli, kohlrabi, kale, and turnip. Solanaceous-feeding flea beetles consume tomato family crops and weeds, including tomato, potato, eggplant, and sometimes pepper; they also sometimes feed on squash and pumpkin.

MINIMAL-IMPACT INTERVENTION

These several options work best if employed together as an integrated management approach.

- Rotate crops and modify your planting schedule. Flea beetles are very mobile and move long

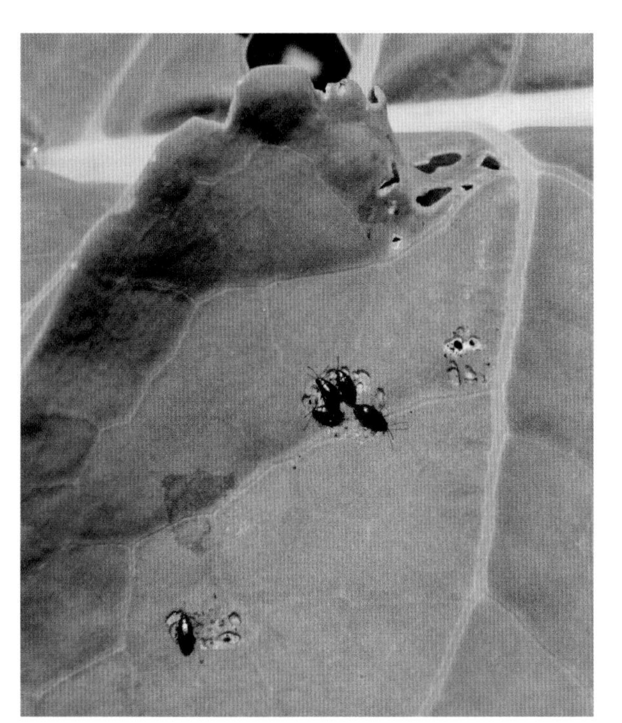

Figure 10.9. Flea beetles feeding on broccoli and flea beetle damage to a potato leaf.

distances; thus rotation helps only in the early season. Beetles emerge in mid- to late spring when temperatures are approximately 50°F (10°C). By planting earlier or later than the swell of emerging beetles, farmers and gardeners can avoid peak adult feeding activity. For example, to minimize damage from potato tuber flea beetle larvae, plant and harvest potatoes earlier than usual in the season, because tuber damage increases as flea beetle populations increase later in the year. In cold climates plant brassica crops later than usual in the season to reduce habitat and food sources for beetles as they emerge.

- Use floating and hooped row cover on brassica and solanaceous crop seedlings and transplants (but only if you are practicing crop rotation, because flea beetle larva pupate in the soil beneath areas where susceptible crops were grown previously).
- Overhead sprinkler irrigation appears to decrease flea beetle populations.
- Trap cropping may tempt flea beetles away from main crops. This can help to build up predators near and also away from your main crop. But unless you manage the beetles in your trap crop, they usually move to main crops eventually. In a Connecticut study, solanaceous-feeding flea beetles preferred the elongated Italian eggplant variety Vittoria to other eggplant and tomato varieties. In California, Chinese southern giant mustard (*Brassica juncea*) is especially preferred by brassica flea beetles. Other studies done at Washington State University concluded that a diverse trap crop containing Pacific Gold mustard Dwarf Essex rape (*B. napus*), and pak choi (*B. rapa*) protected broccoli from brassica flea beetles better than trap crops containing any of these species planted alone.
- Reduce tillage to encourage flea beetle predators. Because flea beetles spend a large portion of their life cycle as larvae in the soil, they may be vulnerable to ground-dwelling predators (such as ground beetles and spiders) whose populations diminish with increased tillage. Reduced tillage and no-till decreased potato flea beetle populations and damage while also increasing ground-dwelling beetle predator populations during a 2005 study in Oregon. In another Oregon study, flea beetle incidence and damage to broccoli foliage was lowest in strip-till/living mulch plots compared with conventionally tilled plots. Populations of brassica-feeding flea beetles were greater in weed-free collard monocultures than in polycultures intercropped with beans and left weedy for 2 or 4 weeks after transplanting.[13] In a 2013 Michigan State University study, fewer flea beetles were observed in rye living mulch plots compared with plots with vetch and bare soil row middles. However, I have noticed that after several years of minimal soil disturbance, the high-residue living mulch in row middles seems to provide overwintering sites for adult flea beetles (as well as predators), and flea beetle populations are then not as well suppressed by the predators.

MODERATE-IMPACT INTERVENTION

Kaolin clay has worked well to deter flea beetle feeding in some studies but not others. Surround provided some efficacy against flea beetles on eggplant in one study, but fruit quality was negatively affected.

HEAVIEST-IMPACT INTERVENTION

Spinosad has consistently given good control of flea beetles.

Grasshoppers

Most species of grasshoppers (*Melanoplus* spp.) overwinter as eggs in the soil. Nymphs begin emerging in early spring. Researchers have found that grasshopper hatch often corresponds with the time lilacs are in full to late bloom. Nymphs feed on vegetation for 40 to 60 days before molting. Adults disperse to suitable hosts during the summer and can do serious damage to crops when populations are high. Adults mate in late summer and lay the overwintering eggs. In most areas eggs are laid in grassy areas along roadsides and around field margins, especially in south-facing areas.

Heavy mortality occurs in the spring if cool, wet weather follows warm weather, leading to premature hatching of eggs. In late spring short periods of hot weather increase the incidence of fungus and bacterial diseases. When grasshopper adults migrate into gardens and vegetable fields, they are leaving areas that are drying out and looking for fresh new irrigated vegetation.

HOSTS

Grasshoppers are general feeders, and some will feed on most any green plant.

MINIMAL-IMPACT INTERVENTION

The best time to suppress grasshoppers is during the early nymph stage in the spring when they are most vulnerable. Optimal management is possible when grasshoppers are still immature and restricted to their breeding areas. Adult grasshoppers are difficult to control. An integrated approach is necessary:

- Grasshoppers are drawn to monocultures and dislike nitrogen-fixing crops like peas and sweet clover. Clover living mulches tend to discourage adult migration into crop fields.
- Create habitat for predators and parasites that control localized grasshopper infestations. Flies, blister beetles, and ground beetles feed on grasshopper egg pods.
- Try trap cropping. Undisturbed strips of vegetation left after spring tilling may serve to attract nymphs that are mobile enough to search for food. An irrigated greenbelt along garden and field perimeters can act as a trap crop when the surrounding vegetation begins to dry up in late summer.
- Domesticated livestock, such as chickens, turkeys, guinea fowl, geese, and ducks help keep grasshopper populations in check.
- A well-known biological control for grasshoppers is *Nosema locustae*, a naturally occurring protozoan that causes disease and death in crickets and grasshoppers. Spores of the parasite are impregnated into wheat bran flakes and applied by hand. It takes between 1 and 5 weeks for the grasshoppers to be infected. Infected individuals are lethargic and slow, making them easy prey for birds. *Nosema locustae* is not toxic to birds, animals, or other insects. When using *Nosema*, growers should locate spring hatching areas. Bait broadcast over these locations will sicken and kill the nymphs. *Nosema* is effective against adults, too, but most effective against the second- and third-instar nymphs. Reports on the success of *N. locustae* are mixed. It is *not* a good "rescue" treatment and will *not* result in instant adult suppression.
- The fungus *Beauveria bassiana* is registered for grasshopper control; it requires moist, humid conditions for effectiveness.

HEAVIEST-IMPACT INTERVENTION

In one study in Colorado, spinosad reduced adult grasshopper populations. Canola oil added to spinosad sprays increases mortality because canola oil attracts grasshoppers.

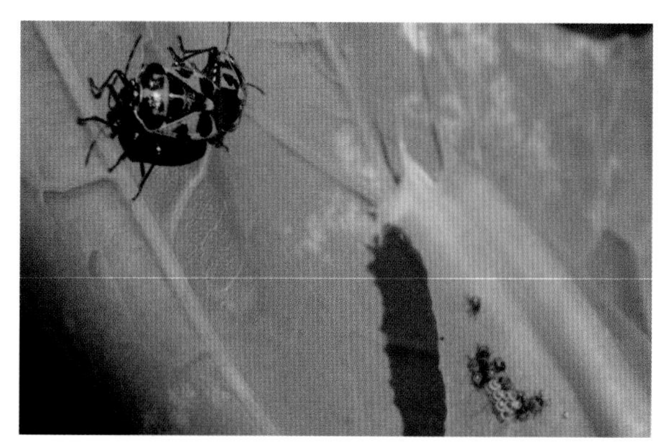

Figure 10.10. Adult harlequin bugs and newly hatched nymphs on Brussels sprouts.

Harlequin Bugs

Adult harlequin bugs (*Margantia histrionica*) are flat, 3⁄8-inch-long (1 cm), shield-shaped stink bugs with red and black spots; nymphs look like adults, but are smaller and rounder. Harlequin bug nymphs can be mistaken for beneficial predator bugs.

Adults overwinter under debris. Cold winter temperatures reduce survival of this subtropical native. They emerge in midspring and lay eggs by early summer. After hatching, nymphs feed for 2 months on leaves, causing them to wilt, brown, and die. There can be one to three generations per year, depending on your climate.

HOSTS

Cabbage family plants (cabbage, cauliflower, Brussels sprouts, broccoli, kohlrabi, kale, mustard, turnip, and Chinese cabbage).

MINIMAL-IMPACT INTERVENTION

Create habitat for predators and parasites. Harlequin bug egg predators include lady beetles, green lacewings, nabid bugs, minute pirate bugs, big-eyed bugs, crab and wolf spiders, and ants.[14] Predaceous stink bugs and praying mantis feed on young nymphs.

In one study of broccoli, mulch was shown to increase spider predation and also to reduce damage from harlequin bugs by 60 percent.[15]

Trap crops of kale planted in the fall and mustard or radishes in the spring have been used to encourage harlequin bugs to congregate. Once the bugs congregate in the trap crop, they can be handpicked, mowed, or sprayed. Cleome flowers are a favorite of harlequin bug and can also be used as a trap crop.

HEAVIEST-IMPACT INTERVENTION

Insecticidal soap plus neem or spinosad can be sprayed on nymphs and adults, but adults are not very susceptible to these sprays.

Leafhoppers

Individual species of leafhoppers feed on different species of plants. For example, **carrot (aster) leafhopper** (*Macrosteles quadrilineatus*) adults are smoky green, wedge-shaped, 3⁄16 inch (5 mm) long, and six-spotted. **Potato leafhopper** (*Empoasca fabae*) adults are yellow green, spindle-shaped, and 1⁄8 inch (3 mm) long. Leafhoppers fly up from underneath leaves when disturbed and are sometimes mistaken for whiteflies. Nymphs are wingless, 1⁄8 inch (3 mm) long, pale green to whitish, and resemble fat aphids. Several generations occur per season, and populations can build to high levels on some crops later in the growing season.

HOSTS

Potato, carrot, flowers, and fruit crops.

MINIMAL-IMPACT INTERVENTION

Yellow sticky traps can attract leafhoppers sufficiently to decrease populations, but the traps also catch beneficial insects. Floating row covers over emerging seedlings can help protect from this pest. Shading plants helps to avoid leafhopper damage.

HEAVIEST-IMPACT INTERVENTION

Insecticidal soap plus neem sprayed on leaf undersides in the early morning or later evenings, when leafhopper adults are moving more slowly, can provide control if applied every 5 to 7 days. Best control occurs if you discover an infestation early and treat when nymphs are present, before adults emerge.

Leafminers

Several species of leafminers attack vegetable crops. Their feeding usually does not seriously affect plant growth but can damage edible leaves to the point of making them unsalable. Hence it is rarely necessary to treat leafminers when they are attacking the leaves of a root crop such as beets. If you can tolerate the cosmetic damage, it is best to ignore leafminers or use only minimal-ecological-impact methods.

Larvae are tiny, white, and maggot-like. Adults are tiny flies. Larvae burrow underneath leaf surfaces leaving a visible trail as they eat their way through the leaf. The top and bottom surface of leaves damaged by leafminers can be pulled apart at the tan-colored blotch or serpentine trail. Larvae or black, sawdusty droppings may be present inside trails or blotches. Larvae overwinter in the soil under host plants. As temperatures warm, larvae pupate into adult flies in mid- to late spring.

HOSTS

Spinach leafminer (*Pergomya hyoscyami*) feeds on spinach, chard, beet, and lambsquarters. **Serpentine leafminer** (*Liriomyza brassicae*) feeds on beet, cabbage, radish, spinach, and turnip. **Onion leafminer** (*Phytomyza gymnostoma*), also called allium leafminer, feeds on onion, garlic, leek, and chives. **Vegetable leafminer** (*Liriomyza sativae*) feeds on many crops including beans, peas, cabbage, lettuce, pepper, eggplant, potato, tomato, squash, watermelon, cucumber, beet, onion, lettuce, celery, and a variety of flowers.

MINIMAL-IMPACT INTERVENTION

Native parasitoids keep a check on leafminer populations when a farm or garden ecosystem has reached a dynamic equilibrium. Vegetable leafminers, for example, are usually suppressed well as long as their natural enemies are present, and intervention is not required. If you find damage on plants, remove and destroy infested leaves when the mines are small. Yellow or blue sticky traps catch egg-laying adults, but also catch beneficial insects.

Floating row covers over emerging seedlings can help protect from this pest. In a Pennsylvania study, row covers were tested for periods ranging from 0 to 45 days. The row cover was found to protect onion seedlings from damage during the adult leafminer flight and egg-laying periods. But row cover also served as a warm microclimate that resulted in increased onion yield when covered for 30 or 45 days. In the same study, plastic mulch also increased onion yield and reduced onion leafminer damage compared with bare soil.

In research done at the Rodale Institute in Pennsylvania, two cover crop mixes were evaluated for control of onion leafminer. One mixture included hairy vetch, rye, oats, sunflower, and white Dutch clover. The other mixture included brassicas—mustard, rape, and daikon radish. Onion plants grown in plots preceded by the first cover crop mix were 50 percent more damaged by onion leafminer than onions preceded by brassica cover crops. As discussed in chapters 7 and 8, when brassica cover crops are winterkilled in the fall or tilled in the spring, they decompose quickly, releasing chemical compounds such as thiocyanates into the soil, where they temporarily suppress soilborne pests and weeds.[16]

HEAVIEST-IMPACT INTERVENTION

According to USDA studies in Maryland, neem has some systemic activity against leafminers when applied as a soil drench; neem soil drenches are an option for vegetable leafminers if applied early. Both spinosad and neem inhibited leafminer feeding and egg laying up to 24 hours after application in one study. In another study, larval mortality was greater than 99 percent when 1 percent neem seed extract was sprayed to the point of runoff on tree foliage containing either eggs or early instars of leafminers.

Mexican Bean Beetles

Mexican bean beetles (*Epilachna varivestis*) are oval, 1/3 inch (8 mm) long, and coppery brown with 16 black spots on their wing covers. Yellow, elliptical eggs are found on leaf undersides. Larvae are yellow to bright yellow and spined.

Adults overwinter under plant debris, in open fields, or in wooded areas. In the spring adults emerge and lay eggs on leaf undersides. Larvae hatch within 2 to 3 weeks and join adults feeding on foliage. Both adults and larvae cause characteristic lacy damage on leaves. Adults are strong fliers.

HOSTS

Lima and snap beans; cowpeas; soybeans; pinto, navy, kidney, and fava beans; and lentils.

MINIMAL-IMPACT INTERVENTION

Pediobius foveolatus is a very effective wasp parasite of Mexican bean beetle. It has shown good control in commercial bean fields in warmer climates and in the mid-Atlantic states and in New England but will not survive in colder climates. This wasp parasite is mass-reared and sold by the New Jersey Department of Agriculture. It is also available from other beneficial insect suppliers. *Pediobius* wasps are killed by frost, so annual releases are required except in warm climates. Lady beetles, green lacewings, and minute pirate bugs also feed on both the egg and young larval stages.

Bush bean varieties seem to be attacked more readily than pole bean varieties. Plant early-maturing bean varieties to outpace a buildup of bean beetles.

Use row cover, putting it in place after planting. It can be left on until beans begin to flower and form pods.

You can sacrifice a small trap crop of beans by planting a small area with bean seeds in early spring. When overwintered adult beetles infest these plants, pull them up and destroy the beetles.

Till crops under after harvest to destroy overwintering beetles.

HEAVIEST-IMPACT INTERVENTION

Spinosad, neem, insecticidal soap and neem in combination, and pyrethrin have all shown some efficacy against bean beetles, especially larval stages.

Pea Weevils

Pea weevils (*Bruchus pisorum*) overwinter in plant debris, then emerge in the spring when peas are blooming and feed on flowers, leaves, or pods. Adults are small beetles, 1/5 inch (5 mm) long, that have a distinctive "snout." They are brown and flecked with white, black, and gray patches. Cream-colored larvae burrow into pea pods where they feed and develop in the pea seed.

HOSTS

Peas and fava beans.

MINIMAL-IMPACT INTERVENTION

Early planting and harvest may help avoid this pest. Spiders, ground beetles, and birds prey on pea weevils.

HEAVIEST-IMPACT INTERVENTION

Pyrethrin has shown some efficacy against these weevils.

Root Maggots

Root maggots (*Delia* spp.) are cream to white in color, legless, and about 3/8 inch (1 cm) long when mature. The adult fly is gray and resembles a housefly, but is only 3/16 inch (5 mm) long.

Root maggots spend the winter in a pupal stage buried 1 to 5 inches (2.5–13 cm) deep in the soil. In early spring adult flies emerge from pupae and rise to the soil surface; they lay very small, white, oblong eggs on or just below the soil surface near the base

Figure 10.11. Root maggot damage to turnip roots.

of host plants. Maggots hatch in 3 to 7 days, then migrate through the soil and feed on underground plant parts. The insect causes damage only during the maggot stage, which lasts from 3 to 5 weeks. Mature maggots leave the plant and pupate in the soil nearby. In 2 to 4 weeks the adult fly emerges. Several generations occur in a growing season.

HOSTS

Cabbage root maggots feed on cabbage, broccoli, cauliflower, kohlrabi, and radish. **Onion root maggots** feed on onion and related bulb plants. **Corn root (or seed corn) maggots** are pests of corn.

MINIMAL-IMPACT INTERVENTION

An integrated approach is best for root maggots.

- Numerous parasites and predators attack cabbage maggots, but usually do not reduce populations quickly enough to avoid damage in the short run. Among these parasites are wasps in the Braconidae and Ichneumonidae families. Among the predators are several ground beetles, including rove beetles (*Aleochara* spp).
- Avoid adding uncomposted manure or plant residues, such as fresh grass clippings, to soils. Root maggot adults are attracted to decaying raw organic matter to lay their eggs. There is some indication that well-composted materials may reduce maggot problems. Avoid planting root-maggot-susceptible crops after a rotation of legume crops, especially alfalfa. Adult flies are attracted to decaying legume residues. Thoroughly incorporate organic residues and cover crops into the soil well in advance of planting (3 to 4 weeks prior) so decaying resides are not attractive to root maggots at planting. Plant in conditions that favor rapid emergence of seeds for crops susceptible to root maggot.
- Delaying planting is an important practice to reduce root maggots, though this may be less feasible in cool climates with short growing seasons. Cabbage maggots do not develop at temperatures below 43°F (6°C). Plant before or after peak adult maggot flight in the spring (300 degree days at base 43°F after soils thaw). For onion maggots, wait to plant until after the first generation of adult flies has emerged in the

spring (792 to 812 degree days at base 39°F [3°C] after soils thaw). Refer to the "Degree-Day Models" section on page 128 for details on how to calculate degree days.

- Row cover may be placed over seeds or transplants at planting time. Be sure there are no gaps in the cover through which the tiny flies could enter. Washington State University scientists tested other non-chemical techniques on experimental plots. Two popular treatments, garlic sprays and wood ashes, were not effective at repelling or preventing maggot flies from laying eggs.

Spider Mites

Mites are more closely related to spiders than insects. Like spiders, mites have two body regions and four pairs of legs. Their very small size makes them difficult to detect, identify, and monitor. The most common vegetable leaf mites are two-spotted spider mites; they can complete a generation in as little as 10 days at temperatures above 80°F (27°C). Mites prefer warm temperatures, dry and dusty conditions, and low humidity. They thrive with excess nitrogen fertilization.

HOSTS

Tomato, eggplant, potato, melon, and cucumber.

MINIMAL-IMPACT INTERVENTION

Sprinkler irrigation can diminish mite populations as they increase in dry, dusty conditions. Providing adequate water for plant growth needs is also important in managing spider mites. Drought and fluctuating wet/dry soil conditions can stress plants in a manner that can cause spider mite populations to increase. Excessive nitrogen fertilization can enhance mite populations.

Many organisms prey on spider mites. Minute pirate bugs and predatory mites are among the most important. Mites usually do not become a problem unless your farm/garden agroecosystem is out of balance as a result of spraying insecticides that kill beneficial insects.

MODERATE-IMPACT INTERVENTION

Spray a 2 percent solution of insecticidal soap at a time when temperatures are less than 85°F (29°C) and the plants are not in direct sun or drought-stressed. Mix 1 tablespoon sulfur per gallon (4 L) of soap spray for higher populations; spray in the evening, and repeat applications every 5 to 7 days in hot weather.

Squash Bugs

Squash bugs (*Anasa tristus*) are flat, brown, ⅝-inch-long (16 mm) true bugs with long legs and a triangle shape on the body below the head. Nymphs, which are small, gray-colored, and spider-like, are usually found in groups on leaf undersides. Both adults and larvae release an unpleasant odor when crushed. Eggs are bright orange and are laid on the underside of leaves. Adults overwinter near their food source and emerge in early summer to feed on squash vines. Their feeding causes pale yellow spots that eventually turn brown; ultimately whole plants wilt and die. Squash bugs are more common in warm climates. In cold climates they tend to be worse following mild winters.

HOSTS

Cucurbits (pumpkin, squash, melon, cucumber, crookneck squash, and watermelon).

MINIMAL-IMPACT INTERVENTION

An integrated approach using several of the techniques below is necessary for this insect.

- Create undisturbed ground cover for predators such as ground beetles (*Carabidae*) and damsel bugs close to or within squash plants. Other predators include spiders, predatory mites, disease organisms, and robber flies. In one study at the University of Kentucky, ground beetles preyed on squash bugs enough to increase fruit yield by 33 percent. However, wolf spiders preyed upon minute pirate bugs and nabid bugs (both good predators of squash bugs) and actually increased squash bug density. There are pros and cons to mulch. It provides habitat for ground beetles and spiders, but may also provide protective cover for squash bugs. Since squash bugs often seek shelter around the base of plants, keep those areas clear of mulch.

- Floating row covers are the most effective preventive control.[17] They must be put in place as seedlings emerge or immediately after transplanting with secure edges so bugs can't crawl under them. Remove row covers when the plants start to bloom because good pollination cannot occur otherwise. Researchers from Iowa State University Organics Research Program found that mulching with hay combined with tightly secured row covers provided the best control of squash bugs in pumpkin.
- Another study that delayed row cover removal until 50 percent of the female squash plants had flowers showed no negative impact on yield.[18]
- Growing vining types of squash, cucumbers, and melons on a trellis seems to make them less vulnerable to squash bug infestation.
- In the field or garden beds, plant transplants rather than seeds to outgrow squash bugs and provide vigorous plants as early as possible. Vigorous, older plants tolerate squash bug damage without a reduction in yield.
- Certain squash varieties, especially butternut and acorn squash, are more resistant to squash bugs. Kabocha winter squash, crookneck summer squash, and watermelon are particularly susceptible to squash bugs. They might be used as perimeter trap crops for squash bugs. Kansas researchers ranked the following types and varieties from least to most squash-bug-susceptible: butternut, Royal Acorn, Sweet Cheese, Green Striped Cushaw, Pink Banana, and Black Zucchini. I rotate squash varieties within the row, planting 5 to 10 plants of butternut in between more-susceptible kabocha squash varieties.
- Since squash bugs overwinter near the previous year's cucurbit crops, plant cucurbits as far away from last year's crops as possible. However, squash bugs are strong fliers, so crop rotation alone does not fully control them.
- Handpick adult squash bugs as soon as they appear. They are large and easy to see and remove. Look for them as soon as squash plants begin to vine.

HEAVIEST-IMPACT INTERVENTION

Spraying insecticidal soap plus neem or pyrethrin on leaf undersides in the early morning or evening is effective on the smallest nymphs, but not as effective on adults or even larger nymphs. Good penetration throughout foliage is essential so that nymphs under the leaves and deep within plants will be covered. However, field trials have reported only marginal success using organic certified insecticides to control squash bug adults.

Figure 10.12. Two adult squash bugs mating.

Squash Vine Borers

The larvae of squash vine borer (*Melitta satyriniformis*) are white and thick worms with brown heads; adults are moths that fly during the day rather than at night. Moths lay eggs on the stems of the host plants. Larvae tunnel into the vine and feed. There is usually only one generation per year in most areas.

HOSTS

Cucurbits (pumpkin, squash, melon, and cucumber).

MINIMAL-IMPACT INTERVENTION

An integrated approach is necessary for this insect.

- Locate the point of injury, split the vine, and puncture the larva. Cover the cut stem with moist soil to encourage roots to sprout.
- Vigorous plant growth is a vital part of borer suppression. Supplemental fertilizer may be necessary to promote vigorous growth. Vigorous plants tolerate the presence of one or two borers and still produce a crop by growing additional roots along the stem.
- Floating row covers are a commonly recommended method of vine borer suppression. In one study, researchers delayed row cover removal until 50 percent of the female squash plants had flowers with no negative impact on yield.
- The University of Illinois rated the susceptibility of 12 squash varieties to squash vine borer attack. The most resistant types and varieties were blue Hubbard, Boston Marrow, Golden Delicious, Connecticut Field Pumpkin, Small Sugar Pumpkin, and zucchini. White Bush Scallop and acorn squash were moderately resistant. The most susceptible varieties were Summer Crookneck, Green Striped Cushaw, and butternut. These susceptible varieties might be used as a perimeter trap crop.
- Perimeter trap cropping is reportedly effective at preventing crop damage. Research at the University of Connecticut reports that squash vine borer was reduced by 88 percent in main crops of zucchini or summer squash when a perimeter trap crop of blue Hubbard squash was planted, infested by borers, and then tilled in. Another study found that zucchini planted early in the season and left unharvested was an effective trap crop for a cash crop of summer squash planted later. Unharvested zucchini had a 53 percent rate of infestation, whereas harvested zucchini's was below 15 percent.[19]
- In South Carolina trials the insect-parasitic nematodes *Steinernema carpocapsae* and *S. feltiae* applied to squash stems and to soil provided good borer suppression.
- Crop rotation is most effective if fields are far apart. Squash vine borer adults are strong fliers and reportedly travel up to a mile from where cucurbits were grown the previous year. Waiting at least a year between plantings can also be helpful.
- Though avoiding tillage is our goal, with this insect, disrupting mature larvae or pupae inside squash plants or in the soil by tilling in infested crop debris soon after harvest prevents a buildup of the pest.

HEAVIEST-IMPACT INTERVENTION

Scout for moths and monitor their presence (by using pheromone traps), and then spray spinosad to control the adult moths.

Stink Bugs

Stink bug adults are shield-shaped, ½ to ⅔ inch (13–17 mm) long, brown or green in color, with a triangle on the body below the head. Nymphs look like smaller versions of adults, with more rounded, brightly patterned black, red, white, and green bodies.

Stink bugs overwinter as adults and become active in the spring. Damage first appears on green fruit, as dark pinpricks surrounded by light-colored areas that remain green or turn yellow and become hard spots just under the fruit skin when fruits ripen. The spots result in a pithy area and a mealy texture.

Stink bugs are strong fliers and readily move in and out of vegetable fields and gardens. Brown stink bugs are usually the most abundant species in the spring. Green stink bugs become more abundant in late summer. There are several species of stink bugs, but one of the most damaging to crops is the brown marmorated stink bug. The highest concentrations of brown marmorated stink bugs occur in the mid-Atlantic region, but they have been found in 38 states as well as all over the world.

HOSTS

A wide range of vegetables, including tomato, pepper, eggplant, peas, corn, and beans.

MINIMAL-IMPACT INTERVENTION

- Handpick adults and nymphs from plants. Natural enemies of stink bugs include birds, spiders, daddy longlegs, assassin bugs, predatory stink bugs, earwigs, green lacewing larvae, ground beetles, praying mantises, and parasitic wasps. Researchers have observed stink bug egg mortality as high as 50 to 60 percent due to predators eating the eggs. Specific birds that feed on stink bugs are wrens, bluebirds, and cardinals.
- Covering vegetables with row cover material can prevent feeding by stink bugs.
- Stink bug pheromone traps are available, but a study in Maryland home gardens showed that single pheromone traps placed 1 yard (1 m) from garden rows resulted in more stink bugs and damage on the plants near the traps.

HEAVIEST-IMPACT INTERVENTION

Pyrethrin or neem, or a combination of the two, may provide some suppression of young stink bug nymphs. In a field study on peppers, a mixture of pyrethrin and neem resulted in less adult stink bug injury than the no-spray control, but most of the organically acceptable sprays tested were not effective.

Figure 10.13. At first glance this dark-colored bug might look like a plant-damaging stink bug, but this is a predatory rough stink bug. Note the spines projecting from the corners of its shoulders; plant-damaging stink bugs have rounded shoulders.

Tarnished Plant Bugs

Tarnished plant bugs (*Lygus pratensis*) are oval, brown, ¼-inch-long (6 mm) true bugs with a white triangle on their back; nymphs are smaller, yellow-green, and have black spots on their backs. The tarnished plant bug overwinters as an adult. It is active as soon as the weather is warm. Eggs are laid on plant stems. Nymphs hatch in the spring, feed on leaves till maturity, and then disperse. Development occurs as long as temperatures exceed about 50°F (10°C), but slows or stops when temperatures exceed 94°F (34°C). In beans, feeding on flowers causes them to drop, and young bean pods may look pitted and blemished. Fruits may also be attacked, leading to indentations, bumps, or yellowing of the flesh. This damage might be confused with stink bug damage, but tarnished plant bugs do not cause the white pithy areas beneath the skin that is typical of stink bug damage (which is more common).

Tarnished plant bugs also feed on many weeds, flowers, forage crops, and orchard crops. Once weeds flower, tarnished plant bugs tend to remain in the weeds unless they start to dry or are mowed. Weed hosts include wild carrots and other umbelliferous plants, redroot pigweed (and other amaranths), lambsquarters, mustards, shepherd's purse, goldenrod, and mullein. When these are mowed, tarnished plant bugs move into crops. Alfalfa is a favorite of tarnished plant bug, and harvesting or mowing alfalfa often causes them to move into crops.

HOSTS

Potato, celery, beans, lettuce, endive, escarole, cauliflower, broccoli, tomato, eggplant, and pepper.

MINIMAL-IMPACT INTERVENTION

Predators such as lady beetles and spined soldier bugs prey on nymphs. A parasitic wasp (*Peristenus digoneutis*) reportedly can cause up to 50 percent mortality.

White sticky traps may be used to suppress low infestations in small gardens.

HEAVIEST-IMPACT INTERVENTION

If populations are high and damaging, consider spraying PyGanic, which is a commercial form of concentrated pyrethrin allowed for control of tarnished plant bugs on certified organic farms. PyGanic is hard on predators and parasites. Before choosing to spray, assess whether the economic loss due to bug damage will really outweigh the impact of spraying on overall farm ecosystem health.

Thrips

Among the many species of thrips, certain ones visit a wide range of plants but damage only a few of the crops on which they are found. The distinguishing signs of thrips damage are tiny black dots (frass) associated with white, stippled leaves. Leaves look like the chlorophyll has been sucked out. Thrips are tiny, elongate, and slender, ranging from 1⁄32 to ⅛ inch (0.8–3 mm) in length, and can be translucent white, yellow, brown, or black. Several species of thrips are common in vegetable crops. **Western flower thrips** (*Frankliniella occidentalis*) are responsible for vectoring tomato spotted wilt virus on solanaceous crops and also appear to transmit fungal disease microorganisms (*Fusarium* and *Botrytis*) and viruses in greenhouse crops. Western flower thrips pupate in growing media, leaf litter, and/or flowers. Adults emerge and lay eggs in leaves or flowers. Optimal temperatures are 80 to 85°F (27–29°C). Thrips can readily move long distances, floating with the wind. Other species found on vegetable crops include **melon thrips** (*Thrips palmi*), **onion thrips** (*T. tabaci*), and **bean thrips** (*Caliothrips fasciatus*).

HOSTS

Many kinds of flowers and vegetable crops.

MINIMAL-IMPACT INTERVENTION

An integrated approach is necessary for this insect.

- Keep plants well irrigated, and avoid excessive applications of nitrogen fertilizers, which may promote higher populations of thrips.
- Create blooming habitat for predators and parasites that feed on adult thrips, including minute pirate bugs, predatory thrips, green lacewings, mites, and certain parasitic wasps. Avoid creating dust, because thrips like dusty conditions. It may help to rinse dust off small plants, because thrips predators do not like dust. Predatory thrips can be distinguished from pest species because predators are seldom seen at the high levels that are common with certain pest thrips. The mite predators *Amblyseius cucmeris* and *A. swirskii* are very effective at controlling thrips larvae in greenhouses, but not adult thrips. The timing of mowing living mulches is very important with thrips pests. Thrips tend to live in flowering living mulches or buckwheat strips and move to crops when these are mowed. Hence I don't mow the living mulch beneath peach and nectarine trees in the spring when thrips are present and damaging crops.
- Insect-parasitic nematodes in the genus *Thripinema* have been shown to naturally parasitize some thrips species.
- White sticky traps are best for monitoring thrips because yellow traps attract a large number of beneficial insects that prey on thrips.
- Set up hooped row cover while plants are young and susceptible to thrips damage.
- Ultraviolet, reflective mulches interfere with thrips' ability to locate plants. In vegetable crops that are especially sensitive to insect-vectored viruses, reflective mulch can be effective in preventing or delaying infection of small plants. Silver or gray is the most effective color for reflective mulch, but white also works. Apply reflective mulch before plants emerge from the soil, and leave a narrow, mulch-free strip along the planting row.

MODERATE-IMPACT INTERVENTION

Beauveria bassiana has been used for thrips, but it is not effective at relative humidity below 50 percent and performs best in greenhouse conditions. Efficacy of this fungal parasite increases with increasing humidity.

Insecticidal soap can be used for thrips, especially nymph stages.

HEAVIEST-IMPACT INTERVENTION

Both neem oil and pyrethrin suppress thrips. Spinosad performs well against most thrips species, and was most effective against western flower thrips in one study.

Figure 10.14. White stipples on onion leaves from onion thrips damage.

Wireworms

Wireworm (*Limonius canus*) larvae are cylindrical, brown to yellowish, ⅓ to ½ inch (8–13 mm) long, and wormlike, with a shiny, hard skin. Adults are black to grayish or brown click beetles. Larvae burrow into potato tubers and crop roots, creating shallow to deep holes, as they feed.

Adult beetles overwinter in the soil (preferably heavy, wet soil) or in rotten wood and plant debris. Eggs are laid underground in early spring or summer in damp soil several inches beneath the surface. Larva move into the top 6 inches (15 cm) when soil temperatures reach 50°F (10°C), but then move lower at 80°F (27°C). Wireworm populations tend to build up on grass and sod and in fields that have been in continuous cereal production. Manure and high-nitrogen compost amendments also encourage wireworms.

Figure 10.15. Wireworm damage to potato tubers.

HOSTS

Beans, beet, carrot, corn, lettuce, onion, peas, brassica crops, and potato.

MINIMAL-IMPACT INTERVENTION

The presence of grass sod within a rotation results in a population increase of overwintering wireworm adults.[20] But rotations that include non-susceptible crops such as onions, lettuce, sunflower, and buckwheat in the spring appear to suppress wireworm egg laying. Besides the host crops named above, wireworm-susceptible crops include brassicas, grains (especially barley), clover, cowpeas, melons, and sweet potatoes.

Soil tillage in late spring or early fall when larvae or eggs are in the upper soil layers can decrease wireworm populations.

Planting a black mustard cover crop may be beneficial. Black mustard incorporated when soil temperatures reached 50°F (10°C) decreased wireworm larva populations in one study. To avoid allelopathy to vegetable crops, wait 2 weeks after tilling black mustard and then seed the crop.

The soil fungus *Metarhizium anisopliae* infects and kills wireworms in nature.

HEAVIEST-IMPACT INTERVENTION

Spinosad interacts synergistically with the insect antagonistic fungus *Metarhizium anisopliae* to suppress several wireworm species.

CHAPTER 11

Vegetable Crop Diseases and Interventions

Most of the diseases described in this chapter are common throughout North America. Many, such as late blight, are found worldwide. As with insect pests, I categorize interventions for plant disease into minimal impact, moderate impact, and heaviest impact, based on their potential to disrupt ecological systems. I include information and research about suppressive soil microorganisms because these microorganisms are a natural part of a functioning and healthy soil organic matter system. These beneficial microorganisms suppress disease in several ways: They improve plant health so plants can better resist disease, and they directly compete with and attack disease microorganisms. I also focus on suppressive soil microorganisms because many researchers and farmers are experimenting with commercial formulations of these microorganisms, which are now or will become available for farmers and gardeners.

Note that for some diseases, I list only minimal- and moderate-impact interventions. This is because the heaviest-impact interventions available to organic growers, such as copper fungicides, are not really worth the ecological impact risk. For other diseases, there are no moderate-impact options available. The "Physiological Disorders and Nutrient Deficiencies" section beginning on page 260 covers deficiencies and other growth problems that may sometimes be mistaken for disease.

Alternaria Diseases:

Early Blight, Leaf Blight, Purple Blotch

Alternaria is a common fungal genus with many species; individual species attack different plant families. For example, *Alternaria cucumerina* attacks melon and cucumber, *A. solani* attacks tomatoes and potatoes, and *A. porri* prefers onion and garlic. *Alternaria* causes leaf spotting and wilting. Leaf spots progress from older leaves to newer ones. Circular spots start as small, sunken, water-soaked areas and progress to purple to brown spots surrounded by yellow rings that create a diagnostic bull's-eye pattern seen in figure 11.1. Entire leaves may die and drop from the plant. Spots are also found around stem ends on fruits and on the plant stem. This fungus is spread by splashing water and by walking through plants when wet. *Alternaria* survives in plant debris and may be spread by insects, or wind, and by rain/irrigation. Spores germinate in several hours during wet, high-humidity conditions. Infection is encouraged if water remains on plant tissue for longer than a couple of hours. Optimal temperatures for infection are 60 to 80°F (16–27°C) (depending on the

Figure 11.1. Early blight symptoms on tomato leaves.

Alternaria species) plus high humidity. The disease is encouraged by low soil fertility.

HOSTS

Depending on the fungus species, cucurbits (cucumber, squash, melon, and pumpkin), solanaceous crops (eggplant, pepper, potato, and tomato), onion, and other vegetables such as peas and cabbage.

MINIMAL-IMPACT INTERVENTION

Keep water off foliage; irrigate with soaker hoses or drip irrigation. Follow 3- to 4-year rotations between susceptible crops in the same plant family. Remove infected leaves. Maintain balanced nutrient levels and high soil organic matter levels to enhance native soil biological control microorganisms.

SUPPRESSIVE SOIL MICROORGANISMS

Some specific soil microorganisms have shown suppression of *Alternaria* species.

- Early blight on tomato: *Bacillus amyloliquefaciens* strain F727 has shown some suppression of *A. solani*. Plant-growth-promoting rhizobacteria reportedly improve tomato growth and suppress early blight disease.[1]
- Fruit rot on field-grown chili peppers: Application of *Trichoderma harzianum* IMI 392432 to pepper seeds before planting significantly suppressed disease caused by *A. tenuis*.[2]
- Tests at Cornell University report suppression of *Alternaria* on basil using an organic product containing the beneficial bacterium *Streptomyces lydicus*.
- Tests at Cornell University report suppression of *Alternaria* on brassica crops using an organic product containing the beneficial bacteria *Bacillus amyloliquefaciens* strain D747.

HEAVIEST-IMPACT INTERVENTION

Sulfur (or as a very last resort, copper) can be sprayed when temperatures are between 55 and 85°F (13–29°C) and weather is wet, to protect leaves from infection.

Angular Leaf Spot

Angular leaf spot (*Pseudomonas syringae*) is a bacterial disease causing dark, angular spots between leaf veins. One good diagnostic is that tear-shaped droplets ooze from infected tissue. As the leaf dries, tissue tears and shrinks. Fruit may have circular spots or rotted areas. The bacteria overwinters in plant debris and can persist for more than 2 years on dry leaves.

HOSTS

Cucurbits (cucumber, squash); a similar strain infects beans.

MINIMAL-IMPACT INTERVENTION

Follow 2- to 3-year rotations between cucurbit crops. Avoid wetting foliage with irrigation water. Use resistant cultivars when possible. Plant on raised beds. Excessive nitrogen fertilizer increases disease severity. Angular leaf spot is also seed-borne; treat seed for 20 minutes with 120°F (49°C) water.

Maintain balanced nutrient levels and high soil organic matter levels to enhance native soil biological control microorganisms.

SUPPRESSIVE SOIL MICROORGANISMS

Beneficial soil bacteria have been reported to induce systemic resistance against angular leaf spot disease of cucumber.[3] (See the section "Epiphytic Bacteria and Fungi" on page 141 for a discussion of induced systemic resistance.)

HEAVIEST-IMPACT INTERVENTION

As a very last resort, if a crop is already showing symptoms and wet weather is expected, copper can be sprayed to protect leaves.

Anthracnose

Small, depressed, and circular fruit spots are the most common symptom of anthracnose, which is caused by *Colletotrichum* species fungi. Spots become much larger, with centers ranging from tan or orange to brown or black. Anthracnose prefers temperatures of around 80°F (27°C) and wet conditions. The disease is seed-borne and can also overwinter in the soil. Rain splash spreads anthracnose spores from infected soil, plant debris, and/or fruits onto susceptible plants.

HOSTS

Vegetables, members of the cucurbitaceae family, and peppers.

MINIMAL-IMPACT INTERVENTION

Rotate crops, allowing a 2- to 3-year gap period between susceptible crops (including strawberry). Avoid wetting foliage with splashing irrigation water. Use resistant cultivars when possible. Plant on raised beds. Use straw or living mulch or plastic mulch to avoid splashing of soil onto plants. Avoid excessive nitrogen fertilization.

HEAVIEST-IMPACT INTERVENTION

During wet, warm weather, copper can be sprayed to protect leaves as a very last resort if anthracnose is a common problem for your crops.

Aster Yellows

Aster yellows is a disease caused by an organism called a phytoplasma. It is spread only by leafhopper feeding. This disease causes twisted distorted new growth (including leaflike petals) and a yellowing/reddening of leaf tissue; it also causes hairy roots in carrots. Leafhoppers that spread the disease overwinter in warmer regions. After feeding on infected plants, they migrate north and can then transmit the disease through continued feeding on uninfected plants. Peak infection periods are in late summer and early fall.

HOSTS

Carrot, celery, and various flowers, including asters and zinnia.

MINIMAL-IMPACT INTERVENTION

Plant resistant cultivars when possible. At the Frontier Herb Company Experimental Farm in Iowa, planting several crop species together in growing beds greatly reduced the incidence of aster yellows.

HEAVIEST-IMPACT INTERVENTION

Control leafhoppers that spread this disease by applying insecticidal soap plus neem.

Bacterial Spot or Bacterial Blight

Leaves, stems, and fruit infected with bacterial spot (*Xanthomonas campestris*) develop brown, circular spots or specks, but without concentric zones (as for fungal leaf spot diseases). On beans there is a diagnostic yellow ring around the brown spots. The bacterium overwinters on plant debris and is seed-borne. It is favored by wet conditions, high humidity, and temperatures of 75 to 86°F (24–30°C).

HOSTS

Tomato and pepper. Another race of the bacteria also affects beans.

MINIMAL-IMPACT INTERVENTION

Keep water off foliage, stake tomatoes and peppers to keep them off the ground, and follow a 3-year rotation. Treat seed with hot (122°F [50°C]) water or bleach.

SUPPRESSIVE SOIL MICROORGANISMS

Tests at Cornell University report suppression of bacterial spot on tomato using an organic product containing the beneficial bacterium *Bacillus amyloliquefaciens* strain D747.

HEAVIEST-IMPACT INTERVENTION

Copper can be sprayed to protect leaves as a very last resort, during wet weather when temperatures exceed 75°F (24°C). Copper is only moderately effective on bacterial spot, however, and may have phototoxic effects on tomato plants.

Bacterial Wilt of Cucurbits

Infection by *Erwinia tracheiphila* bacteria causes wilting and death of most cucurbits. White ooze is visible at any cut in the plant tissue. Bacterial wilt is spread by and depends upon the presence of cucumber beetles and overwinters in their guts. In fact, the bacterium survives the winter only in the digestive tract of cucumber beetles. In spring, when beetles begin to feed on new cucumbers and melons, the overwintered bacteria are spread to new cucurbit leaves from the beetles' fecal droppings. The bacteria can then infect plants through wounds caused by beetle feeding. Bacteria are unable to infect plants through normal plant openings (such as stomata) and the disease is not carried on seed. The disease can be more severe with excessive levels of nitrogen, but it also seems to be promoted by very low levels of nitrogen and potassium in unbalanced nutrient situations.

HOSTS

Cucumber, squash, muskmelon, pumpkin, and gourd. Watermelon is immune.

MINIMAL-IMPACT INTERVENTION

Use tightly secured row covers over seedlings and transplants to prevent cucumber beetles from gaining access.

HEAVIEST-IMPACT INTERVENTION

Control populations of cucumber beetles (see the "Cucumber Beetles" entry on page 226). When the first cucumber beetles are observed, spray with neem and pyrethrum.

Black Canker

Black canker fungus (*Tersonilia perplexans*) causes brown, black, or purple black cankers to form mainly on the surface of root crowns and shoulders. The disease organism overwinters in infected parsnip and carrot roots or as spores in the soil. Disease development is enhanced by cool, wet weather.

HOSTS

Carrot and parsnip.

MINIMAL-IMPACT INTERVENTION

Use of resistant varieties is the best suppression practice. Rotate parsnips and carrots with non-host crops.

Black Rot

Black rot (*Xanthomonas campestris*) turns young plants yellow, then brown, and eventually kills them. Yellowing begins first on older leaf edges and progresses inward in a wedge-shaped pattern. Leaf veins turn black. Black rings and yellow ooze are present in cut stems. The most common source of the infection is infested seeds or infected transplants. The bacteria can remain infectious for up to 2 years. Infection occurs during wet weather. Optimal temperatures for disease development are 80 to 86°F (27–30°C) when dew, rain, or sprinkler irrigation are present. Transmission occurs via insects, water, and gardening or farming tools.

HOSTS

A wide range of brassica crops and some brassica weeds (weeds can be a source of inoculum).

MINIMAL-IMPACT INTERVENTION

Keep water off leaves or water in the morning to allow maximum rapid drying. Follow 2- to 3-year rotations and destroy infected plants. Grow resistant varieties. Treat seed with hot (120°F [49°C]) water for 15 to 20 minutes before planting. Note that mustards are susceptible to heat-induced seed damage and require shorter duration hot water treatment.

Low nutrient content in soils and plants aggravates black rot disease. In one study, disease severity diminished significantly when organic sources of foliar nitrogen and potassium (from manure-based compost) were added to kale plants, and lignin cotent increased. The compost appears to have activated defense mechanisms in kale plant tissues.[4]

SUPPRESSIVE SOIL MICROORGANISMS

An antagonistic, endophytic strain of *Bacillus subtilis* suppressed three strains of black rot bacteria in cabbage, cauliflower, rape, and broccoli. However, when encouraging or applying these microorganisms, details matter; this beneficial bacteria was less effective when crops were grown on heavy soils during heavy rainfall periods.[5] Tests at Cornell University report suppression of black rot on brassica crops using an organic product containing the beneficial bacteria *Bacillus amyloliquefaciens* strain D747.

HEAVIEST-IMPACT INTERVENTION

Spray copper only if you have had severe problems with black rot in the past and weather is wet. Copper is ineffective once the disease is well established and works only to limit spread to uninfected plants.

Botrytis

Senescent leaves, fruits, and petals are susceptible to botrytis (*Botrytis cinerea*), which is also commonly called gray mold. Under cool, moist conditions a soft, brown decay develops, covered by a dense gray to light brown mass of spores. Germination of spores and infection requires moisture for 8 to 12 hours, relative humidity of 85 percent or greater, and temperatures 55 to 75°F (13–24°C). Growth of this fungus is inhibited at temperatures above 89°F (32°C). Healthy, actively growing green plant parts are seldom infected directly by *Botrytis* fungi.

HOSTS

Many vegetable, fruit, and flower crops.

MINIMAL-IMPACT INTERVENTION

Plant on raised beds, maintain low humidity, and keep water off leaves or maximize drying conditions when you irrigate. Organic biofungicides containing the beneficial fungus *Ulocladium oudemansii* (U3 strain) have shown some suppression of *Botrytis*.

HEAVIEST-IMPACT INTERVENTION

If botrytis has been a past problem, spray sulfur or potassium bicarbonate when humid, cloudy weather persists.

Cercospora Leaf Spot

Leaf spots typical of cercospora leaf spot are pale, round circles with dark margins (see figure 11.2). Symptoms appear on older leaves first and can cause defoliation. *Cercospora* fungi overwinter on plant residue. Spores are carried on the wind relatively long distances. Infection requires free water on leaf surfaces and warm nights combined with high humidity. Optimal daytime temperatures for disease development are 77 to 95°F (25–35°C) with night temperatures above 61°F (16°C) and a relative humidity of 90 to 95 percent. Inoculum occurs on residue from a previously infected crop, but the fungus can be carried on seed.

HOSTS

Beet, Swiss chard, spinach, carrot, celery, cucurbits (cucumber, squash), tomato, and lettuce (different species on each vegetable). Also infects several weed species, such as lambsquarters, pigweed, mallow, and bindweed.

Figure 11.2. Cercospora leaf spot on Swiss chard.

MINIMAL-IMPACT INTERVENTION

Grow resistant varieties. Soak seed in 122°F (50°C) water for 25 minutes before sowing. Rotate crops, allowing a 3-year break between susceptible vegetables. Keep water off leaves. Increase air movement by staking crops if possible.

SUPPRESSIVE SOIL MICROORGANISMS

An antagonistic strain of *Bacillus subtilis* has been reported to suppress cercospora leaf spot.[6] Repeated foliar applications of a liquid culture of two *Trichoderma* fungi isolates also suppressed cercospora in one study.[7] Tests at Cornell University report suppression of this disease using an organic product containing the beneficial bacteria *Streptomyces lydicus.*

HEAVIEST-IMPACT INTERVENTION

Spray sulfur at first sign of the disease if it has been a serious problem in your crops in the past.

Clubroot

Clubroot (*Plasmodiophora brassicae*) is not seedborne, but the main means of spread is contaminated transplants. The disease causes wilting as a result of deformed and constricted roots. It is favored by wet, heavy soils and rain events that cause temporarily flooded soils.

HOSTS

All brassica crops.

MINIMAL-IMPACT INTERVENTION

Choose transplants carefully or grow your own. Adjust soil pH to 6.8 with lime. Some research indicates that in acidic soils, annual application of

Figure 11.3. The deformed kohlrabi roots on the right plant are caused by clubroot disease. The kohlrabi roots on the left are normal, without disease.

1,500 pounds per acre (1,680 kg/ha) of hydrated lime (calcium hydroxide) before planting helps to diminish clubroot. A rotation of at least 7 years out of susceptible brassicas is also necessary.

Curly Top (Virus)

Infection by the curly top virus causes dwarfing, leaf vein swelling, curling, puckering, distortion, and yellowing; death occurs if very young plants are infected. The causal viruses are curtoviruses (family Geminiviridae), which are spread by leafhopper insects. Tomato, melon, and cucurbit fruits in general appear to ripen prematurely, but have an odd taste. Younger plants seem to be more susceptible to damage and develop more symptoms from the virus compared with adult plants. High light intensity, prolonged summer heat, and high evaporation lead to severe infection. Relative humidity above 50 percent reduces curly top; humidity below 35 percent increases it. High humidity may also delay visits of leafhoppers. The virus is *not* seed-borne; it overwinters in crop debris and weed hosts, such as lambsquarters, Russian thistle, and four-wing saltbush.

HOSTS

Chard, beet, spinach, watermelon, tomato, cucurbits (cucumber, squash, melon).

MINIMAL-IMPACT INTERVENTION

Grow resistant varieties.

Remove diseased plants immediately. There are reports that crops grown in high tunnels or under plastic tunnels may resist this virus better than those grown unprotected. Increasing humidity may help avoid this virus.

HEAVIEST-IMPACT INTERVENTION

Some recommend spraying for leafhoppers, but this has not been effective for control of curly top and is probably not worth the ecological impact.

Damping Off

Damping off is caused primarily by *Pythium* fungi, but also *Fusarium* and *Rhizoctonia* species. Only young seedlings are affected. Once plants have a well-developed root system and mature leaves, they naturally resist the fungus or mold that causes damping off. Infected seedling stems look water-soaked and thin, and roots are stunted or have dark, sunken spots. Young leaves wilt and turn green-gray to brown. The microorganisms that cause damping off thrive in cool wet conditions and are encouraged by slow plant growth due to low light, overwatering, and high salt from overfertilizing. See also the "Rhizoctonia Diseases" entry on page 255.

HOSTS

Many vegetables and flowers.

MINIMAL-IMPACT INTERVENTION

Plant seeds under temperature conditions that favor rapid germination, or pre-sprout seeds before planting outside. For indoor plant production, use a heating pad under trays to warm the soil to 65 to 75°F (18–24°C). Wait until the outside soil has reached optimal temperatures for specific crop germination. Instead of a sterile growing media, use a compost-based growing media with good drainage and water to keep it moist but not so wet that you can squeeze water out of it. Provide good light sources to your seedlings at least 12 hours a day. In some cases, ambient light from windows does not provide enough light. Wait until late morning to water plants to maximize drying conditions.

SUPPRESSIVE SOIL MICROORGANISMS

Tests at Cornell University report suppression of damping off using an organic seed treatment containing the beneficial bacteria *Streptomyces lydicus.*

Downy Mildew

Symptoms of downy mildew begin as early as seedling stage, and it can be mistaken for other leaf diseases. Downy mildew has darkly colored, angular spores that cause blackish to gray spots on leaves compared with the white, diffuse spores of powdery mildew disease that cause white patches or a white film on foliage.

Different species of fungi cause downy mildew on different crops. Downy mildew on spinach is caused by *Peronospora effuse,* which does not colonize other crops in the same family (beets and Swiss chard) or weedy relatives, such as lambsquarters, and nettleleaf goosefoot. On lettuce, downy mildew is caused by *Bremia lactucae.* On cucurbits, it is caused by *Pseudoperonospora cubensis.* The story is further complicated by downy mildew's ability to evolve different races. For example, in cucurbits, there are two main races, one that infects cucumbers and melons, and another that infects pumpkins and squash. Hence it is possible to have a healthy pumpkin field alongside a diseased cucumber field. Downy mildew is ecologically adaptable! New races continue to occur as the disease changes.

The optimal temperature range for this disease is 55 to 70°F (13–21°C). However, spores of the downy mildew pathogen have been observed on plants over a very wide temperature range, from less than 40°F up to 118°F (4–48°C). The disease needs moisture on the leaf surface in order to germinate and start a new infection. It is favored when temperatures are low and there are long periods of leaf wetness caused by overnight dew. Since these conditions are common in late fall and in cool-season greenhouses, that is often where we find downy mildew. It is a seed-borne disease.

HOSTS

Cucurbit crops, spinach, basil.

MINIMAL-IMPACT INTERVENTION

An integrated approach is vital to suppressing this disease.

- The use of resistant cultivars is the most effective way to suppress spinach downy mildew. There are many races, however, and most varieties are not resistant to all races.
- Nitrogen fertilizer seems to encourage downy mildew by boosting lush vegetative growth, which encourages higher downy mildew infection. Nitrogen reportedly increased the disease in both grape and pearl millet.[8]
- Soaking seeds in hot water (122°F [50°C]) for 25 minutes is useful for minimizing the downy mildew inoculum in spinach seeds.[9]
- Keep water off leaves. Drip irrigation and wide row spacing help to dry leaves and encourage good air movement around plants. You can also trellis vining plants to improve air circulation and avoid wet foliage.
- Follow a 3-year crop rotation. The only place I ever see downy mildew is in my high tunnel where I insist on growing spinach every late fall and winter. I need to forgo spinach for a winter, or build another high tunnel.

SUPPRESSIVE SOIL MICROORGANISMS

Tests at Cornell University report suppression of downy mildew on basil using an organic product containing the beneficial bacteria *Streptomyces lydicus* and on other vegetable crops in products containing *Bacillus amyloliquefaciens*.

HEAVIEST-IMPACT INTERVENTION

Potassium-bicarbonate-based fungicides have been used on this disease. Potassium bicarbonate has been reported to kill downy mildew on contact by pulling water from spores and their growing strands.

Figure 11.4. Downy mildew on spinach.

Fusarium Wilt or Yellows

Fusarium oxysporum, and crop-specific subspecies, are fungi that affect both seedlings and mature plants. Infection causes top growth to wilt, yellow, and die. Lesions form at the plant base or slightly below the soil line. Reddish brown streaks are present internally in the root, stems, and leaf petioles. The earliest symptom is the yellowing of old leaves, often on only one side of the plant. The symptoms can be mistaken for verticillium wilt (see the "Verticillium Wilt" entry on page 259).

Fusarium is most prevalent on acidic sandy soils. It can survive several years in soil and is favored by warmer weather (80 to 85°F [27–30°C]), high levels of phosphorus, certain micronutrients, and ammonium forms of nitrogen.

HOSTS

Broccoli and other brassicas, celery, tomato, eggplant, pepper, potato, cucumber, and cantaloupe; occasionally in pumpkin, squash, and watermelon and weeds such as pigweed, mallow, and crabgrass.

MINIMAL-IMPACT INTERVENTION

Use of resistant varieties is the best suppression practice. Follow a 5- to 7-year rotation. In general, raise the pH to 6.5 to 7.0 if the soil is acid. In studies on cucumber, higher levels of nitrate-nitrogen seemed to protect cucumber plants against fusarium wilt disease by suppressing colonization of cucumber *Fusarium oxysporum* ssp. *cucumerinum*. *Remove and destroy* infected plants.

SUPPRESSIVE SOIL MICROORGANISMS

Many microbial species, including *Bacillus* spp., *Pseudomonas* spp., *Trichoderma* spp., *Streptomyces* spp., and *Acinetobacter* spp. have been shown to effectively suppress plant-disease-causing *Fusarium* species.[10] Commercially available biocontrol products containing *Gliocladium virens* and *Trichoderma harzianum* can reduce disease when granules are incorporated into potting medium at 0.2 percent (weight/volume). Co-inoculation of an antagonistic bacteria strain, *Bacillus amyloliquefaciens*, and the fungus *Pleurotus ostreatus* suppressed cucumber fusarium wilt in a greenhouse pot study.[11] Organic products containing *Bacillus amyloliquefaciens* strain F727 and *Streptomyces lydicus* can be applied as a foliar spray or root drench. A biofungicide with extract of giant knotweed (*Reynoutria sachalinensis*) can also be applied as a foliar spray or root drench.

Late Blight

Late blight (*Phytophthora infestans*) first manifests as water-soaked spots on older leaves; spots enlarge into brown blotches. Leaf undersides may be covered with a gray to white moldy growth. Infected leaves, petioles, and stems shrivel and die. Tomato fruit develops dark, greasy-looking spots that enlarge until the fruit rots. Potato tubers show irregular, slightly depressed areas of brown to purplish skin.

The fungus overwinters on crop debris. It thrives at 91 to 100 percent humidity and optimal temperatures of 65 to 72°F (18–22°C) with cool days plus warm nights. Inoculum carried by wind and water infects young plants. When weather is favorable, infection moves so fast that plants appear to have been damaged by frost. Hot, dry days with temperatures above 86°F (30°C) decrease infection.

HOSTS

Potato, tomato, occasionally eggplant.

MINIMAL-IMPACT INTERVENTION

Use of resistant varieties is the best suppression practice. For potatoes, keep tubers covered by hilling with soil or mulch throughout the growing season. A biofungicide with extract of giant knotweed (*Reynoutria sachalinensis*) can also be applied as a foliar spray or root drench. In tests at Cornell University, organic products containing *Bacillus amyloliquefaciens* strain F727 applied as a foliar spray were somewhat effective against late blight. Researchers in Germany reported that compost extracts provide some control against late blight on tomato leaves.

Figure 11.5. Water-soaked spots and brown blotches on older tomato leaves at summer's end caused by late blight.

HEAVIEST-IMPACT INTERVENTION

Spray copper only if you have had severe problems with the disease in the past and weather is wet and humid with temperatures between 60 to 75°F (16–24°C).

Mosaic Virus: Tobacco Mosaic Virus, Cucumber Mosaic Virus

Specific virus diseases are difficult to distinguish, but generally they cause stunted, slow-growing plants, twisted, crinkled, cupped, or deformed leaves, and yellow/green mottling, puckering, and distortion of leaves. Viruses can also cause colored circles or mottling and streaks on fruits. There are many species of mosaic virus, and they are very persistent. The virus is spread either through insects (aphids, thrips, and leafhopper feeding) or mechanically (infected tools, hands and clothing). The virus reproduces within plant cells and disrupts the cell's normal function. Viral diseases are systemic and symptoms usually progress and worsen through the growing season.

HOSTS

Cucurbits (cucumber, squash), solanaceous crops (eggplant, pepper, potato, tomato), celery, corn.

MINIMAL-IMPACT INTERVENTION

Use of resistant varieties is the best suppression practice. Control aphids and cucumber beetles, which help spread the virus. Pre-soak seeds in a 0.5 percent bleach solution. Remove and destroy infected plants.

Phytophthora Root Rot

Several species of soilborne pathogens in the genus *Phytophthora* cause root rot diseases. Stunted, yellowing, wilting leaves are the first sign of this disease. Leaves may also turn dull green or in some cases red or purplish.

Phytophthera fungi overwinter as spores in soil or diseased plant material. Species that cause root and crown rots enter host plants near the root collar via wounds or the succulent parts of small roots. Fungal spores move in water and are attracted to

Figure 11.6. Wilted and yellowing pepper leaves and fruit due to infection by *Phytophthora*. Photo by Doug O'Brien.

the root exudates from stressed plants. These spores can survive for years in moist soil without host plant roots. However, if the soil is completely dried out, these spores are less likely to survive for more than a few months. *Phytophthera* can be spread in splashing rain or irrigation water. Flooded and saturated soils favor the spread. Different *Phytophthora* species are favored by different temperatures and conditions. Root rot of tomato is favored by warm conditions.

HOSTS

Many vegetable crops, including asparagus, tomato, pepper, eggplant, beans, and brassica crops.

MINIMAL-IMPACT INTERVENTION

The most important factor in reducing *Phytophthora* disease is proper water management. Plant in raised beds to provide for good drainage. Group your crop plants according to their specific irrigation needs. Separate those needing frequent, light irrigations, such as potatoes, from those needing less frequent, but deep irrigations, such as tomatoes. Avoid prolonged saturation of the soil or standing water. Keep the soil pH above 6.0. *Remove and destroy* infected plants, including the roots. Follow a 2-year rotation that includes a resistant crop such as corn.

Potato Scab

The bacterium that causes potato scab, *Streptomyces scabies*, is inhibited at soil pH higher than 7.4, and disease severity is reduced in soils with pH levels of 5.2 and below. There are usually no aboveground symptoms, but potato tubers have roughened, russeted areas and scab-like protuberances with corky tissue.

HOSTS

Potato mainly, but also other root crops, including beet, radish, carrot, and parsnip.

MINIMAL-IMPACT INTERVENTION

Use resistant varieties. Maintaining soil moisture near field capacity during the 2 to 6 weeks following tuber initiation will inhibit infection by potato scab. Bacteria that flourish at high soil moisture outcompete potato scab on the tuber surface. Mulching with straw (as long as the soil has good drainage) may help to maintain higher moisture levels and discourage scab. Applying manure to potato fields has been shown to cause an increase in scab infection. This is probably because *Streptomyces* bacteria are involved in the decomposition of soil organic residues and hence stimulated by its presence. Rotation with small grains, corn, or alfalfa help to decrease potato scab, but red clover stimulates it and should be avoided where potato scab has been a problem. Also do not rotate with alternate hosts of potato scab, such as radish, beet, and carrot. Limit the use of soil amendments, such as lime and manure that raise soil pH. Light-textured soils that dry out easily and those with high levels of organic matter are favorable to scab infection.

Powdery Mildew

Powdery mildew is a common occurrence on plants in the squash family at the end of the season. It is a good reason to rotate crops, but not to panic. There are many species of fungi that cause powdery mildew; each attacks specific plant families. White, powdery spots form on both upper and lower surfaces of leaves and on shoots, flowers, and fruit. All powdery mildew species can germinate and grow without the presence of water. In fact, spores of some powdery mildew fungi are inhibited when plant surfaces remain wet for extended periods. Temperatures of 60 to 80°F (16–27°C) and shady conditions favor powdery mildew development. The disease is inhibited at temperatures above 90°F (32°C) and in extended direct sunlight.

HOSTS

Beans, beet, carrot, cucumber, eggplant, lettuce, melon, parsnip, peas, pepper, pumpkin, radish, squash, tomato, and turnip.

MINIMAL-IMPACT INTERVENTION

Plant in full sun and avoid shade for susceptible crops. Avoid excess fertilizer. Overhead sprinkling can reduce powdery mildew because spores are washed off the plant (but it could encourage other diseases!).

SUPPRESSIVE SOIL MICROORGANISMS

Bacillus subtilis may suppress the microorganisms that cause powdery mildew.

MODERATE-IMPACT INTERVENTION

Potassium-bicarbonate-based fungicides have been used on this disease. Potassium bicarbonate has been reported to kill powdery mildew on contact by pulling water from spores and their growing strands. Many organic materials have been tested for powdery mildew over the years, but in comparative studies on zucchini at Purdue University, only the potassium bicarbonate treatment had significantly less powdery mildew than the untreated control. JMS Stylet-Oil, insecticidal soap and sulfur were effective in tests at Cornell University.

HEAVIEST-IMPACT INTERVENTION

If powdery mildew is a consistent, serious problem in your fields or garden, consider use of sulfur or copper sprays. Sulfur was very effective in tests at Cornell University. Copper sprays were most effective in tests at Purdue University.

Rhizoctonia Diseases

The *Rhizoctonia* fungus can be subdivided into strains based on the plant family preferred and optimal temperature to cause disease. For example, strains of *Rhizoctonia* that attack potato do not attack

Figure 11.7. White powdery mildew patches on both surfaces of cucumber leaves.

brassica crops. The strain of *Rhizoctonia* that causes bottom rot and head rot of cabbage grows at temperatures ranging between 48 and 91°F (9–33°C), while lettuce *Rhizoctonia* is favored by warm temperatures between 77 and 81°F (25–27°C). Generally, *Rhizoctonia* causes damping-off-like symptoms on seedlings, root rots, and aboveground stem cankers and leaf and fruit rots. *Rhizoctonia* persists in soil and in plant debris. It is persistent over very long periods. Cool, moist soils favor disease development; dry or waterlogged soil discourages it.

HOSTS

Many vegetable crops, including lettuce and brassica crops as well as beet, beans, carrot, celery, cucumber, eggplant, onion, peas, pepper, rhubarb, spinach, tomato, and sweet potato.

MINIMAL-IMPACT INTERVENTION

Practice a 3-year or longer rotation to non-susceptible crops such as sweet corn or onions. Plant crops like lettuce on raised beds to promote air movement and drainage and to minimize foliage contact with the soil. Keep irrigation water off foliage and, for lettuce heads, avoid irrigation near harvest, when crops are most susceptible to *Rhizoctonia*.

SUPPRESSIVE SOIL MICROORGANISMS

Treat seed with organic products containing *Bacillus subtilis* or *Pseudomonas fluorescens*. Organic products containing *Bacillus amyloliquefaciens* strain D747 can be applied to soil on 14- to 28-day intervals right up until harvest.

Root Knot Nematodes

Nematodes are microscopic roundworms. They can move only short distances on their own but may be spread in transported soil or plant debris. They persist over a fairly long time in gelatinous, sacklike structures when conditions are unfavorable. Females, eggs, and juveniles survive in plant roots. Eggs and juveniles are released into the soil when plants decompose. Nematodes are active when soil is moist and warm. Plants affected by root knot nematodes (*Meloidogyne* spp.) show symptoms of stunting, wilting, and yellowing. Closer inspection of roots reveals swollen, distorted areas, called galls or knots.

HOSTS

Many kinds of vegetables and flowers, woody shrubs, and weeds.

MINIMAL-IMPACT INTERVENTION

An integrated approach is best for root knot nematodes.

- Practice rotations that include a summer fallow and/or winter cover crops of grains, such as wheat. Winter grains planted when soil temperatures are below 65°F (18°C) help to decrease nematode populations.
- In hot weather heat the soil with solarization techniques (covering the soil with clear plastic for 3 to 5 weeks), then uncover the soil and leave it to dry.
- Apply parasitic nematodes. (See the "Parasitic Nematodes" section on page 125.)
- Several bacterial genera, namely *Pasteuria, Pseudomonas, Burkholderia, Arthrobacter, Serratia, Achromobacter*, and *Rhizobium*, are known to suppress nematodes. Application of *Bacillus cereus* strain BCM2 in tomato decreased nematode populations by 60 percent and reduced nematode damage.[12] A study testing a commercial biocontrol product containing *Bacillus subtilis, Bacillus licheniformis*, and *Trichoderma longibrachiatum* also inhibited nematode reproduction on tomato.[13]
- *Trichoderma longibrachiatum* decreased nematodes by 88 percent during in vitro experiments and reduced nematodes in cucumber in a greenhouse study.[14] *Trichoderma* strains have also been proven effective both as plant growth promoters and to suppress nematodes in pepper.[15]
- Cover crops of French marigolds can reduce the number of root knot nematodes. Marigolds release a chemical that is highly toxic to root knot nematodes and prevents egg hatching. Also, root knot nematodes do not seem to be able to develop properly in marigold roots. Other cover crops increase the diversity of microorganisms in the soil and encourage the growth of certain bacteria and fungi that feed on root knot nematodes and parasitize their eggs.
- Several varieties of nematode-resistant tomatoes are available, and some resistant varieties of peppers, peas, and beans are on the market.

Rust Diseases

Though different rust diseases infect different crops, most can be identified by the minute, circular to elongate, golden or reddish brown pustules that form on the upper and/or lower leaf surfaces (as seen in figure 11.8). White rust (*Albugo occidentalis*) on spinach shows up as small yellow spots on upper leaf surfaces and white pustules on lower leaf surfaces. Asparagus rust (*Puccinia asparagi*) and corn rust (*P. sorghi*) disease are favored by temperatures near 80°F (27°C) with high humidity and frequent dews. Optimal conditions for infection by *P. porri*, which causes rust in garlic, onion, and leek, occur around 59°F (15°C) with 100 percent relative humidity for at least 4 hours. Spinach white rust is favored by warm (72°F [22°C]), sunny days followed by cool nights with dew.

HOSTS

Asparagus, corn, onion, garlic, chives, spring onion, leek, spinach.

MINIMAL-IMPACT INTERVENTION

Plant resistant varieties. Rotate away from susceptible crops for 2 to 3 years. Plant in well-drained soils. Till under infected plant residues or remove and destroy infected plants in smaller areas. Organic products containing the beneficial bacterium *Bacillus amyloliquefaciens* strain D747 have been used on rust disease with variable results.

Sclerotinia Disease:

White Rot, Lettuce Drop

Initial symptoms of sclerotinia disease or white mold (*Sclerotinia sclerotiorum*) are small, circular, light green spots that appear water-soaked. Affected plant parts dry, turn brown, and become covered with a white, cottony fungal growth, hence the common name. Infected fruits and foliage rapidly disintegrate with a watery rot, giving rise to the common name of lettuce drop.

The disease is favored by moist conditions and temperatures of 68 to 77°F (20–25°C). *Sclerotinia* overwinters in the soil as resting structures that can persist there for 5 to 8 years. The microorganism infects susceptible crops in the spring. Then the cycle continues as the resting structures, called sclerotia, fall from infected crops onto the soil surface, and become incorporated by soil cultivation and tillage. The sclerotia can germinate after 20 to 100 days, depending on soil moisture and temperature. In the soil, sclerotia can be degraded by

Figure 11.8. Rust disease on garlic.

Figure 11.9. Sclerotinia disease is just starting to infect the bottom leaves of this lettuce plant that are touching the soil.

bacterial decomposition and fungal parasitism and also eaten by mites, collembolans, and earthworms.[16]

HOSTS

A wide range of plants, including lettuce (lettuce drop), beans (white mold), and tomato (white mold). Other crops are susceptible to one of several *Sclerotinia* species, including Brussels sprouts, cauliflower, cabbage, carrot, collards, eggplant, pepper, potato, squash, and melon.

MINIMAL-IMPACT INTERVENTION

Standard crop rotation practice is not effective because of the wide host range of this disease, but a very long rotation of up to 8 years with non-hosts, such as grasses and grains, may be effective. Avoid excessive irrigation and irrigate in mid- to late morning to allow plants to dry as quickly as possible. Mulch or living mulch can decrease this disease by preventing soil from splashing up on plants. *Sclerotinia*'s ability to persist in the soil for extended periods makes deep plowing ineffective. However, some brassica cover crops are reported to suppress this disease. Indian mustard (*Brassica juncea*) reduced potato stem rot caused by *Sclerotinia sclerotiorum* by more than 50 percent in one study.[17] Field mustard (*B. campestris*) and rapeseed (*B. napus*) also reduced *Sclerotinia* in the same study.

Creating a soil organic matter system with an active microbial community and undisturbed habitat for earthworms is the best preventive against *Sclerotinia*.

SUPPRESSIVE SOIL MICROORGANISMS

Tests at Cornell University report suppression of *Sclerotinia* on lettuce using an organic product containing the beneficial bacteria *Coniothyrium minitans* applied once as soil spray. These beneficial bacteria can also be incorporated into the soil in the fall to help protect crops the next season. Tests at Cornell report suppression of white mold on bean crops using an organic product containing the beneficial bacteria *Bacillus amyloliquefaciens* strain D747.

Septoria Leaf Spot

Septoria leaf spot (*Septoria* spp.) appears initially on lower leaves after the first tomato fruit forms. The small circular leaf spots have dark brown borders and tan to gray centers. This disease is favored by wet weather and temperatures of 72 to 79°F (22–26°C). Irrigation water, rain, high humidity, and dew on leaves lead to rapid disease development. *Septoria* may overwinter on solanaceous weeds such as groundcherry. Septoria leaf spot can be confused with early blight (see the "Alternaria Diseases" entry on page 243). Early blight causes fewer, larger circular spots with concentric rings surrounded by a yellow halo.

HOSTS

Tomato, eggplant, potato.

MINIMAL-IMPACT INTERVENTION

Use resistant cultivars. Iron Lady tomato has some tolerance to *Septoria*. Keep water off leaves. Avoid irrigating in late afternoon or evening. Stake plants and space plants well to reduce the amount of time leaves remain wet.

SUPPRESSIVE SOIL MICROORGANISMS

Tests at Cornell University report moderate suppression of *Septoria* on tomato using organic products containing the beneficial bacteria *Streptomyces lydicus* and *Bacillus subtilis* GB03. Another organic product containing extract of giant knotweed (*Reynoutria sachalinensis*) was also moderately effective.

HEAVIEST-IMPACT INTERVENTION

Copper sprays have been recommended for septoria leaf spot.

Smut

The corn smut fungus (*Ustilago maydis*) causes swelling in aboveground plant tissues (ears of corn). Plant cells become spongy gray, then black as the spores mature. Galls can be up to 4 inches (10 cm) in diameter. Smut overwinters in the soil in corn residue and is seen after damage to the plant, such as hail damage.

HOSTS

Corn. Pasture grasses are susceptible to a similar smut fungus.

MINIMAL-IMPACT INTERVENTION

There is no control for this disease after infection.

Remove and destroy all infected plants. *Do not* compost these plants, unless your composting practice creates high enough temperatures to kill this disease. Soil applications of raw manure favor infection. Plants grown with high nitrogen levels or with high rates of manure are more susceptible to the disease. Use resistant corn varieties.

Verticillium Wilt

Verticillium wilt (*Verticillium* spp.) symptoms vary depending upon the host crop, but foliar symptoms typically include wilting, curling, yellowing, marginal or interveinal browning, and death. Overall, these symptoms may resemble water stress and can occur on only one side of a plant. Inside stems, you may see discolored streaks or bands that range in color from light tan to grayish olive to brownish black. Yellowing and defoliation usually progress upward.

Verticillium survives in plant debris and in the roots and trunks of killed trees (for as long as several years). As a resting structure (microsclerotia), the fungus can persist in the soil for 10 years or longer. Water-stressed or wounded plants are most susceptible. Wet, warm (65 to 72°F [18–22°C]) weather encourages this fungus.

HOSTS

Solanaceous crops (eggplant, pepper, potato, tomato) are especially susceptible. Several other vegetable crops can be hosts, including cantaloupe, pumpkin, watermelon, mint, spinach, and strawberry. Some cover crops can be hosts, including common vetch. Weed species such as dandelion, groundsel, lambsquarters, nightshade, pigweed, sagebrush, and shepherd's purse are also hosts to *Verticillium* species.

MINIMAL-IMPACT INTERVENTION

There is no cure for verticillium wilt. Use resistant cultivars. Resistant varieties are available for alfalfa, mint, potato, strawberry, sunflower, tomato, and other crops. Soak seeds in hot water or a 0.5 percent bleach solution before planting. Do not plant out transplants until soil temperatures are 65 to 70°F (18–21°C). Rotate crops on a 4- to 5-year basis with non-susceptible plants such as sweet corn, spinach, beans, peas, grasses, asparagus, carrot, and sweet potato. Rotation and incorporation of cover crops or other organic amendments prior to planting has been shown to reduce verticillium wilt in some crops. Several cover crops or crops, including broccoli, Sudan grass, and various mustards, may suppress verticillium wilt. High-nitrogen fertilizers can increase wilt severity. *Remove and destroy* infected plants. Soil solarization for 4 to 6 weeks in midsummer can reduce but does not completely eliminate *Verticillium* inoculum.

White Rot

White rot (*Stromatinia cepivora*) is favored by cool, moist soil conditions. The soil temperature range for infection is 50 to 75°F (10–24°C), with an optimum of 60 to 65°F (16–18°C). As soil temperatures rise above 78°F (26°C), the disease is naturally suppressed. Unfortunately, the temperature and soil moisture conditions that are good for onion and garlic growth also favor white rot development. Leaves of infected plants turn yellow, wilt, and die back. Older leaves and bulbs collapse with a watery decay.

HOSTS

Onion, garlic, leek.

MINIMAL-IMPACT INTERVENTION

Long rotations may help, but since white rot has long-lived resting structures (sclerotia), once it is present in a field, it is very difficult to grow *Allium* species there successfully. Some work is being done with sclerotial germination stimulants, which are stimulants to get the disease's resting structures to germinate rapidly rather than over a long period of time. In one study, onion oil, garlic oil, and *Allium* crop waste were the most effective treatments. Within 6 months of treatment, more than 70 percent of the sclerotia germinated and then died in the treated plots. But onions and garlic planted in those plots the following year still became infected with white rot.

Physiological Disorders and Nutrient Deficiencies

Sometimes symptoms such as leaf and fruit spots and discoloring are caused not by disease microorganisms, but instead by a plant's physiological response to environmental stress (like too much heat or cold) and to nutrient deficiencies. Refer to "Deciding When a Nutrient Intervention Is Needed" on page 82 for descriptions of crop nutrient deficiency signs in detail and in a larger ecosystem context. Here I explain how to fix the most common nutrient deficiencies and physiological disorders.

Figure 11.10. Blossom-end rot on tomato.

BLOSSOM-END ROT

Scientists used to think that calcium deficiency directly caused the physiological disorder known as blossom-end rot. Calcium can play a part in causing blossom-end rot, but the situation is a bit more complicated than that. A review of the recent literature concluded that calcium deficiency is not the cause but rather a result of blossom-end rot in tomato and pepper fruit.[18] Blossom-end rot is actually a physiological disorder aggravated by several interacting conditions of environmental stress, including drought, high light intensity, heat, ammonium-nitrogen nutrition, excessive nitrogen fertilization, uneven soil moisture, cold stress in the spring, and poor root development caused by an inactive soil organic matter system. Often blossom-end rot occurs even when soil calcium levels are high. If that is the case in your garden or farm soil, adding calcium won't prevent symptoms. The way to manage blossom-end rot is to create an effective soil organic matter system and avoid tomato and pepper crop environmental stress. Blossom-end rot has virtually disappeared on my farm the past 7 years since I began to grow my own fertilizers that are high in carbon. If soil calcium levels are actually deficient, however, calcium addition can help to alleviate blossom-end rot. For example, research indicates that spray applications of calcium chloride solution on a weekly basis reduced blossom-end rot symptoms by 50 percent.[19] Note that calcium chloride is not considered an acceptable fertilizer amendment for use by certified organic farms, with the caveat that "calcium chloride, from brine process is natural and prohibited for use except as a foliar spray to treat a physiological disorder associated with calcium uptake."[20] Unless your fields or garden beds suffer from severe soil calcium deficiency, it does not make much sense to spray calcium chloride on tomato or pepper foliage (see "Calcium Deficiency" below).

BORON DEFICIENCY

Some vegetables, such as brassicas, are heavy boron feeders. You can correct a deficiency with the careful addition of borax to the soil in liquid form (1 tablespoon per 100 square feet [9.3 m^2] in 1 gallon [4 L] water) or 1 tablespoon Solubor in 1 gallon water. Or add boron more slowly by adding kelp (10 pounds per 100 square feet [49 kg/100 m^2]) or wood ashes (note that wood ashes also increase soil pH and soil potassium, sometimes excessively). Foliar seaweed sprays as soon as transplants are set, and repeated several times until head or fruit formation, may help to prevent deficiency problems. While boron is essential for root growth and fruit development, it can become toxic if overapplied. Always test the soil and apply only the recommended amount.

CALCIUM DEFICIENCY

Add gypsum if your pH is above 6.2, or calcitic lime if your soil pH is below 6.2 and a soil test shows that soil calcium is low.

Figure 11.11. Nitrogen deficiency causing yellow, pale leaves on newly transplanted heirloom tomatoes.

IRON DEFICIENCY

In high-pH soils, iron forms chemical compounds that are unavailable for uptake by most plants. High pH-induced iron deficiency is common in high-pH western US soils. You can lower soil pH by adding sulfur. For an emergency intervention, apply a root drench or spray leaves with iron chelate. Iron that has been treated with lignin derived from woody plants is usually allowed by organic certification, but check with your certifier first. Iron that has been chelated with synthetic materials, such as EDTA, is not allowed. Apply 1 tablespoon of chelated iron in 1 gallon (4 L) of water as a soil drench or as a spray on new foliage, in the evening when humidity levels are high.

NITROGEN DEFICIENCY

Nitrogen is the most commonly found nutrient deficiency in vegetable crops. Be careful not to overdo nitrogen, though, because too much nitrogen causes rapid leaf tissue growth with little or no root growth as well as low flower and fruit production. Apply liquid nitrogen sources as a root drench, such as ¼ pound (113 g) of alfalfa meal plus

¼ pound of soybean meal mixed in 20 gallons (76 L) of water. Let sit at 65 to 70°F (18–21°C) for 12 to 36 hours before application. Covering transplants with row cover also helps to warm the soil and allows the plants better uptake of slowly available organic residue sources of nitrogen (see "Helen's Magic Mix Liquid Fertilizer Intervention" on page 87).

PHOSPHORUS DEFICIENCY

Cool spring soils and high or low soil pH can make phosphorus unavailable to plant roots. Sometimes this does not affect yield if it occurs in older plants. But leaf purpling of young plants can decrease yields. To improve phosphorus availability, warm soil with plastic mulch before planting and use row cover after planting. Phosphorus can be added in liquid form as a root drench, such as 2 cups (0.5 L) of rock phosphate mixed in 10 gallons (38 L) of water. Let the mixture sit at 65 to 70°F (18–21°C) for 12 to 36 hours before application. (See "Helen's Magic Mix Liquid Fertilizer Intervention" on page 87.)

SUNSCALD

Sunscald is a physiological disorder caused by sudden exposure of fruits to intense direct sunlight. Rapid growth also encourages sunscald. On hot, sunny days, fruit exposed to the sun becomes extremely hot, in contrast with fruit protected by a dense canopy of foliage. Optimal crop fertilizer and irrigation help develop a thick foliage canopy that protects fruit from direct sunlight. On a small scale, shade cloths that provide 35 percent shade reduce sunscald significantly, especially when plants are in the midst of a quick growth spurt and it suddenly turns very sunny and hot. In a study at the University of Georgia, bell pepper fruit levels of nitrogen, phosphorus, and potassium increased with increasing shade level. And, phytophthora blight in plants and fruit sunscald decreased with shade. However, fruit soluble solids and fruit weight decreased with increasing shade level, too. Shading bell peppers reduces light intensity, both air and soil temperatures, and heat stress in the plants. Balanced nutrition and avoidance of excessive nitrogen also seem to reduce sunscald.

Figure 11.12. Minor phosphorus deficiency shows up as purpling in older leaves of mature fall-planted broccoli.

Figure 11.13. Sunscald on pepper fruits.

CHAPTER 12

Fruit Tree Growing and Troubleshooting Guide

It's no small undertaking to plant fruit trees, but they are more forgiving than annual crops once established. During the establishment phase and as they grow into a garden or farm ecosystem, each type of fruit tree has its own cultural requirements and is susceptible to specific pests. In this chapter I cover the fruit tree crops that I have the most experience with: apples, apricots, cherries, peaches, nectarines, pears, pluots, and plums. Peaches and nectarines are mostly grown in the same ways as pluots and plums, so they will sometimes be lumped together. For beginning growers and to help everyone who is managing an orchard or planning to add or expand one, the chapter begins with general fruit tree management details. I also include a summary of Woodleaf Farm's successful ecological orchard management system, which started in California and evolved when we moved to Oregon. Woodleaf's system in California was developed over 35 years of farming and experimenting, with both mistakes and successes.

In the second half of the chapter, individual crop entries cover temperature, light, soil and fertility, water, and special ecological requirements for optimal growth, as a well as a troubleshooting section that describes symptoms of pests and problems and the probable causes. Use the troubleshooting guide to key out specific problems with your trees, and then refer to chapters 13 and 14, which provide a range of interventions, from minimal impact to heaviest impact, for managing specific pest and disease problems.

General Fruit Tree Management

Fruit trees have specific environmental needs that help them to thrive and suppress pests. Managing these fiddly details may seem like a lot of mental work, but it helps to later avoid all of the extra inputs to manage diseases, insects, and fertility problems.

DECIDING WHAT TO GROW

Consult with other growers, state universities, and local extension agents to learn what species and varieties of fruit trees do well in your area. Species and varieties that thrive in cold northern climates will be very different from those that thrive in hot southern climates.

As you plan where to locate plants, first do enough detective work so that you know what the basic preferences of your fruit trees are. Research the light, moisture, and soil preferences (see below) as well as the disease susceptibility (or resistance) for the trees and rootstocks you plan to grow. Match species, varieties, and rootstocks to the soil, sun, and shade microclimates in your garden or fields.

- Species that require full sun, such as cherry and plum trees, will not grow well in shady locations.
- Avoid planting species that are susceptible to foliar fungal diseases or bacterial canker diseases in lawn areas or near crops that are regularly irrigated. Regular watering of lawn grass or crops provides the higher humidity that many disease-causing microbes require to infect fruit trees.
- Avoid heavy, wet soils for plant species susceptible to root rot diseases.

POLLINATION

Most fruit trees need cross-pollination in order to produce fruit or to produce well. This means that two or more individual varieties with compatible pollen and similar bloom times must be planted near one another. Some fruit species are self-fruitful and do not require cross-pollination. Sour cherry, most apricots, European-type plums (such as Stanley and

Table 12.1. Rootstock Comparisons

Rootstock Name	Size: % of Normal Seedling Tree Height	Suggested Spacing in Feet (Meters)	Pest and Problem Resistance
APPLE			
M.7	40–50	12 × 20 (3.7 × 6.1)	Fire blight, phytophthora root rot
M.106	50–70	12 × 25 (3. × 7.6)	Woolly apple aphid
M.111	70–90	1 × 26 (5. × 7.9)	Woolly apple aphid
Apple seedling or standard-sized clonal rootstock	100	2 × 28 (6.1 × 8.5)	Wet feet and drought, low fertility, phytophthora root rot
Ottawa 3	30–40	12 × 16 (3.7 × 4.9)	Collar rot
PEAR			
Old Home × Farmingdale (OH×F)	70–90	10–1 × 18–20 (3.0–3.7 × 5.5–6.1)	Fire blight
Pear seedlings	100	10 × 28 (3.0 × 8.5)	Fire blight (depending on the seedling source)
PEACH, PLUM, APRICOT			
Lovell	100	12–14 × 20 (3.7–4.3 × 6.1)	Bacterial canker, wet soils
Bailey	90	12–14 × 20 (3.7–4.3 × 6.1)	
Nemaguard	100	12–14 × 20 (3.7–4.3 × 6.1)	Root knot nematodes
Guardian	100	12–14 × 20 (3.7–4.3 × 6.1)	Root knot nematodes
Citation	90–100 for apricots, but 75 for plums	12 × 20 (3.7 × 6.1)	Root knot nematodes
Krymsk 86	75	12 × 20 (3.7 × 6.1)	Phytophthora root rot, root knot nematodes
CHERRY			
Mazzard	90–100	Sweet: 20 × 30 (6.1 × 9.1) Sour: 14 × 20 (4.3 × 6.1)	Root knot nematodes, bacterial canker

Italian Prune), and peach/nectarine are self-fruitful. Apple, pear, sweet cherry, some apricots, and Japanese and American plums are not; they require companion pollinator trees of another compatible variety to produce good yields. Companion pollinator trees should be sited within 100 feet (30 m).

Insects make pollination happen for fruit trees. Honeybees and native bees do the majority of this pollination work, but a diversity of insects visiting fruit tree blossoms helps to transfer pollen throughout the orchard. (Chapter 5 describes ways to create habitat for a diversity of pollinator insects.)

ROOTSTOCKS

For commercial production, most fruit tree varieties are grown as grafted plants. This means that the fruitful variety is "attached" to a rootstock variety through a process called budding. The rootstock variety imbues the trees with characteristics that the fruiting variety may not possess, such as size control, hardiness, and insect and disease resistance. Careful rootstock selection for your soils, climate, and conditions is another way to avoid potential pest and disease problems.

Pest and Problem Susceptibility	Comments
Woolly apple aphid	
Phytophthora root rot	Avoid heavy soils with poor drainage.
Phytophthora root rot	
Fire blight, woolly apple aphid	
Fire blight, woolly apple aphid, apple mosaic virus	Very cold-hardy.
	Cold-hardy, vigorous. Old Home, OH×F 40, OH×F 333, and OH×F 97 are fire-blight-resistant.
	The most common rootstock worldwide is some selection of a Bartlett seedling. Most are cold-hardy.
Root knot nematodes, peach tree short life, bacterial canker	Cold-hardy, best to promote longevity.
Bacterial canker	Cold-hardy; promotes tree longevity.
Bacterial canker, cold conditions, zinc deficiency	
	Moderately cold-hardy.
Bacterial canker	Cold-hardy, tolerant of wet soils.
	Tolerant of wet soils, cold-hardy.
	Cold-hardy.

Apple tree size as mediated by rootstocks is generally divided into three categories: *standard* (also called seedling), *semi-dwarf*, and *dwarf*. Standard trees are budded onto apple seedlings or standard-sized clonally produced rootstocks; they produce large trees that are 20 feet (6 m) tall or more. The most common semi-dwarf apple rootstocks are M.7 or M.7a, M.26, and MM.106 or MM.111. *M* refers to rootstocks that were developed at the East Malling Research Station in England; *MM* rootstocks developed jointly at the East Malling and Merton stations. The MM series are not hardy in zones colder than USDA hardiness zone 4. It is important to remember that shallow, low-fertility, and/or high-pH soils generally result in smaller-than-normal fruit trees, regardless of the rootstock. If your soil is not in great condition, it may not be a good idea to choose a dwarfing rootstock.

For pear, some rootstocks are available for size control, but I recommend choosing fire-blight-resistant pear rootstocks as top priority! Many peach, plum, apricot, and cherry rootstocks are available, but long-term evaluations of these rootstock varieties are still ongoing. In table 12.1, I have included the rootstocks for which the most information is available regarding soil preferences, disease resistance and susceptibly, and cold tolerance.

SPACING

Appropriate spacing for fruit trees depends on the rootstock and the vigor of the variety. It also depends on your farming system. Trees on dwarfing and semi-dwarfing rootstocks can be planted more closely. Generally, it is best for tree health to plan on and provide more than adequate space per tree. If you plan to grow your own fertilizer within your orchard, as described in chapter 4, be sure to leave enough room between trees so that ground cover "fertilizer" and "habitat" plants can grow successfully. When we began to grow our own fertilizer in our Oregon orchard, we increased the spacing both between trees in the row and also created wider row middles. We added 3 to 5 feet (0.9–1.5 m) to row middles and 1 to 3 feet (0.3–0.9 m) between trees in crop rows. We have plenty of land and space, so now I wish we had added even more space to our row middles. Tree health and fruit quality is generally better with more space.

PRUNING

Shape fruit trees early in their lives to establish a central leader, a modified central leader, or an open center framework system. For a central leader system, the main trunk is encouraged to develop, and lateral branches called scaffolds extend out from the leader in three or four well-spaced layers (see figure 12.1). An open center system is a good choice for most stone fruit that want to grow naturally in a vase shape. After heading back the central leader, allow three to five lateral scaffold

Figure 12.1. This cherry tree has naturally upright growth and is pruned to a central leader system.

branches between 2 to 3 feet (0.6–1.2 m) above ground level to develop. (See figure 12.2.) Another choice is a modified central leader system, which is a combination. The central leader is allowed to grow, and lateral branches are also encouraged to grow upright in a vase shape around the leader.

These pruning system frameworks provide light to all parts of the tree and help to support the weight of fruit. Choose well-spaced main scaffolds that branch out at a 45-degree angle from the trunk to provide a strong skeleton for each tree. Then prune lightly annually to remove upright branches, diseased branches, or those too shaded by other branches, and encourage lateral branches to grow out away from the inner space of a tree. Apples and pears prefer a central leader or modified central leader pruning system. Stone fruit species do best with an open center or modified central leader pruning system. How much to prune depends on the species. Ranking in order from most to least vigorously pruned: apple, pear, peach, apricot, cherry, Japanese plum, pluot, and European plum.

FERTILIZATION AND SOIL PH

Fruit trees generally require more nitrogen and phosphorus when they are young and developing their woody skeleton (1 to 8 years old). Fruit-bearing trees require less nitrogen and more potassium, calcium, and micronutrients (especially boron, iron, manganese, and zinc). Most fruit tree species covered in this book prefer a soil pH of 6.0 to 7.5. See the crop entries later in this chapter for more details on specific nutrient needs. In general, peaches and nectarines use the most nitrogen, followed by pears, while apples, apricots, and plums/pluots and cherries use the least amount of nitrogen per year.

Figure 12.2. When this peach tree was young, its central leader was headed back to the top of a whorl of branches to stimulate an open center system. Growth now continues outward along the scaffolds each year.

Woodleaf Farm's Ecological Orchard Management Techniques

At Woodleaf Farm in both California and Oregon the success of the agroecosystem that Carl and I developed depended on managing the details of overlapping ecological relationships:

- A diverse mix of tree species and varieties; year-round soil cover for soil health.
- Sequential ground cover bloom to support natural enemies.
- Regular surface application of plant residues to feed the soil organic matter system.
- Selective, minimal application of minerals to recharge the reservoir of soil nutrients.
- Removal of sources of disease inoculum.
- Pruning to maintain tree shape.
- Bloom sprays of minerals to boost tree health and suppress disease.

The overall goal of our management system is to build the soil organic matter system through the natural process of nutrient cycling and recycling, and also to enhance pest insect and disease suppression. Let's look at each aspect of the management program.

OPTIMIZING DIVERSITY

Our main reason for diversification in the orchard is to support disease and insect suppression while still arranging crop species with the same cultural needs together. Species and varieties are arranged in fields, areas, and rows depending on light, water, and fertility needs and soil preferences. Hence we generally block stone fruits together and apples and pears together into separate areas or fields. The fields at Woodleaf California are small: 1 to 2 acres (0.4–0.8 ha) planted in one genus or similar genera, but with a mixture of species and/or of different varieties (including disease-resistant varieties) within each species. For example, we grow peaches

Figure 12.3. The range of bloom colors shows both in-row and in-field variety diversity at Woodleaf Farm, California. There are several varieties of pink-blooming peaches and nectarines as well as white-blooming cherries and pluots in this field.

Table 12.2. Ripening Sequence for Varieties Grown at Woodleaf Farm (OR)

July: 1st–3rd weeks	July: 3rd–4th weeks	August: 1st–2nd weeks	August: 2nd week	August: 3rd–4th weeks
Springcrest peach Wescot apricot Leahcot apricot Flavor Royale pluot Flavorella pluot Honey May peach	Bella Gold peacotum Zee Pride peach Tomcot apricot Chinese apricot Flavor Delight apricot	Rich Lady peach Harglow apricot Snow Queen nectarine Sugar May peach	Eva's Pride peach Independence nectarine Hermosa aprium	June Pride peach Harken peach Splash pluot Spice Zee nectarine Harko nectarine Sweet Cap Donut peach
August: 4th week	**September: 1st week**	**September: 2nd week**	**September: 3rd week**	**September: 4th week**
Dapple Jack pluot Suncrest peach Reliance peach Halehaven peach Vetern peach Mericrest nectarine Geo Pride pluot	Emerald Gem pluot Flavor Queen pluot Elegant Lady peach Red Globe peach Flavortop nectarine Fantasia nectarine Loring peach Contender peach	Dapple Dandy pluot Crimson Royale pluot Zee Lady peach Baby Crawford peach Red Gold nectarine Mount Royal plum	Autumn Sprite aprium Flavor Grenade pluot Flavor King pluot Honey Punch pluot Ginger Gold apple O'Henry peach	Kaweah peach Emerald Beaut plum Superior plum Buckeye Gala apple McIntosh apple Candy Stripe pluot
October: 1st week	**October: 2nd week**	**October: 3rd week**	**October: 4th week**	**November: 1st week**
Macoun apple Fuji apple Empire apple Mutsu apple	Red Cameo apple Jonathan apple	Red Delicious apple Golden Delicious apple Crimson Crisp apple	Idared apple Spartan apple Rome apple	Warren pear Pink Lady apple Gold Rush apple

(our main market crop) together, but to ensure genetic diversity and disease suppression, we plant many different peach varieties, mixing them up within the rows, so there are no more than 5 to 20 trees of the same variety in a row within any of the small fields. The usefulness of in-row variety mixtures for disease management has been well demonstrated for several diseases (including apple scab) and is becoming more widely practiced as a disease suppression tool.[1] We also add a row or two of cherries, plums, or pluots to a field of peaches (see figure 12.3), but generally we block together the species of tree fruit crops that have the same water, light, and soil fertility needs.

Woodleaf Oregon has fruit from three different genera with 85 varieties of 10 different species: sweet cherry (*Prunus avium*), sour cherry (*P. cerasus*), apricot (*P. armeniaca*), European plum (*P. domestica*), Japanese plum (*P. salicina*), pluots (*P. salicina* × *P. armeniaca*), peach/nectarine (*P. persica*), European pear (*Pyrus communis*), Asian pear (*Pyrus pyrifolia*), and apple (*Malus domestica*) arranged with in-row and in-field diversity, including disease-resistant varieties and a succession of varieties with different ripening dates to extend the harvest season (see table 12.2). It took Carl and me a long time to work out the timing precisely so that we had fruit ripening continuously over the longest

Figure 12.4. Year-round ground cover builds both soil health and habitat for natural enemies like this praying mantis hunting in a peach tree at Woodleaf Farm, Oregon.

period of time, and not too much ripening all at one time. The ripening times listed in table 12.2 are unique to eastern Oregon and change with annual weather conditions, but the ripening sequence of these varieties in relation to one another should be similar no matter where they are grown.

KEEPING SOIL COVERED

Keeping the soil covered year-round is essential to maintain orchard soil health and ecosystem function. The perennial living mulch ground cover at Woodleaf in California and Oregon consists of grasses, clover, and weeds beneath tree fruit crops to ensure the presence of growing roots at all times.

Table 12.3. Woodleaf Farm (CA) Living Mulch Bloom Sequence*

Plant Name	Month and Week								
	FEB W3	FEB W4	MAR W1	MAR W2	MAR W3	MAR W4	APR W1	APR W2	APR W3
Oxalis				B1	B1	FB	FB	FB	FB
Purple deadnettle	B1	B1	FB	FB	FB	FB	FB	FB	S
Chickweed	B1	B1	FB	FB	FB	FB	FB	FB	S
Wild mustard					B1	B1	FB	FB	FB
Bindweed									B1
Canada thistle									
Ragweed				B1	FB	FB	FB	FB	S
Dandelion			B1	FB	FB	FB	FB	FB	FB
Speedwell	B1	B1	FB	FB	FB	FB	FB	FB	FB
Pennycress	B1	FB	S	S	S	S	S	S	S
Yellow dock						B1	B1	B1	FB
Sow thistle			B1	B1	B1	FB	FB	FB	FB
Buttercup		B1	B1	FB	FB	FB	FB	FB	FB
Common plantain						B1	B1	B1	B1
Wild strawberry	B1	FB	FB	FB	FB	S	S	S	S
Shepherd's purse	B1	FB	FB	FB	S	S	S	S	S
Vetch						B1	B1	FB	FB
Clover			B1	B1	FB	FB	FB	FB	FB
Perennial grasses						B1	FB	FB	FB

* B1 = First Bloom. FB = Full Bloom. S = Seeding.

The ground cover has a diversity of species, growth forms, root architecture, root depth, and root exudates to entice a wide range of microbial species.

YEAR-ROUND HABITAT

As described earlier in the book, we have succeeded in relying on biological controls rather than pesticidal sprays in our orchard. Table 12.3 details the living mulch plants that provided shelter, pollen, nectar, seeds, and sap for beneficial birds, bats, predators, and parasites during nearly every week of the year at the Woodleaf orchard in California. In the Oregon orchard we have many similar, but also some different living mulch plants. (Chapter 4 provides more details about which plants encourage which natural enemies.)

Some orchardists worry that allowing vegetation to grow in and between tree rows may reduce tree fruit yields. That has not been my experience in Montana, California, or my Oregon orchard. And in a Michigan State University study over a 5-year period, fruit yields were not significantly lower in orchards with irrigated ground covers than in bare soil orchards using herbicides to manage weeds. Research also supports Woodleaf Farms' finding that blooming ground cover builds habitat for beneficial insects and decreases pest damage. In a 3-year study in peach orchards in two provinces in

APR W4	MAY W1	MAY W2	MAY W3	MAY W4	JUNE W1	JUNE W2	JUNE W3	JUNE W4	JULY, AUG, SEPT
FB	FB	FB	FB	FB	FB	FB	FB	FB	S
S									
S									
S	S	S	S						
B1	FB	FB	FB	FB	FB	FB			
B1	FB	FB	FB	FB	FB	FB	S		
S	S	S	S						
FB	FB	FB	S	S	S				
FB	S	S	S						
S	S								
FB	FB	FB	FB	FB	S	S	S	S	S
FB	FB	FB	FB	S	S	S	S		
FB	FB	FB	FB	FB	FB	S	S		
FB	FB	FB	FB	FB	FB	FB	FB	S	S
S									
S	S	S	S						
FB	FB	FB	FB	S	S				
FB	FB	FB	FB	S	S	S	S		
FB	FB	FB	FB	FB	FB	FB	FB	FB	S

Figure 12.5. In early July the living mulch in our no-till orchard at Woodleaf Oregon is a lush mass of blooming grasses, clovers, and weeds providing habitat for beneficial organisms.

China, a white clover ground cover promoted an abundance and diversity of predators and reduced the number of peach pests in tree canopies compared with bare soil row middles.[2] Ground cover has also been reported to increase orchard spider populations and decrease apple aphid abundance.[3] Mowing selectively or minimally is also vital. Reduced mowing increases the areas of undisturbed orchard ground cover that spiders need.[4]

REDUCING TILLAGE

Fruit trees hate root disturbance, so they love the perennial living mulch in the Woodleaf California and Oregon orchards. I don't do any cultivation to control weeds, and I do not till in the organic residues that result from mowing. Hence, the soil microbial community is able to establish root associations with both the fruit trees and the ground cover plants, providing an ecosystem that is reminiscent of that in a natural forest ecosystem, where the lack of disturbance allows co-evolution of

Figure 12.6. In the fall Carl is applying hay mulch from another Woodleaf Oregon field to the spot where each fruit tree will be planted the following spring.

Figure 12.7. At Woodleaf Oregon we mow selectively. Here, the living mulch is cut short near the trees, with early-summer grasses and weeds in the row middle left unmowed.

the plants, soil microbes, insects, birds, and animals in the system. We used to till when it was time to replace old peach trees (after about 20 years). But our new orchard in Oregon was planted using no-till methods: hay mulch was applied to the tree-planting holes to be for 6 months over the winter instead of tillage.

ADDING CARBON

As discussed in earlier chapters, surface application of plant residues is very important in the Woodleaf system. A carbon-rich soil organic matter system is even more important for perennial tree fruit than it is for vegetable crops. One reason is that regular carbon additions enhance microbial communities, especially fungal communities, including mycorrhizae. The orchard microbial community is dominated by fungi, which flourish during early stages of organic residue breakdown. Hence I maintain some orchard areas with decomposing residues at all times. To maintain nutrient balance, I regularly apply both heavy-carbon and light-carbon organic residues. This prevents nitrogen and phosphorus immobilization, which could be a problem if too much heavy carbon was applied all at once.

Mowed Living Mulch

At Woodleaf California, the living mulch understory seed mix originally included low-growing, shade- and drought-tolerant grasses as well as 5 percent New Zealand white clover (*Trifolium repens*). Over time we observed that the ground cover morphs naturally into a mix of grasses, clover, and weeds. After 5 to 10 years the living mulch becomes approximately 70 percent grass species (some taller grasses move in) and 30 percent broadleaf weeds and clover. This is part of the natural movement of biological systems discussed in chapter 1. Dominant grass species in our California orchard's dynamic equilibrium stage of development include orchard grass (*Dactylis glomerata*), California brome (*Bromus carinatus*), Blando brome (*Bromus mollis*), and tall fescue

Figure 12.8. In this late-summer scene, what was a tall living mulch between rows of pear trees has become mowed surface residue, equivalent to 1 to 2 tons of hay mulch per acre (907–1814 kg/0.4 ha).

(*Festuca arundinacea*). Table 12.3 lists the dominant broadleaf species in the dynamic equilibrium stage.

The living mulch is selectively mowed two to four times annually, leaving at least 30 percent unmowed and undisturbed at any given time. The height of the living mulch before each mowing ranges from 8 inches to 3 feet (20–91 cm). On average, the mowed living mulch provides 2 to 3 tons (1.8–2.7 metric tons) of material each year (dry weight) added to the soil surface. The estimated nutrient content of 2 tons (1.8 metric tons) of this type of hay mulch is as follows: 50 pounds (22.7 kg) total nitrogen, 22 pounds (10 kg) phosphorus, 76 pounds (34.5 kg) potassium, 8 pounds (3.6 kg) sulfur, 8 pounds calcium, 3 pounds (1.4 kg) magnesium, 0.2 pound (91 g) manganese, and 0.12 pound (54 g) zinc. Beneath the ground surface, the mulch also contributes some nitrogen through nitrogen fixation, and supplies carbon through the sloughing off of root material over time.

A 2001 study at Washington State University found that mowing clover living mulch growing beneath young apple trees resulted in 46 percent of the total clover nitrogen released into and cycled through the orchard soil each year. The mowed clover living mulch also reportedly contributed to enhanced tree growth and fruit yield. Other species contribute different average nitrogen amounts; for example, 47 pounds per acre (52.6 kg/ha) per year for alfalfa but only 24 pounds per acre (26.9 kg/ha) per year for a mixed grass mulch.

Chipped Branch Prunings

Carl and I pruned orchard trees once or twice per year, and added the high-carbon pruned branches to the soil. In our California orchard we left the young branches in row middles after pruning. Branches were ½ inch (13 mm) to a little more than 1 inch (25 mm) in diameter. Tree branches have a higher carbon-to-nitrogen (C:N) ratio and degrade more slowly than succulent mowed living mulch.

In my Oregon orchard I prune once, usually in late winter. I place the pruned branches in the row middles. I use a rotary tractor-mounted mower to break up the branches and then mow again with a small riding mower and blow the chipped material beneath the trees. Researchers have found that fallen tree leaves and pruning wood contain significant amounts of nitrogen that is slowly recycled back to the fruit trees.[5] Thus this practice is an important part of maintaining the orchard's soil organic matter system.

In my Oregon orchard soil organic matter system, nutrient cycling and recycling occurs through mowing and blowing into tree rows: fallen leaves, tree prunings, and living mulches. Fruit tree leaves, pruned-off branches, and mowed ground cover contain significant amounts of nitrogen, potassium, phosphorus, and micronutrients. Recycling plant residue within the orchard makes mineral nutrients available again for tree root uptake and also maintains them in the soil microbial community.[6] Maintaining an active soil microbial community in turn recycles nutrients back to fruit trees. In fact, recent evidence suggests that mycorrhizal fungi associated with tree roots can help trees take up both inorganic and organic forms of nitrogen directly. However, both tillage and nitrogen fertilization limits this ecological process.[7]

Compost

At Woodleaf in California, we also applied chipped branch wood from off the farm as yard-waste compost. The compost consisted of grass clippings and branch/leaf prunings that had been aerobically composted with a bed turner. We applied compost one to three times per year, usually in spring and/or fall. The rate in the early years of the farm was 4 to 6 tons (dry weight) per acre (3.6–5.4 metric tons/0.4 ha) per year, reduced in later years to 2 tons (dry weight) per acre (1.8 metric tons/0.4 ha). We used a compost spreader to apply the material to row middles. Some was blown under trees then irrigated into the living mulch immediately following application. On my farm now, I am watching my trees carefully to see how long I can get away without adding off-farm fertilizer except the annual Mineral Mix Bloom Spray (described in table 7.3). As of this writing, it has been 6 years.

BALANCING SOIL MINERALS

Soil mineral (plant nutrient) balancing was a foundation of Woodleaf's soil management system in California. It was based on 20 years of Carl's farming experience and on research and recommendations by soil fertility specialist Neal Kinsey. Of course, the basis of the system was application of plant-based organic residues, but Carl also supplemented with off-farm, purchased minerals when soil

Figure 12.9. Fallen tree leaves and pruned branches contain significant nitrogen and other nutrients. I use a rotary tractor-mounted mower to chip the branches, and all the nutrients are slowly recycled back to the trees.

test results indicated they were needed. Carl considered two factors when interpreting soil test results: *levels* of specific minerals (such as total ppm soil phosphorus) and *ratios* of minerals to one another.

Mineral Balancing

Mineral balancing is a soil test interpretation and fertilization theory based on the work of Dr. William Albrecht. The focus is on the importance of a balanced plant nutrient equation; it is the ratio of one element to another that counts, rather than the absolute amount of any particular nutrient. For example, in a soil with calcium levels of 1,700 ppm, calcium may be rated as high. But if potassium levels are excessive in this same soil (900 ppm), a calcium level of 1,700 ppm may actually be only medium, and can even result in calcium deficiency signs in crops, such as blossom-end rot, tipburn, or cork spot in apple. Dr. William Albrecht proposed a general standard in 1940 for agronomic crops. He theorized that that there is an approximate ratio of basic cations that must occupy a soil's cation exchange capacity or plant growth will be limited. (Cation exchange capacity is a measure of how many positively charged cations can be held by the negatively charged particles in a soil.) Dr. Albrecht suggests that potassium should be 3 to 5 percent of the total soil cation exchange capacity, magnesium 10 to 20 percent, calcium 60 to 70 percent, and hydrogen 10 to 15 percent.

Carl also paid particular attention to the ratio of base cations (positively changed mineral ions) to one another, especially the ratios of calcium to magnesium to potassium. Sodium and hydrogen are other base cations present in soil, but they are not as important to focus on for soil and plant health. Soil test results report the quantity of these individual positively charged soil minerals as *cation base saturation*, which is the sum of all the base cations held on to the soil particles divided by the total cation exchange capacity and expressed as a percentage. The level of any individual base cation, such as calcium, can be expressed as a percentage of the total cation base saturation. Carl believed that there is an approximate ideal ratio of basic cations in soil that results in optimal plant growth (see the sidebar). Hence he applied minerals based on ratios of calcium, magnesium, and potassium in Woodleaf's soil. Carl also considered the ratio of other mineral nutrients to one another. For micronutrients such as boron, the ideal amounts are expressed as parts per million, because they are needed in such minute quantities in comparison with the macronutrients like calcium and magnesium.

Carl surface-applied off-farm minerals when soil analysis showed nutrient levels below his target levels. Carl's general mineral targets were:

Potassium	4–7 percent of cation base saturation
Magnesium	15 percent of cation base saturation
Calcium	68–75 percent of cation base saturation
Boron	0.8 ppm
Sulfur	20 ppm
Manganese	15 ppm

When Carl and I started our Oregon orchard and began to experiment with growing our own fertilizer, our goal was to reduce the addition of mined minerals—to see how little supplementation we could get away with. This experiment is ongoing. At the time of this writing, we applied minerals only once: a pre-plant application to the fruit tree planting hole area. Overall, only 8 percent of land in the orchard has been covered with supplemental minerals, within a 9-square-foot (0.8 m^2) area for each newly planted tree. Before we planted trees, we took soil samples from the orchard site, and then Carl formulated a mineral blend based on the mineral levels already present in the soil. One pound (0.45 kg) of Carl's soil mineral blend was applied to each tree, consisting of:

Soft rock phosphate	0.40 pound (181 g)
Elemental sulfur	0.083 pound (38 g)
Sulfate of potash	0.083 pound (38 g)
Manganese sulfate	0.041 pound (19 g)
Iron sulfate	0.083 pound (38 g)
Zinc sulfate	0.0082 pound (4 g)
Boron	0.0041 pound (2 g)
Azomite*	0.21 pound (95 g)
Nutramin*	0.041 pound (19 g)

Since 2016, I have not added any further minerals, except a little manganese in 2022 to peaches only, and the orchard is still producing good yields, great fruit quality and color, and excellent taste. This system takes precise monitoring, however. Here is an example of how closely I monitor the trees in order to practice my minimal fertilizer input system. In July 2022, during an unusual and early heat wave, some of the peach trees developed minor manganese deficiency symptoms in early summer. I immediately applied manganese in a granular form to the soil beneath each peach tree and also made a liquid drench of manganese plus seaweed to apply to the trees showing foliar manganese deficiency symptoms. By harvest in early August, the trees that had been showing symptoms looked healthy and produced a high yield of tasty peaches. It is difficult to test empirically whether mineral recycling is actually occurring in the orchard, but the tangible results are a good sign that the system is working.

* Azomite and Nutramin are natural products that contain a broad spectrum of minerals and trace elements.

SANITATION

Removing diseased fruit/mummies and pruning out diseased branches is vital. During the growing season, we squish diseased fruit into the ground cover and rely on our active microbial community and ground-dwelling insects to "digest" the disease microorganisms and thus decrease disease inoculum. Carl used to say that when the soil fungal webs are active and happy, they eat up the diseased fruits quickly. In October, when soil microbes and decomposer organisms slow down, we also remove mummy fruits from the trees and destroy them. We do a late-fall mowing of the entire orchard floor to grind up any leaves with disease inoculum and recycle nutrients and also to discourage voles from overwintering in the orchard.

PRUNING

Regular pruning is important to maintain the best tree shape for each species. In California we did light pruning twice per year, in the summer after fruiting and in late winter/early spring. In Oregon I prune once per year in late winter/early spring. Timing is vital. Apples and pears are pruned first each winter, followed by European plums, pluots, Japanese plums, apricots, peaches, and nectarines. Cherries are pruned last to avoid increased susceptibility to diseases, because pruning wounds that provide an entry point for disease organisms will heal more quickly as the weather warms in early spring.

FOLIAR SPRAYS

In chapter 7 I discussed the use of Carl's Mineral Mix Bloom Spray, a blend of minerals, kelp, and rock powders, for disease suppression and to shift the balance from disease-causing foliar microorganisms to beneficial leaf microorganisms. It was part of a slow evolution from a belief in the necessity of killing disease organisms in the orchard to our systems approach of suppressing disease organisms while simultaneously encouraging beneficial soil, leaf, and bloom microorganisms and optimal tree health.

Carl's interest in phasing out use of fungicides dates back to the early 1990s. At that time, brown rot was a serious problem at Woodleaf California and for all organic peach and stone fruit farmers. The only allowable options for organic disease management were copper and sulfur fungicides. Carl tested his soil every year and was concerned by an increase in soil copper levels. He worried about how the increased copper might affect soil mineral balance and fungal food webs. He also wondered what might be happening to general orchard ecology as a result of so much copper and sulfur spraying.

Carl and Woodleaf California received funding in 1993 and 1994 from the Organic Farming Research Foundation to test materials for brown rot disease control. He tested individual materials and mixes of materials he thought might help to suppress brown rot fungi, comparing their effectiveness with that of copper and sulfur, the organic brown rot control standards.

Results of 1993 Trials

Figure 12.10 shows the results and gives the details of Carl's 1993 spray evaluations. The best three treatments in order of effectiveness were:

1. A mix of kelp and basalt rock dust.
2. Kelp sprayed alone.
3. A mix of kelp, basalt rock dust, compost tea, sugar, white wine vinegar, and hydrogen peroxide plus *Aureobasidium pullulans* yeast that was found on the leaves of Woodleaf Farm's peach trees and cultured by a microbiologist at California State University–Chico.

Carl tested 10 materials: the ones listed above and also copper hydroxide, wettable sulfur, and Farewell (a microbial organism). To make the compost tea,

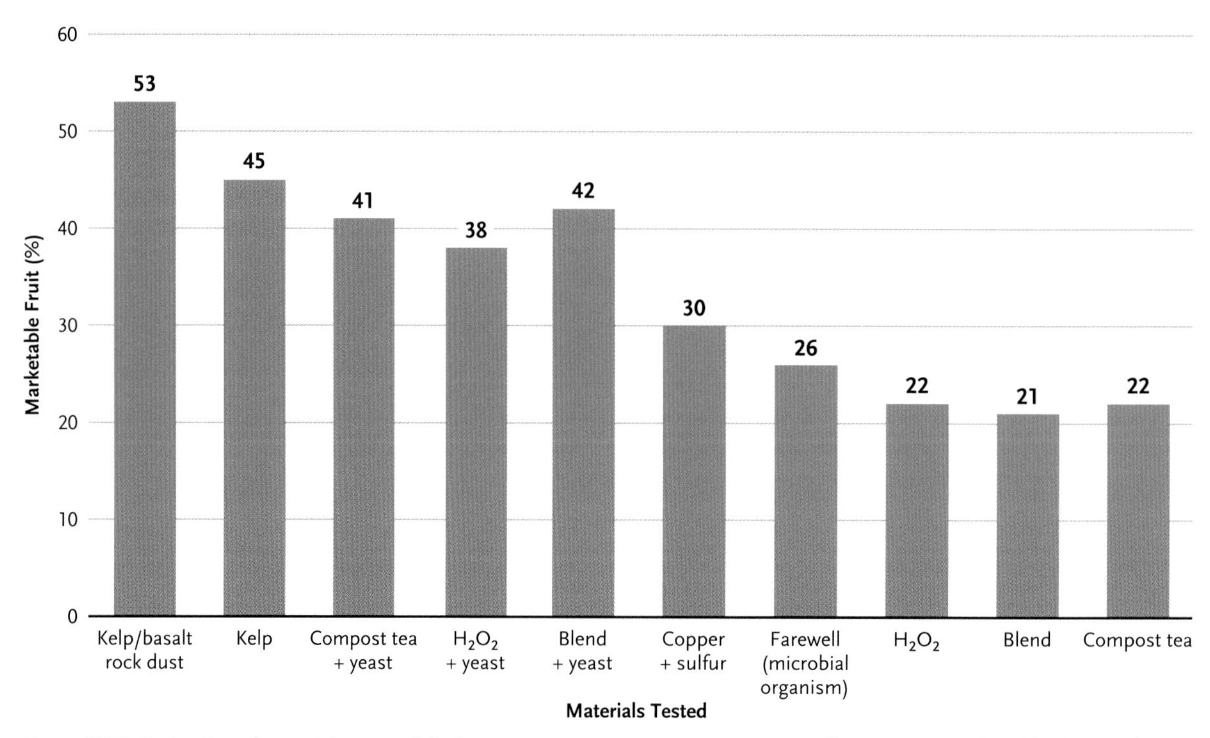

Figure 12.10. Evaluation of materials sprayed for brown rot management on O'Henry peaches: percent marketable, disease-free fruit (percent of total harvest) with no disease per each tested material(s).

10 gallons (38 L) of finished compost was stirred into 55 gallons (208 L) of water for 5 minutes, left to settle for 15 minutes, and applied at 80 gallons per acre (748 L/ha). Except for copper, sulfur, and yeast, each material was applied three times per week at bloom and four times per week prior to harvest. Copper hydroxide was applied once just prior to bloom (15 pounds per acre [17 kg/ha]). Sulfur was applied three times during bloom and three times in June following three rainfall events (15 pounds per acre). The copper and sulfur schedule was Woodleaf California's standard spray program in the 1990s. Yeast was sprayed three times during bloom. Applications were made on 15 contiguous (non-randomized) trees per treatment, for a total of 150 trees.

Carl was surprised by two revelations revealed by the study data: First the *amount* of fruit he was actually losing to brown rot disease during rainy years, and second that the copper/sulfur sprays did *not* provide the best results for managing brown rot disease.

Results of 1994 Trials

Figure 12.11 shows the results of Carl's 1994 spray evaluations, in which he tested the top-performing materials from 1993 individually and in a mix against the standard control, sulfur. By this time he had decided that there was no good reason ever to spray copper in his orchard. In the 1994 trial, the best treatment produced almost 90 percent disease-free, marketable fruit. The best treatments in order of effectiveness were:

1. A mixture of basalt rock dust, *Aureobasidium pullulans* yeast, and kelp.
2. Basalt rock dust alone; also, sulfur alone.
3. *Aureobasidium pullulans* yeast alone.
4. Kelp alone.

The mixture of basalt rock dust, yeast, and kelp, with other additions, was to become Carl's Mineral Mix Bloom Spray that we used for 20 years and that I still use (with some modifications).

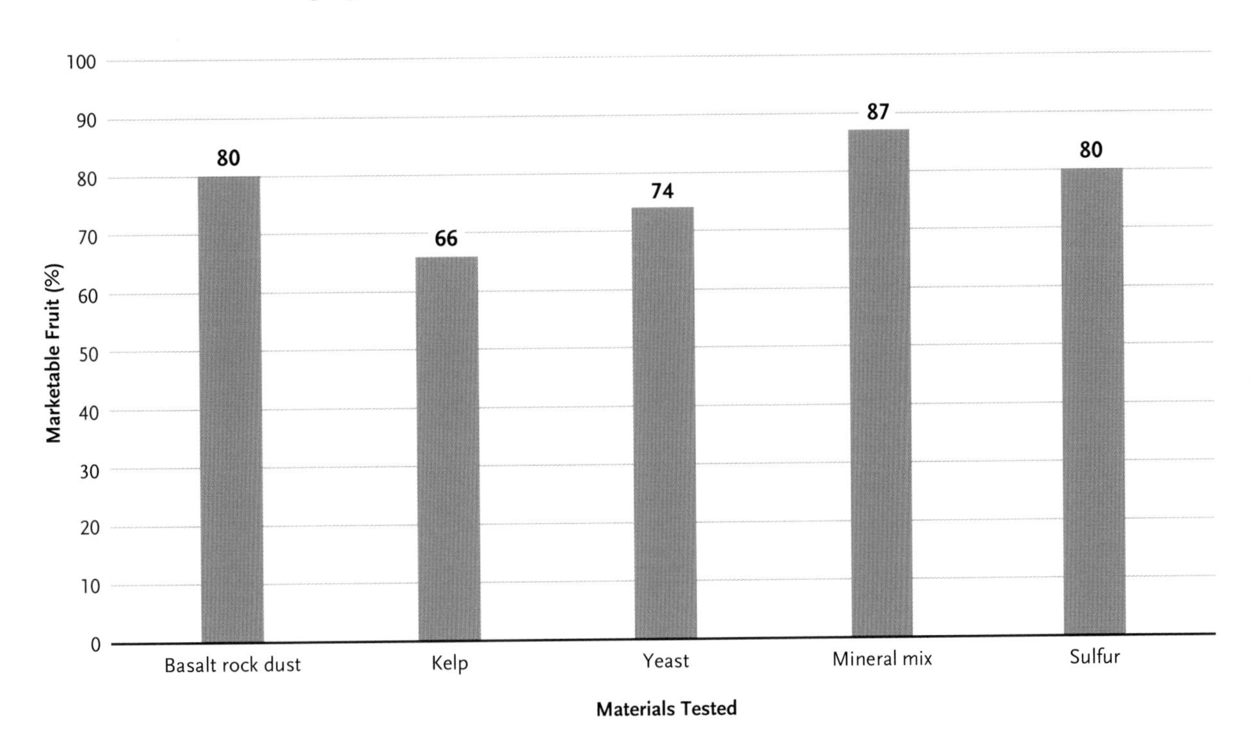

Figure 12.11. Evaluation of materials sprayed for brown rot management on Royal Glo peaches in 1994: quantity of fruit (percent of total harvest) with no disease per each tested material(s).

Figure 12.12. These peaches at Woodleaf Farm are already deeply colored, even though they are still ripening and gaining in size. A lab analysis of tissue from these fruits showed that they were high in macro- and micronutrients.

The Power of Bloom Sprays

It took us nearly two decades to fully understand the implications of the data Carl collected in his 1993 and 1994 trials and how to apply those findings by adjusting our methods to shift the ecological balance away from disease-causing microorganisms and toward beneficial disease-suppressing microorganisms.

In the 1994 study sulfur sprays resulted in almost the same percentage disease-free fruit as the mixture treatment and the same percentage as the basalt rock dust treatment. Carl wanted to stop using high rates of sulfur to manage brown rot disease. Based on that data, Carl decided to add sulfur at lower rates to his Mineral Mix Spray, particularly during years when it was wet during peach bloom. Now I use very low rates of sulfur in my Mineral Mix Bloom Spray (one-third to one-half the rate per acre/hectare recommended to manage fungal diseases). By the late 1990s researchers had learned more about how brown rot disease develops, and they began recommending bloom sprays to control it. So Carl began to spray his Mineral Mix one to four times only at bloom, depending on the weather (more frequently when conditions were rainy or cloudy and humid). Once Carl stopped spraying copper and higher levels of sulfur, he found naturally high levels of *Aureobasidium pullulans* yeast on the leaves in the orchard, so he stopped adding this beneficial microorganism to his Mineral Mix Spray. Our theory was: Create habitat for beneficial leaf organisms, refrain from spraying copper and high rates of sulfur to avoid killing them, and the organisms will come to flourish naturally on foliage and blossoms.

Before Carl gave up using copper and higher levels of sulfur, he tested the effect of the spray materials he was evaluating on the native *Aureobasidium pullulans* yeast he had found on his peach tree leaves, reporting the result as number of

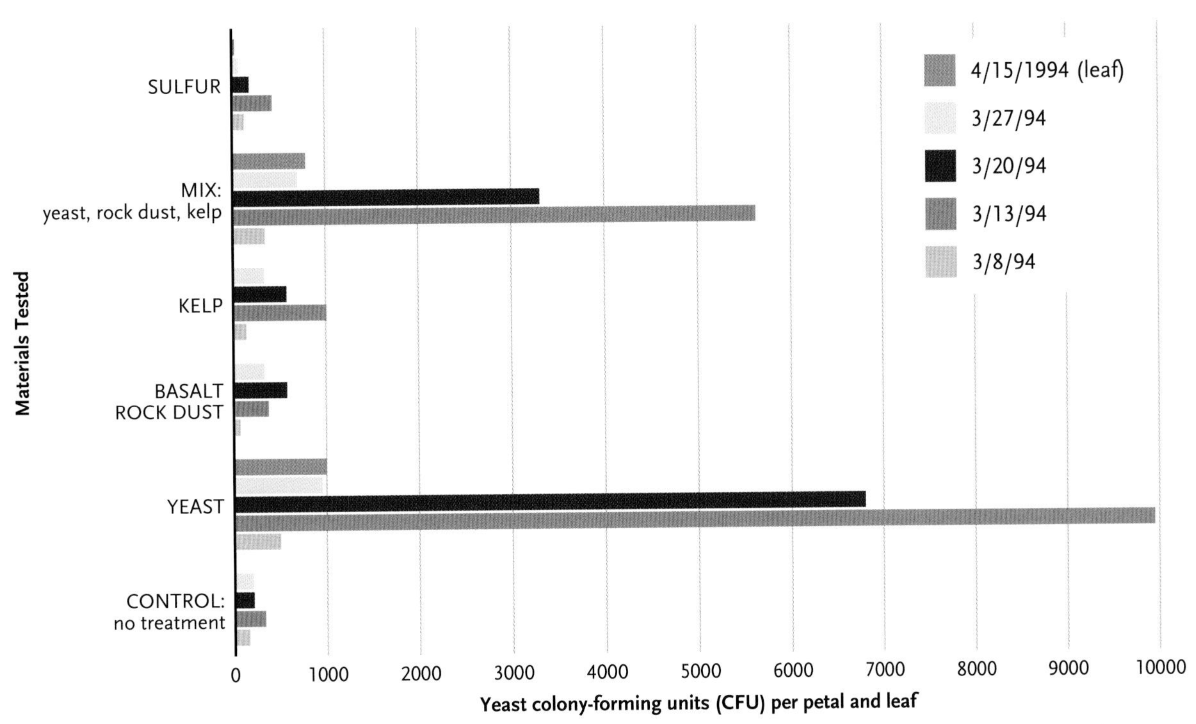

Figure 12.13. Abundance of yeast colony-forming units (CFU) per petal and leaf in an evaluation of spray materials tested at Woodleaf Farm in 1994 for their effect on levels of blossom and leaf *Aureobasidium pullulans* yeast.

yeast-colony-forming units (see figure 12.13). With help from the microbiologist at Chico State University, Carl found that spraying sulfur suppressed the yeast on peach blossoms, compared with the no-spray control, on three of the four sample dates. (In other words, the sulfur was killing the yeast.) The treatments that had yeast added resulted in many more yeast colonies on both blossoms and leaves for all sample dates. But the kelp treatment and even the basalt rock dust also resulted in higher numbers of yeast colonies than the no-spray control (and the sulfur treatment) on three of the four sample dates. Hence Carl was able to scientifically prove that the materials that suppressed the brown rot fungi better than or as well as the sulfur did not suppress the beneficial *Aureobasidium pullulans*. This finding helped Carl to develop his present-day version of the Mineral Mix Bloom Spray and gave us the courage to try to stop depending on sprays to preventively manage disease.

Our farm practices over the past 20 years continue to support Carl's data, and the evidence is especially strong from the past few years, when I have stopped spraying all materials (including a dormant-season lime sulfur spray) except Carl's Mineral Mix Bloom Spray. Both this data and the 20 years of field experience support our ecological theory about suppressing rather than killing disease-causing microorganisms in order to maintain a community of beneficial leaf and soil microorganisms. Just as I said in relation to insect pest management in chapter 6, when we kill something, we inherit its job.

We also wondered how the Mineral Mix Bloom Spray might affect the nutrient levels in peach leaves. Fortunately, Carl had taken extensive leaf tissues samples for analysis as part of his on-farm research in 1994. In 2014 we took another set of extensive leaf tissue samples from the same variety of peaches in the same field (some of the same trees even) and compared 1994 to 2014 samples for macro- and micronutrient content (see figure 12.14). All leaf macronutrient levels increased and all but sulfur were in the normal sufficiency range. Even leaf nitrogen levels increased and were in the normal range, though we had not added any nitrogen fertilizers, except mowed living mulch and yard-waste compost, since the early 1990s.

Leaf micronutrient levels had also increased, except iron, copper, and zinc (see figure 12.15). We were pleased that foliar copper levels had decreased and hoped that was partially due to no copper spraying after 1993. We were a bit disappointed to learn that boron, manganese, and zinc were still below normal levels for peach leaf tissue, despite mineral sprays and mineral applications to the soil. Clearly our peaches use high quantities of these micronutrients. And still today, the intricacies of managing these kinds of relationships continues to keep me questioning.

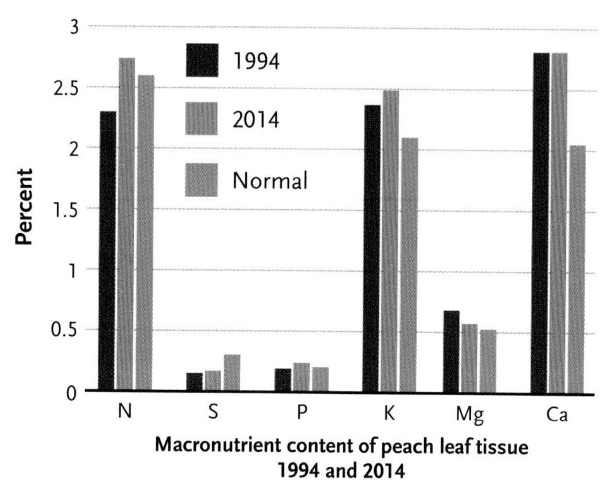

Figure 12.14. Woodleaf Farm O'Henry peach leaf analysis in 1994 and 2014: Macronutrients (percent) compared with normal sufficiency levels for peach leaves.

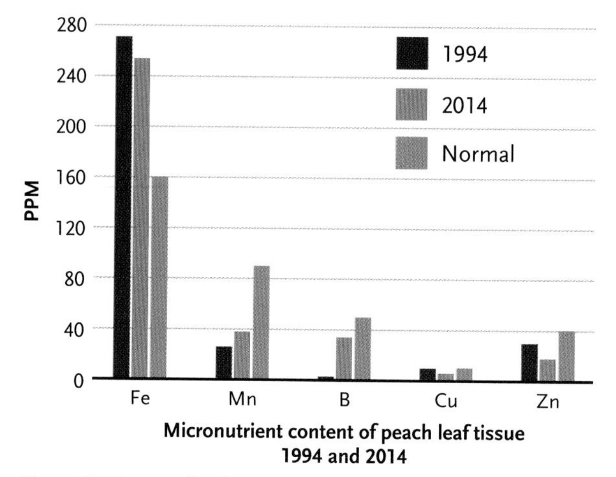

Figure 12.15. Woodleaf Farm O'Henry peach leaf analysis in 1994 and 2014: Micronutrient levels (PPM) compared with normal sufficiency levels for peach leaves.

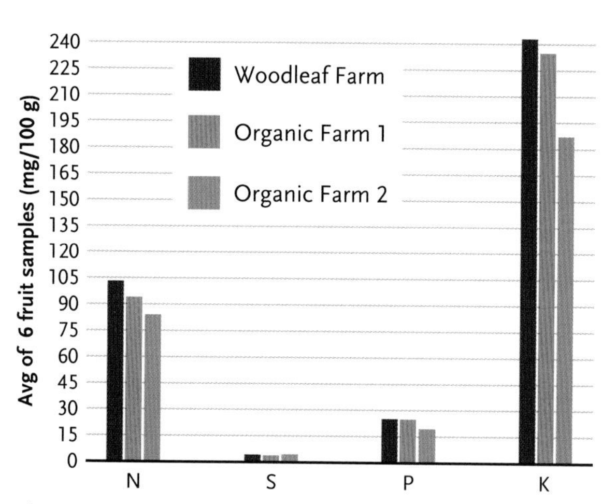

Figure 12.16. Woodleaf Farm O'Henry peach fruit tissue macronutrients levels in 2014 compared with O'Henry fruit from two California certified organic peach orchards. '

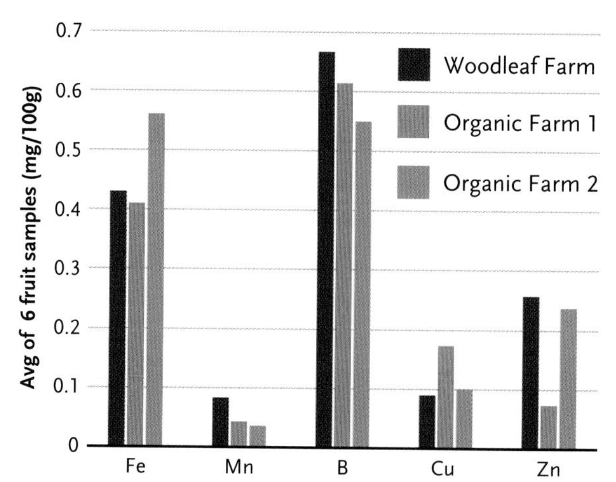

Figure 12.17. Woodleaf Farm O'Henry peach fruit tissue micronutrient levels in 2014 compared to O'Henry fruit from two California certified organic peach orchards.

Figure 12.18. We mow as needed around micro-sprinklers for efficient irrigation, but without disturbing blooming habitat for natural enemies.

Our leaf analysis data led us to wonder how our low-nitrogen, high-carbon soil organic matter system affected fruit mineral nutrient levels. So we did a mini-experiment in August 2014. Carl randomly chose six individual O'Henry fruits from our farmers market pack-out. He also chose six O'Henry fruits from two other long-term certified organic orchards who sell at the same market. We sent samples from the peaches to the lab for analysis. Figure 12.16 shows macronutrient levels in the peach fruits from Woodleaf and the two other organic orchards.

Woodleaf peaches had the highest nitrogen, potassium, and phosphorus levels, despite not adding nitrogen fertilizer for 20 years, except from our mowed living mulch and yard-waste compost. One of the other farms had slightly higher sulfur levels.

Figure 12.17 shows micronutrient levels in the peach fruits. Woodleaf peaches had the highest manganese, boron, and zinc levels, probably due to our mineral mix bloom sprays. One of the other farms had the highest peach tissue iron levels, and both had higher copper levels than Woodleaf. We were pleased that copper levels in our fruit were lowest. It was nice to see that our system was actually showing results inside the fruit as well as on an agroecosystem level. Carl had always assumed the two were inextricably connected!

OTHER RELATIONSHIPS

In California we used micro-sprinklers to irrigate the orchard because these minimize fruit and foliage wetting, which is important for discouraging development and spread of disease. Drip systems and soaker hoses are preferred to overhead sprinklers. To encourage quick drying of foliage, water in the late morning or early afternoon (after 10 A.M., if possible, and before 3 P.M.). Insect-eating birds also like the micro-sprinkler in the summers.

At Woodleaf Oregon, the main form of rodent suppression is my pet rescue cats. Carl's goal was to keep one cat per acre (0.4 ha) of orchard trees, but I often fall short of this ratio, so perhaps the one-cat-per-acre policy is not necessary.

Apple

Apples are one of the easier, less demanding fruit trees to grow. Good sanitation (as described in the "Sanitation" section on page 277), fruit thinning, and maintaining an active soil microbial community at the base of apple trees is vital to better suppress disease and insect pests.

CROP MANAGEMENT

Temperature

Apple trees are one of the most winter-hardy fruits, though some varieties are much hardier than others. Varieties hardy to zone 3 include Honeycrisp, Honeygold, Lodi, Northern Spy, and Frostbite. Varieties hardy to zone 4 include Cortland, Empire, Enterprise, Freedom, Golden Delicious, Red Delicious, Liberty, Paula Red, Red Rome, Arkansas Black, and Spartan.

See table 12.4 for additional apple hardiness information.

Light

Apple trees grow best in full sun (6 to 8 hours of daily sunlight during the growing season). They are more disease-susceptible if grown in partial shade, and fruit is often small and poorly colored with insufficient sunlight.

Soil and Fertility

Apples are less demanding of nutrients, especially nitrogen. In fact, because nitrogen encourages rapid growth, excess nitrogen can increase apple susceptibility to fire blight and aphids.

Water

Apples require good amounts of irrigation, but do best with drip or micro-sprinkler irrigation that keeps water off foliage and fruit. Apples are more tolerant of heavy soils and wet roots than peaches, plums, cherries, or apricots, but they prefer well-drained soil.

Special Ecological Preferences

Prune apples to a central leader or modified central leader. Remember that a lot of heavy pruning encourages excessive shoot growth and fire blight disease. Shape trees when they are young. Then prune young trees (up to 10 years of age) lightly, especially if they are on semi-dwarfing rootstocks. You can prune older trees more vigorously, especially if they are on seedling, standard rootstocks. Birds are much less likely to damage ripe apples than ripe cherries. I encourage birds in the apple and pear areas because they are important codling moth predators.

Codling moths cause the dreaded "wormy apple." With the exception of Japan and part of mainland Asia, codling moth is everywhere and is one of the most challenging pests wherever apples are grown throughout the temperate regions of the world.

Figure 12.19. A robin sitting on her nest in a nectarine tree in the orchard. Robins are voracious insect predators when feeding their babies and rarely damage fruit when nesting.

Table 12.4. Apple Disease Resistance, Ripening Time, and Hardiness

Cultivar	Apple Scab	Fire Blight	Powdery Mildew	Rust	Ripening	USDA Hardiness Zone
Akane	3	3	3	4	S	5–6
Braeburn	3	4	3	3	O	4–10
Carroll	3	2	3	3	S	3–6
Crimson Crisp	1	2, 3	2, 3	3	S	5–8
Cortland	3	2	4	3	O	3–6
Empire	4	2	3	2	O	2–6
Enterprise	1	2	2	2	O	4–5
Florina	1	2	2	4	S, O	5–8
Freedom	1	2	2	1	S	4–5
Goodland	2	2, 3	2	3	S	3–5
Goodmac	2, 3	3	2	2	S	3–5
GoldRush	1	2	2	4	O	4–5
Haralson	3	2	3	3	O	3–5
Jersey Mac	4	3	4	1	A	3–8
Jonafree	1	2, 3	2	3	S	3–8
Liberty	1	2	2, 3	2	O	4–7
Lodi	3	4	2	3	A	3–8
Nova Easygro	1	2	2	3	S	3–6
Novamac	1	1	1	1	S	4–6
Nova Spy	1	3	2	3	O	4–6
Macfree	1	2	3	1	S	3–6
MacIntosh	4	3	3	1	S	4–7
Macoun	3	3	3	3	O	4–7
Paula Red	2	4	3	2	A	4–9
Prima	1	2	2	4	S	4–5
Priscilla	1	1	3	2	S	4–5
Redfree	1	2	2,3	1	A	4–9
Sir Prize	1	3	2	3	S	4–5
Spartan	3	2, 3	2,3	2	O	4–7
Sundance	1	2	2	2	O	5–8
Williams Pride	1	2	3	1	A	4–7
Wealthy	2	3, 4	2	3	S	3–7
Wolf River	2, 3	2, 3	2	3	A	4–5

1 = Very resistant 2 = Resistant 3 = Susceptible 4 = Very susceptible A = August S = September O = October

Figure 12.20. These apple seconds (left), including those with codling moth surface-only stings, are some that Carl and I offered for sale at market with a sign saying FRUIT TO EAT IN THE DARK OR BY CANDLELIGHT. Unblemished apple firsts (right) command a much higher price.

Birds that travel up and down the trunks of trees, searching for insects, pecking and flaking bark with their beaks, seem to provide the best codling moth suppression. The insectivorous birds found in orchards with this feeding habitat include woodpeckers, flickers, oak titmice, nuthatches, and bushtits. In one California study birds and beneficial insects in combination reduced codling moth larvae numbers by close to 50 percent compared with a control, but beneficial insects alone reduced larval numbers by only 11 percent.[8]

It is worthwhile to design beneficial bird habitat near apple and pear fields or areas of your garden where apple trees are growing. On farms, hedgerows of native trees and shrubs encourage beneficial birds. You can also design to encourage beneficial birds and discourage crop-eating birds. For example, in another California study, researchers found that three of the most common crop-eating pest birds (American crow, red-winged blackbird, and Brewer's blackbird) were 10 times more abundant in fields with bare or weedy margins than in fields bordered by hedgerows.

Table 12.5. Woodleaf Farm (CA) Codling Moth Damage Records on Unsprayed Apples, 2012–2015

Year	Apples	Pears
2012	11%	1%
2013	20%	9%
2014	10%	7%
2015	8%	4%

Codling moth was our main apple pest challenge at Woodleaf Farm. We stopped spraying even certified organic insecticides and fungicides on apples in the early 2000s. We thinned off apples injured by the first generation of larvae, and our biological control system kept annual total damage around 10 percent in most years (see table 12.5). About half of the 10 percent damaged fruits were apples with codling moth "stings"—a surface spot where a larva had started to feed but was stopped by the apple's firmness or by a biological control organism on patrol. We were able to sell some of these "stung" apples as seconds. In 2013 total codling moth damage was economically significant, but all other years the money we saved (on materials, equipment/fuel, and labor) by not spraying apples, along with the benefits of maintaining our entire ecological system undisturbed, evened out the economic loss. We called it our annual tithing to ecology. Codling moth damage to apples at the Woodleaf Oregon orchard has also been low enough to justify not spraying (7 to 14 percent depending on the variety). Yellow, softer varieties are more susceptible than the firm red varieties.

TROUBLESHOOTING PESTS AND PROBLEMS

Brown blossoms and shoots.

Other symptoms: Leaves wilt rapidly and turn dark. Wilted shoots bend into the shape of a shepherd's crook. Infected twigs retain dead leaves through winter. In the spring amber-colored droplets of bacteria-filled sap are visible on blighted spurs, shoots, and branch cankers.

Cause: Fire blight.

Comments: Resistant varieties include Melrose, Northwestern Greening, Nova Easygro, Quinte, Winesap, and others listed in table 12.4. Resistant rootstocks include Geneva 11, Geneva 30, Geneva 65, M.7, and M.27.

Yellowing leaves with light brown, velvety spots.

Other symptoms: Leaf spots become dark brown. Dark spots appear on fruit and the spots may be cracked.

Cause: Apple scab disease.

Comments: See table 12.4 for resistant varieties.

New leaves are pale green with yellow in between veins.

Cause: Iron deficiency.

Yellow to brown foliage, branch dieback.

Other symptoms: White mycelial fans in the roots and trunk. Eventual death of the tree.

Cause: Armillaria root rot.

Leaves wilt but remain attached.

Other symptoms: Cankers on the trunk at soil level; a dark discoloration of the bark, which may be slimy with gum or dark sap.

Cause: Phytophthora root and crown rot.

Yellowing of leaves in a distinctive, random pattern.

Cause: Mosaic virus.

Comment: Though these symptoms are distinctive and may look serious, usually mosaic is a cosmetic problem only. No need to treat for it.

Orange leaf spots, defoliation.

Other symptoms: Fruit and twig deformities.

Cause: Rust (cedar apple rust).

Comment: See table 12.4 for resistant varieties.

Distorted, curled leaves at the ends of new shoots.

Other symptoms: White, powdery areas on leaves. Russeting on fruits.

Cause: Powdery mildew.

Comments: Resistant varieties include Fuji, Nittany, and Ambrosia. Red Delicious, Gala, Golden Delicious, and Autumn Glory are moderately resistant. See table 12.4 for other resistant varieties.

Folded leaves with thick silk webbing.

Other symptoms: Sawdust-like frass inside folded leaves; a green to white worm (larva) may be present.

Cause: Leafrollers (several species).

Comments: Honeycrisp, Sunrise, and Yataka are less preferred by leafrollers than other varieties.

Curled and sticky leaves with a black sooty mold.

Other symptoms: Fruit may be small, deformed.

Cause: Aphids (most common are rosy apple and green apple aphids).

Comments: Populations can be very high (50 percent of new shoots infested) before yields are decreased or serious damage is done to fruit. Goldrush is reported to be less susceptible to rosy apple aphids.

Figure 12.21. The hole and surrounding frass are classic codling moth damage to apple, indicating the presence of a worm or worm damage inside.

Curled leaves, cottony masses on twigs.

Other symptoms: Tiny pink-gray insects under the "cotton," galls on roots, and twigs that sometimes crack.

Cause: Woolly apple aphid.

Comments: M.106 and M.111 rootstocks are resistant to woolly apple aphids.

Stippled or bronzed leaves with fine silk webbing.

Cause: Spider mites.

Comments: Some varieties are less preferred by mites, including Enterprise, Pioneer Mac, and Redfree. I did my graduate research on mite resistance of apples at Rutgers University in New Jersey. I was able to confirm that Enterprise and Redfree were less preferred, but was unable to identify the mechanism for resistance. Researchers at Purdue University also found some of their selections, including Redfree, to be mite-resistant and were able to correlate mite resistance with smooth, less hairy apple leaves.[9]

Small yellow to brown raised spots on leaves.

Other symptoms: Spots are blister-like and can be patchy.

Cause: Eriophyid mites.

Sawdust-like excrement on outside of fruits.

Other symptoms: Excrement can form at the calyx end or on the side of fruits; small holes in fruits. In early drops, white larvae with brown heads may be feeding on seeds at the core.

Cause: Codling moth.

Comments: Apple varieties that seemed to be less susceptible to codling moth in on-farm research evaluations at Woodleaf in California and Oregon include: Crimson Crisp, Spartan, Royal Gala, Macoun, Auvil Early Fuji, and Empire. The most susceptible varieties are consistently golden- or yellow-colored, softer-skinned varieties such as Ginger Gold, Golden Delicious, and Hudson's Golden Gem as well as Red Delicious and Fuji. Some of our evaluations are supported by a Pennsylvania State University laboratory bioassay experiment, using a series of no-choice and multiple-choice tests. Golden Delicious, Fuji, and Stayman were the varieties most preferred by codling moths while Arlet, Honeycrisp, Pristine, Gala, and Sunrise were the least preferred varieties.[10]

In a study in New York, Pristine and Sunrise were less susceptible to codling moth than other varieties tested. Very susceptible varieties were Golden Delicious, Cameo, Gala Supreme, Goldrush, and Suncrisp.[11] Other researchers report that later-maturing varieties are more susceptible to codling moth injury. My experience and on-farm research concurs with that finding, except that Ginger Gold, an early yellow variety, is very susceptible and Crimson Crisp, a later red variety, is less preferred.

Small holes in fruits.

Other symptoms: Brown sawdust-like excrement; meandering tunnels and whitish to pinkish tan larvae inside fruits.

Cause: Oriental fruit moth.

Comments: Oriental fruit moth is an uncommon pest of apples, but in the late 1990s began to be seen in apple orchards in mid-Atlantic states. Damage is somewhat like codling moth injury, but codling moth larvae tunnel directly to the core and feed on the seeds rather than making meandering tunnels.

Crescent-shaped scars on fruit surface.

Other symptoms: Tunnels inside fruits; grayish white, legless, C-shaped, brown-headed grubs in tunnels.

Cause: Plum curculio.

Comments: Braeburn, GoldRush, Fuji Red (cv. BC#2), and Suncrisp show some resistance to plum curculio. Honeycrisp and Ginger Gold are very susceptible.[12]

Pitted, dimpled, and misshapen fruit.

Other symptoms: Brownish, irregular, threadlike tunnels and white, legless larvae inside fruits.

Cause: Apple maggot.

Comments: Less-preferred varieties are Goldrush and Pristine.[13]

Deformed fruit and small corky pits on fruit surface.

Cause: Leafrollers (many species) or green fruitworms.

Brown, sunken spots on fruit skin.

Other symptoms: Brown, spongy areas under the skin at the bottom of the fruit.

Cause: Bitterpit (a physiological disorder and nutritional deficiency).

Deformed fruit with deep pits.

Other symptoms: Spongy areas below the skin at the top of the fruit.

Cause: Stink bug; tarnished plant bug.

Comments: Braeburn and Pioneer Mac show some resistance to tarnished plant bug.

Dark spiral scars on fruit skin.

Other symptoms: Tunneling inside fruit and exit holes with reddish brown frass; strong odor.

Cause: European apple sawfly.

Comments: Pristine, Senshu, and Ginger Gold appear to be less susceptible to sawfly than other varieties.

Branches covered with small, wartlike bumps.

Cause: Scale.

Apricot

Apricot trees grow tall and like to spread out wide. Plan to give them plenty of space and light.

CROP MANAGEMENT

Temperature

When completely dormant, apricots tolerate cold temperatures as low as –20°F (–29°C) (USDA zone 5), but they need consistent cold temperatures because they come out of dormancy more easily than other fruit crops. Apricot blossoms can be injured when warm temperatures in late winter are followed by a blast of cold weather. Applying mulch can help keep the trees consistently cold through late-winter warm spells. Since they come out of dormancy easily and can bloom early, apricot blossoms are susceptible to late frosts. On our California orchard, we created a system to cover our most frost-susceptible peach rows with floating row cover. We spread the fabric over the trees and a frame we made out of 10-foot (3 m) T-posts placed in between every other tree in row. Carl welded a 3-foot (0.9 m) section of rebar onto the top of each T-post so it would reach to the top of the fruit trees. The floating row cover was tied down to each tree trunk with heavy-duty twine (see figure 12.22). It was a lot of work, but very effective frost protection. I gave up trying to cover my too-tall apricot trees in the Woodleaf Oregon orchard because wind was injuring blossoms. Floating row cover is a challenge in windy areas! Hardy apricot varieties include Tomcot, Harcot, Harglow, Hargrand, Harogem, Harlayne, Wescot, Canadian White Blenheim, Moongold, Scout, Chinese, Goldcot, Moorpark, Puget Gold, and Tilton.

Light

Apricot trees need full sun (7 to 8 hours of daily sunlight during the growing season). They are more disease-susceptible and usually don't fruit as well if grown in partial shade. However, there are several "escaped" seedling apricots growing in partial forest shade at Woodleaf that produce an annual crop of okay-tasting apricots.

Figure 12.22. These peach tree rows at the Woodleaf California orchard are covered with Agribon for frost protection of tender spring blooms.

Soil and Fertility

Apricots do not do well in heavy, wet, clay soils. They want a well-drained loamy soil with a pH between 6.7 and 7.5. Apricots do not require as much fertilizer as peaches. Several apricot varieties growing very vigorously in my orchard at the time of this writing have not had any nitrogen fertilizer for 7 years other than that from the mowed living mulch.

Water

Regular irrigation is important for apricots, but be sure that the soil is not constantly wet.

Special Ecological Preferences

Good sanitation and pruning for aeration are vital to suppress disease. Never use sulfur products on apricots. Light sulfur injury can cause leaves to cup, and fruit may become brown or develop brown spots. Heavy sulfur injury can scorch or wither leaves and kill buds (or buds may open only slightly).

I do not thin apricots unless a tree is bearing a particularly heavy crop of marble-sized fruits. They definitely will get bigger if you thin some out! To ensure adequate aeration, space trees 15 to 20 feet (4.6–6.1 m) apart. Apricots are self-fertile, but a second variety will increase fruit set of both varieties.

TROUBLESHOOTING PESTS AND PROBLEMS

Small, round, tan-to-purplish leaf spots that drop out.

Other symptoms: Leaves develop a shot-hole appearance. Numerous slightly sunken brown to purplish spots on fruits; spots become scabby.

Cause: Shot-hole disease.

Angular, dark spots on leaf undersides that fall out.

Other symptoms: Spots may be purplish, black, or brown; red halos around holes where spots fell out. Sunken areas and oozing gummy residue on fruit.

Cause: Bacterial leaf spot disease.

Comments: Resistant varieties include Harglow, Harlayne, Haroblush, Harojoy, and Harostar; Tomcot is somewhat resistant.

Leaves curl, thicken, turn yellow.

Other symptoms: Leaves later turn brown.

Cause: Aphids.

Small yellow leaves that curl upward at edges.

Other symptoms: Leaves eventually turn brown and dry.

Cause: Potassium deficiency.

Stunted shoot tips with a rosette appearance.

Other symptoms: Small leaves that are pale yellow or whitish green.

Cause: Zinc deficiency.

Leaves are rolled and webbed together where worms feed.

Cause: Leafrollers.

Bronzed leaves with silken webbing.

Cause: Spider mites.

Small yellow to brown spots on leaves.

Other symptoms: Spots are blister-like and can be patchy.

Cause: Eriophyid mites.

Whitish gray powdery mold on young leaves.

Other symptoms: Mold may also form feltlike patches, and may also appear on buds and twigs.

Cause: Powdery mildew.

Shoot dieback and gummy residue.

Other symptoms: Holes in young fruit with a gummy ooze. Whitish to cream-colored worms in fruit later in the growing season.

Cause: Peach twig borer; Oriental fruit moth.

Comments: Oriental fruit moth larvae usually bore deeper into shoots than peach twig borer larvae do.

Small holes in fruits.

Other symptoms: Brown sawdust-like excrement; meandering tunnels and whitish to pinkish tan larvae inside fruits.

Cause: Oriental fruit moth.

Comment: Usually only an occasional problem on apricots.

Brown, rotted areas on fruit in summer.

Other symptoms: Fuzzy tan-gray spores cover the fruit surface as the disease progresses.

Cause: Brown rot; blossom and twig blight disease.

Comments: Harglow, Tilton, Puget Gold, Harojoy, Harostar, Harcot, Harlayne, Harogem, Tomcot, and Hargrand have some brown rot resistance.

Brown, dead, or wilting blossoms.

Cause: Brown rot; blossom and twig blight disease; winter/frost injury.

Catfacing or dimpling on fruit.

Cause: Stink bugs; tarnished plant bug.

Crescent-shaped scars on fruit surface.

Other symptoms: Tunnels inside fruits with grayish white, legless, C-shaped grubs with brown heads.

Cause: Plum curculio.

Gummy cankers on branches in spring.

Other symptoms: Brown leaf spots later become shot holes. Fruit with dark, sunken, sour-smelling spots.

Cause: Bacterial canker disease.

Yellow to brown foliage.

Other symptoms: Branch dieback and white mycelial fans in roots and trunk; eventual death of the tree.

Cause: Armillaria root rot.

Stunted and slow-growing trees with small leaves.

Other symptoms: Leaves wilt but remain attached to the tree; cankers on the trunk at the soil level; dark discoloration of the bark, which may be slimy with gum or dark sap.

Cause: Phytophthora root and crown rot.

Cherry

There are two species of cherries most commonly grown: tart and sweet cherries. Generally tart cherries are more forgiving and easier to grow, especially in colder climates.

CROP MANAGEMENT

Temperature

Cherries are relatively hardy and are susceptible to fruit loss during frosts at bloom. Tart cherries are much hardier than sweet cherries. Sweet cherries bloom earlier and are more prone to damage from late-spring frosts. Cherries don't like high summer temperatures and are susceptible to fruit sunburn.

Light

Cherry trees need full sun—at least 8 hours of sun each day.

Soil and Fertility

Cherries prefer well-drained soil with a pH between 6.0 and 7.0. Cherry trees are more susceptible to root rot diseases if grown on heavy clay, but they don't do well in very sandy soils, either.

Water

Cherries require regular irrigation, but don't tolerate wet roots. They're sensitive to rain and will crack during periods of heavy rainfall as they begin to ripen. Some varieties are less susceptible to cracking. Kristin, Rainier, Stella, Pacific Red, and Royal Ann are crack-resistant. Black Pearl is tolerant to cracking. Bing has very high rain-cracking susceptibility, so much so that growers in wet climates have decided to plant other varieties in recent years.

Special Ecological Preferences

Allowing sufficient space (15 to 30 feet [4.6–9.1 m]) between trees and pruning regularly is vital to cherry tree health and productivity. Most research indicates that regular, relatively aggressive pruning results in bigger cherries.

Tart cherry trees are small compared with sweet cherries, growing only to about 15 feet (4.6 m) tall. Tart cherries are self-fruitful, but a second compatible variety will ensure even better pollination and fruit set. Evans (Bali) tart cherry has excellent quality and is very cold-hardy. Many sweet cherry varieties require cross-pollination for good fruit set.

Birds love cherries. Covering trees with plastic netting helps to deter birds. Cover just as cherries begin to show the first color.

Cherries, like all fruit trees, respond well to an active microbial community within a functioning soil organic matter system. But a beneficial microbial community may be even more vital for cherry disease suppression. In field studies in Turkey, spraying sweet cherries at bloom with beneficial plant-growth-promoting *Bacillus* species bacteria increased yield, growth, and foliar nutrition.[14]

TROUBLESHOOTING PESTS AND PROBLEMS

Tiny maggots inside cherry fruit.
Cause: Cherry fruit flies.
Comments: Different cherry fruit fly species occur in different parts of the United States. In the West cherry fruit flies are the most damaging cherry pest. Early varieties, like Chelan and Pacific Red, tend to escape cherry fruit fly infestations. Late varieties are usually more heavily infested. From the outside, cherries infested with fruit fly larvae often appear perfect.
At our California orchard, spotted wing drosophila (*Drosophila suzukii*) was the only orchard insect that we were unable to suppress to economic levels without certified organic sprays. In 2013 we tried not spraying cherries at all, and damage was 30 percent. So far, cherry fruit flies have not been a problem in Woodleaf's Oregon orchard, but the cherry trees have not yet begun to bear heavily. One way Carl and I dealt with spotted wing drosophila in California was to pick cherries a bit early and put them in cold storage at 40°F (4°C). Refrigeration stops further development of eggs and larvae inside the fruit and may kill eggs and larvae after longer refrigeration periods for a week or more.

Figure 12.23. Cherry fruit fly damage is due to larvae developing inside fruit and feeding on the flesh.

Whitish gray powdery mold on young leaves.
Other symptoms: Feltlike moldy patches may form. Mold may also appear on buds and twigs.
Cause: Powdery mildew.

Small yellow leaves that curl upward at the edges.
Other symptoms: Leaves eventually turn brown and dry.
Cause: Potassium deficiency.

Stunted shoot tips with a rosette appearance.
Other symptoms: Small leaves that are pale yellow or whitish green.
Cause: Zinc deficiency.

Shoot dieback and gummy residue.
Other symptoms: Holes in young fruit with a gummy ooze. Whitish to cream-colored worms inside fruit.
Cause: Peach twig borer; Oriental fruit moth.
Comments: Oriental fruit moth larvae usually bore deeper into injured shoots than peach twig borer larvae do. These pests only occasionally damage cherry fruit.

Brown, rotted areas on fruit in summer.
Other symptoms: Fuzzy tan-gray spores cover the fruit surface as the disease progresses.
Cause: Brown rot; blossom and twig blight disease.

Brown, dead, or wilting blossoms.
Cause: Brown rot; blossom and twig blight disease; winter/frost injury.

Catfacing or dimpling on fruit.
Cause: Stink bugs; tarnished plant bugs.

Buds chewed; rolled leaves bound with silk webbing.
Other symptoms: Sawdusty frass present inside rolled-up leaves.
Cause: Leafrollers (many species).

Small silvery patches on leaves.
Other symptoms: Yellow to brown blister-like spots or patches.
Cause: Eriophyid mites.

Gummy cankers on branches in spring.
Other symptoms: Brown leaf spots later become shot holes. Fruit with dark, sunken, sour-smelling spots.
Cause: Bacterial canker disease.
Comments: Burgundy Pearl is less susceptible to bacterial canker.

Yellow to brown foliage.
Other symptoms: Branch dieback; white mycelial fans in roots and trunk; eventual death of the tree.
Cause: Armillaria root rot.

Stunted and slow-growing trees with small leaves.
Other symptoms: Leaves wilt but remain attached to the tree; cankers on the trunk at the soil level; dark discoloration of bark, which may be slimy with gum or dark sap.
Cause: Phytophthora root and crown rot.

Branches covered with small, wartlike bumps.
Cause: Scale.

Peach

Peaches and nectarines are more demanding fruit trees to grow, especially in colder climates, but they begin to bear fruit earlier than most other tree fruits.

CROP MANAGEMENT

Temperature

Peach trees are one of the least winter-hardy stone fruits. Some varieties are much hardier than others, however. Hardy varieties include Reliance, Contender, Madison, Veteran, Intrepid, Harken, Harrow Diamond, and Red Haven. Buds and new growth of most varieties die at low temperatures of –15°F (–26°C), and trees can be killed outright if temperatures reach –25°F (–32°C). Peaches are also susceptible to spring frosts that damage fruit buds, resulting in decreased or even no fruit some years.

Light

Peach trees grow best in full sun (6 to 8 hours of daily sunlight during the growing season). When grown in partial shade, peach trees are more disease-susceptible, and their fruit is often small and poorly colored.

Soil and Fertility

Peaches prefer light soil with a pH of 6.4 to 7.0. Some diseases, such as bacterial canker, are more prevalent at lower soil pH.

Peaches respond very rapidly to nitrogen fertilization, but this is not always best for total agroecological function. Peaches require more nitrogen and phosphorus when young (1 to 6 years old), but as they mature and begin to fruit, they require relatively less nitrogen and instead use more potassium and, depending on the soil, more of certain micronutrients (especially boron, iron, manganese, and zinc). Peach trees use more potassium and micronutrients perhaps because their fruits contain high levels of potassium and iron compared with other fruits.

As noted previously, other than mowed living mulch, I have not added any nitrogen to my Oregon orchard for the past 7 years. But I do apply Mineral Mix Bloom Spray to my peach trees one to three times during bloom every year. (See table 7.3 for the contents of this mix.) Too much nitrogen can actually be a problem in peaches. In a California study, high nitrogen levels in peach tree leaves increased the fungal disease brown rot. But if your soil is nutrient-deficient, fertilizing with nitrogen will usually most quickly improve peach tree growth, leaf color, fruit size, and overall yield.

Water

Peaches require regular irrigation, but do best with drip or micro-sprinkler irrigation that does not spread water over foliage and fruit. Peaches do not like wet roots.

Special Ecological Preferences

Good sanitation, fruit thinning, and open-center pruning is vital to maintain peach health. Peaches need light and lots of aeration.

Thin fruit to one peach every 6 to 8 inches (15–20 cm) along fruiting branches when fruit has developed to marble or golf ball size. Fruits that touch are more susceptible to disease. Remove diseased fruits and mummies whenever you spot them, and at least every fall.

Pruning timing is also vital. Wait until trees are completely dormant to prune in the winter, especially if you live in a cold climate. Prune peaches as late in the winter or early spring as possible to avoid encouraging canker diseases. Allow plenty of air and light to reach all peach tree branches for best fruit set and color, and to avoid fruit diseases such as brown rot.

Balancing nutrient cycling and pest suppression by managing the soil organic matter system is the key to healthy peach trees. If the orchard is managed with minimal soil disturbance and nutrient recycling, peaches can be grown with relatively low nitrogen and phosphorus fertilization.[15]

Winter pruning and leaf fall result in the greatest removal of peach tree nutrients, even more than fruit harvest. In fact, peach fruits contain relatively low amounts of nitrogen. Hence it is vital to recycle tree branch prunings and leaf fall back into orchard soil by mowing after leaf drop and pruning. In wetter climates where tree fruit diseases are more problematic, flail mowing of branches and leaves on the ground will break them down more quickly and decrease the possibility that fungal inoculum survives to infect foliage the following spring. Some disease management requires removing diseased branches from the orchard. For example, you should always remove and/or burn branches that show any sign of bacterial or fungal canker.

TROUBLESHOOTING PESTS AND PROBLEMS

Wilting leaves and dying and dead twigs.

Other symptoms: Clear to dark-colored gum or ooze with sawdust-like insect excrement exuding from the base of the tree, lower trunk, and roots.
Cause: Peach tree borer.

Wilting leaves and dying and dead twigs.

Other symptoms: Irregularly shaped, brown, sunken areas on trunk and lower branches with amber-colored gum or ooze.
Cause: Cytospora canker disease.

Wilting leaves and limb dieback.

Other symptoms: Rough cankers and amber-colored gum on branches and on the trunk. Inner bark may be brown, foamy, and sour smelling.
Cause: Bacterial canker.

Wilting shoots.

Other symptoms: Sawdust-like excrement at the base of flagging shoots.
Cause: Oriental fruit moth; peach twig borer.
Comments: Oriental fruit moth larvae usually bore deeper into the shoot than peach twig borer larvae do.

Figure 12.24. The distortion of these peach leaves is due to peach leaf curl disease. Fruit is rarely affected. The parasitic wasp on the small, newly developing peach fruit is looking for Oriental fruit moth or peach twig borer larvae to parasitize.

Leaves thickened, curled, distorted.

Other symptoms: Leaves are red or yellow instead of green.

Cause: Peach leaf curl disease.

Comments: Resistant varieties include Frost, Indian Free, Muir, Autumn Rose, August Etter, Avalon, Avalon Pride, Charlotte, Early Charlotte, Early Crawford, Nanaimo, Autumn Rose, Early Crawford, and Oregon Curl Free.

Yellow to reddish leaves with red spots or patches.

Cause: Nitrogen deficiency.

Small, round, tan to purplish leaf spots that drop out.

Other symptoms: Leaves develop a shot-hole appearance. Numerous slightly sunken brown to purplish spots on fruits; the spots become scabby.

Cause: Shot-hole disease.

Comments: On peach trees, the foliar symptoms of this disease look similar to those of nitrogen deficiency. Tolerant varieties include Lovell and Muir.

Angular, purplish black leaf spots that drop out.

Other symptoms: Leaves develop a shot-hole appearance. Numerous slightly sunken brown to purplish spots on fruits.

Cause: Bacterial leaf spot.

Comments: Resistant varieties include Biscoe, Blazing Star, Candor, Contender, Harrow Beauty, Harrow Diamond, Goldnine, Vulcan, Venture, Vinegold, Virgil, Allgold, GoldJim, Belle of Georgia, Comanche, Dixired, Earliglo, Emery, Encore, Garnet Beauty, Harbelle, Harbinger, Harbrite, Harken, Loring, Madison, Norman, Ranger, Redhaven, Redskin, Sentinel, and Sunhaven.

Yellow, curled, and twisted terminal leaves.

Cause: Aphids.

Small yellow leaves that curl upward at the edges.

Other symptoms: Leaves eventually turn brown and dry.

Cause: Potassium deficiency.

Stunted shoot tips with a rosette appearance.

Other symptoms: Small leaves that are pale yellow or whitish green.

Cause: Zinc deficiency.

Blossoms wilt and turn brown.

Other symptoms: Gumming and oozing sap is often associated with the dead flowers. Wilted shoots and blighted twigs with tan centers with dark margins.

Cause: Brown rot; blossom and twig blight disease.

Comments: See resistant cultivars below.

Brown, rotted areas on fruit in summer.

Other symptoms: Fuzzy tan-gray spores cover the fruit surface as the disease progresses.

Cause: Brown rot; blossom and twig blight disease.

Comments: Cultivars with some brown rot resistance include Venture, Babygold 5, Elberta, Glohaven, Springcrest, Flavorcrest, White Haven, Suncrest, Garnet Beauty, Andross, Harken, Harrow Beauty, Harrow Diamond, Harson, and Harcrest. Hardired nectarine has some resistance. Redhaven, Vivid, Vulcan, Virgil, Vinegold, and Catherina peaches are moderately resistant to brown rot.

Small brownish worms, usually found just under skin.

Cause: Peach twig borer.

Comments: Larvae enter peach fruits at the stem end or along the suture.

Small holes in fruits.

Other symptoms: Brown sawdust-like excrement on fruit and in meandering tunnels inside fruit; sometimes whitish to pinkish tan larvae inside fruits.

Cause: Oriental fruit moth.

Comments: Larvae enter peach fruit at the stem end and immediately bore to the center of the fruit.

Crescent-shaped scars on the fruit surface.

Other symptoms: Gummy exudate at scars; grayish white, legless, C-shaped grubs with brown heads in tunnels inside fruits.

Cause: Plum curculio.

Comments: This weevil is primarily found in the midwestern and eastern fruit-growing regions of North America. I have never seen it on fruit in the West.

Deformed fruit with sunken areas.

Other symptoms: Corky scars and catfacing with oozing on young fruits.

Cause: Thrips.

Comments: Nectarines are much more susceptible to thrips damage than peaches.

Deformed fruit with rough, corky areas.

Other symptoms: Catfacing and dimpling of fruit.

Cause: Tarnished plant bugs; stink bugs.

Yellow to brown foliage.

Other symptoms: Branch dieback, and white mycelial fans in the roots and trunk; eventual death of the tree.

Cause: Armillaria root rot.

Stunted and slow-growing trees with small leaves.

Other symptoms: Leaves wilt but remain attached; cankers on the trunk at the soil level; a dark discoloration of the bark, which may also be slimy with gum or dark sap.

Cause: Phytophthora root and crown rot.

Branches covered with small, wartlike bumps.

Cause: Scale.

Figure 12.25. Thin peaches well to avoid brown rot. I thin the fruits to 6 to 8 inches (15–20 cm) apart, spacing susceptible varieties farther apart than resistant ones.

Figure 12.26. Rough areas and catfacing on these deformed peaches are due to feeding by stink bugs and/or tarnished plant bug.

Pear

It can take pears a long time to come into bearing (3 to 10 years for trees to begin flowering and producing fruit). But unlike most stone fruits, pear trees are (or can be) very long-lived.

Asian pears are the same genus as the more familiar European pears, but they are a different species, are native to East Asia, and are shaped more like apples, with a crunchy texture and sweet taste.

CROP MANAGEMENT

Temperature

Pear trees are one of the most winter-hardy fruit crops, hardy in USDA zones 4 through 8. Some varieties are even hardy to zone 3, such as Early Gold, Loma, Golden Spice, Julienne, John, Patten, Southworth, and Summercrisp. Pears don't do well in hot climates, though the University of Florida recommends a few varieties for hot, humid climates, including Flordahome, Orient, and Pineapple. My favorite pear, Warren, grows well in Arkansas and Mississippi.

Light

Pear trees grow best in full sun (6 to 8 hours sunlight daily during the growing season). If grown in partial shade, they are more disease susceptible and the fruit is often small and poorly colored.

Soil and Fertility

Pears do best at soil pH between 6.0 and 7.5, and they prefer well-drained soil. They are less demanding of nutrients than stone fruits (like peaches), especially nitrogen. Because it encourages rapid growth, excess nitrogen can increase pear susceptibility to fire blight and aphids. If pear trees are generating between 16 and 22 inches (41–56 cm) of new growth each year, there is no need to feed nitrogen.

Water

Roots need to be moist but not waterlogged. Pears can handle wet roots better than stone fruit.

Special Ecological Preferences

Good sanitation, fruit thinning, and pruning out diseased branches, twigs, and cankers suppress disease. Pears are more susceptible to fire blight than apples. In wet climates this disease can be limiting to growth of non-resistant pear varieties. Carl and I pulled out most of our fire-blight-susceptible pears and committed to one of our best-tasting pear varieties that is also fire-blight-resistant, Warren pear. The pear area is the only place in the orchard where we planted a block of one variety only. This is not a good ecological practice, and I am testing other fire-blight-resistant pear varieties.

Prune pears to a central leader or modified central leader system. Excessive pruning encourages excessive shoot growth and fire blight disease and reduces fruit quality. Bacterial biological control agents have been used on foliage to control frost injury and fire blight of pear.[16]

TROUBLESHOOTING PESTS AND PROBLEMS

Leaves wilt rapidly, turn dark, and remain attached.

Other symptoms: Terminal growth bends into the shape of a shepherd's crook. Amber-colored droplets of bacteria-filled sap appear on blighted spurs and shoots. Infected branches may be girdled, resulting in the loss of the entire branch.

Cause: Fire blight.

Comments: Resistant pear varieties include Warren, Honeysweet, Kieffer, Le Conte, Magness, Moonglow, Old Home, Harrow Crisp, Harrow Gold, Harrow Delight, Harrow Sweet, Harvest Queen, Harovin Sundown, Orient, Seckel, Starking Delicious, Sunrise, Gem, Blake's Pride, Shenandoah, Potomac, Comice, and Maxine.

Resistant pear rootstock varieties include Old Home (OH), Old Home Farmingdale (except OHF 51), *Pyrus calleryana, Pyrus betulifolia* seedlings.

Resistant Asian pear varieties include Chojuro, Kosui, Olympic (Korean Giant), Seuri, Shinko, Shinsui, Singo, Tse Li, Ya Li, and Kikusui.

Figure 12.27. Green fruit worms do only minimal damage to pears as long as you have not killed the many predators and parasites that prey upon them.

Figure 12.28. Leafrollers and green fruit worms leave sunken, scabby scars on fruit. White-flecks on leaves is residue from mineral mix bloom sprays.

Small yellow to brown blister-like spots or patches on leaves.
Cause: Eriophyid mites.

Yellow, curled, and twisted terminal leaves.
Cause: Aphids.

Chewed buds and rolled-up leaves.
Other symptoms: Leaves rolled and tied together with silk webbing, with sawdusty frass present inside.
Cause: Leaf rollers (many species) and green fruit worm.

Crescent-shaped scars on the fruit surface.
Other symptoms: Grayish white, legless, C-shaped grubs with brown heads in tunnels inside fruits.
Cause: Plum curculio.

Catfacing or dimpling on fruit.
Cause: Tarnished plant bugs; stink bugs.

Branches covered with small, wartlike bumps.
Cause: Scale.

Plum

There are many varieties of plums and new plum crosses with apricot (called pluots and plumcots) that produce fruit with different colors, shapes, and flavors than traditional plums. There are two types of traditional plums, European and Japanese. In general, European plums (also called prune plums) are easier to grow, more frost-tolerant, and more disease- and pest-free than Japanese varieties. The most trouble-free plum trees in my orchard are two old-fashioned Stanley prune plums.

CROP MANAGEMENT

Temperature

Plums, along with tart cherries, are the hardiest of the stone fruits, but Japanese types are less hardy than European (prune) types. The European types Mount Royal, Pembina, and Black Ice are hardy to zone 3 or 4.

Light

Plums generally need 6 to 8 hours of direct sunlight each day. But if you have a less sunny spot, European plums can handle less sun than Japanese plums.

Soil and Fertility

Plums prefer loamy, well-drained soil with a pH between 6.0 and 6.5. They are susceptible to zinc deficiencies, especially if heavy applications of manure have been added to the soil. Prune plums are also susceptible to potassium deficiency.

Water

Plums need plenty of water, but they do not tolerate wet roots.

Special Ecological Preferences

Good sanitation, fruit thinning, and open center pruning are necessary to suppress disease.

All plums are susceptible to wind and heavy rain injury that cause minor scabby areas and spots on fruit surfaces. Sometimes gardeners and farmers worry that this environmental injury is disease-related. Don't worry! Usually environmental damage to fruit is on the predominant wind side of the tree. Plum leaves also sometimes develop shot-hole-like spots on older leaves due to environmental conditions and genetic disorders. To tell the difference between disease-caused leaf spots and environmentally caused spots, look closely at the spots. Environmentally caused spots do not have the light centers with dark dots characteristic of fungal shot-hole disease.

Most plums require a second compatible variety to ensure best pollination and good fruit set. Wild American plums (*Prunus americana*) are good pollen sources for hybrid plum varieties.

TROUBLESHOOTING PESTS AND PROBLEMS

Crescent-shaped scars on the fruit surface.

Other symptoms: Gummy exudate near scars; grayish white, legless, C-shaped grubs with brown heads in tunnels inside fruits.

Cause: Plum curculio.

Catfacing or dimpling on fruit.

Cause: Tarnished plant bugs; stink bugs.

Buds chewed; rolled leaves bound with silk webbing.

Other symptoms: Sawdusty frass present inside rolled-up leaves.

Cause: Leafrollers (many species) and green fruitworm.

Angular, purplish black leaf spots drop out.

Other symptoms: Leaves develop a shot-hole appearance. Numerous slightly sunken brown to purplish spots on fruit.

Cause: Bacterial leaf spot.

Comments: Resistant varieties include Bruce, Stanley, Burbank, D'Agen, America, Obilinya, and Vanier. In general, European plums are more tolerant than Japanese varieties.

Figure 12.29. Candy Stripe pluots are as tasty as they are beautiful.

Small yellow to brown blister-like spots or patches on leaves.
Cause: Eriophyid mites.

Yellow, curled, and twisted terminal leaves.
Cause: Aphids.
Comments: Plums and pluots are especially susceptible to early-spring aphid infestations that cause terminal new leaves to curl. But plums can tolerate 10 to 20 percent of their terminal shoots curled by aphids; the trees almost always grow out of the damage easily, and natural biological control organisms take care of the aphids.

Small yellow leaves that curl upward at the edges.
Other symptoms: Leaves eventually turn brown and dry.
Cause: Potassium deficiency.

Stunted shoot tips with a rosette appearance.
Other symptoms: Small leaves that are pale yellow or whitish green.
Cause: Zinc deficiency.

Flagging, wilting new shoots.
Other symptoms: Small holes in fruits with brown sawdust-like excrement; pinkish white worms bore from the stem end into the center to feed around the pit.
Cause: Oriental fruit moth or peach twig borer.

Wilting leaves and dead or dying twigs.
Other symptoms: Irregularly shaped, brown, sunken areas on the trunk and lower branches with amber-colored gum or ooze.
Cause: Cytospora canker disease.

Wilting leaves and limb dieback.
Other symptoms: Rough cankers and amber-colored gum on the branches and trunk. Inner bark may be brown, foamy, and sour smelling.
Cause: Bacterial canker.

Branches covered with small, wartlike bumps.
Cause: Scale.

CHAPTER 13

Fruit Tree Insect Pests and Interventions

This chapter provides biological and intervention information on the most common tree fruit insect pests in North America. However, many pests covered here are found on fruit trees worldwide, such as codling moth, Oriental fruit moth, and western flower thrips.

I describe each pest and the damage it causes and then offer interventions, focusing first on those with minimal ecological impact, then moderate-impact options, and the heaviest-impact materials as a last resort. For some pests, I list only minimal- and moderate-impact interventions because heaviest-impact interventions are never really appropriate. For many of the pests, there are no effective moderate-impact interventions available, and thus I offer only minimal interventions and heaviest interventions.

Aphids

Many species of aphids, including green peach, rosy apple, green apple, and woolly apple aphids, commonly feed on fruit tree terminal leaves in the

Figure 13.1. Aphids feeding on new spring growth of pluot are causing leaves to curl and pucker. Lady beetles and other predators usually provide sufficient control.

spring and move to vegetable crops in the summer. These yellowish green to blackish pink aphids are not considered a serious fruit tree pest except occasionally on plums and pluots.

HOSTS

Plum and pluot particularly, and peach, nectarine, cherry, apricot, apple, and pear as well as a very wide range of vegetable crops (see the "Aphids" entry on page 219 for information on aphid management in vegetables).

MINIMAL-IMPACT INTERVENTION

Avoid rapid and excessive foliage growth by balancing mineral nutrient levels, specifically avoiding too much rapid-release nitrogen fertilizer. Encourage predators and parasites. Specific predators and parasites, such as syrphids, spiders, lady beetles, lacewings, soldier beetles, and parasitoid wasps prey on green peach aphid. Mowing selectively and creating greater orchard vegetation complexity by allowing flowering weeds and ground cover to grow and bloom under and between fruit trees diminishes aphid populations, according to Washington State University and USDA studies.

MODERATE-IMPACT INTERVENTION

Spray infested areas with insecticidal soap in the evening or early morning. Repeat applications are often required.

Neem oil can be added to an insecticidal soap solution, but doing so raises the impact level from moderate to heavy.

Apple Maggots

Apple maggot (*Rhagoletis pomonella*) is mostly a pest of the eastern US and the mid-Atlantic states, but it has found its way to apple growing regions in certain parts of the western US. It causes misshapen fruit and pitted, sunken areas on fruit surfaces. Inside, the fruit flesh is brown and rotting. Adults are black clear-winged flies with distinctive black bands on the wings in the shape of a W. The larva, which is found inside fruit, is a white, legless maggot without a distinct head. Apple maggots overwinter in the soil as pupae, and emerge as adult flies in early to midsummer. They migrate to apple trees and can travel for at least ½ mile (0.8 km) to find susceptible apples. There is usually only one generation per year. Varieties most susceptible to maggot attack include Wealthy, Cortland, Gravenstein, Red Delicious, Golden Delicious, and early sweet apple varieties. Honeycrisp and Ginger Gold are very susceptible to apple maggot. Less-preferred varieties are GoldRush and Pristine.[1] Spartan, McIntosh, and Macoun also appear to be less preferred.

HOSTS

Apple. Also, quince and the fruits of native and ornamental hawthorn.

MINIMAL-IMPACT INTERVENTION

Larvae and pupae may be preyed upon in the soil by nematodes, ground-dwelling beetles, and insect-eating fungi. In a lab study, seven species of parasitic nematodes were tested on apple maggot pupae. *Steinernema riobrave* was the most virulent, reducing apple maggot pupae by 71 percent and 77 percent, followed by *Steinernema carpocapsae* with a 68 percent reduction.[2]

Kaolin clay (see chapter 6) applied as a spray to leaves, stems, and fruit acts as a repellent if applied by early summer.

In Oregon researchers recommend trapping with red sphere sticky traps for small home orchards, but only to manage low apple maggot density. The red spheres can be purchased and then coated with a sticky product such as Tangle-Trap. You can also buy red sphere traps that come with an adhesive to apply. Spheres baited with butyl hexanoate (fruit odor) and ammonium carbonate (food odor) are more effective than unbaited spheres at capturing flies and preventing them from laying eggs. The recommendation is to hang one apple maggot trap in each small tree, two to four traps per tree in medium-sized trees, and six to eight traps per tree on trees that are 20 to 25 feet (6–8 m) tall. Clean traps weekly or when they fill up, and replace or reapply adhesive every 3 weeks.

MODERATE-IMPACT INTERVENTION

A predictive model of adult apple maggot emergence is available in many states. Ask your local extension agent if one is available for yours.

At a commercial scale, it's not economically or logistically feasible to hang (and then clean) enough plain sticky traps to control the apple maggot flies. Thus, the strategy has to be intensified by also adding a bait and a certified organic insecticide to the traps. Effective suppression has been reported by late-spring/early-summer placement of spinosad-treated and odor-baited red sphere traps on perimeter-row apple trees. In 2003 Ron Prokopy, an entomologist from the University of Massachusetts, reported on ongoing experiments with sticky-coated red spheres baited with both a five-component blend of attractive odor and with spinosad. In Prokopy's study red spheres were placed at varying distances in perimeter-row apple trees; distances were based on both tree size and pruning regimen. Small trees on M.9 rootstocks, well-pruned trees, and an apple-maggot-tolerant variety (McIntosh) received traps 50 feet (15 m) apart (55 traps per 10-acre [4 ha] block). Orchard blocks with larger trees on M.7 rootstocks, trees in need of pruning, apple-maggot-susceptible varieties (such as Gala), and trees with hedgerow or woods bordering them received traps 16 feet (5 m) apart (160 traps per 10-acre [4 ha] block). Orchard blocks with intermediate size and pruning received traps 32 feet (10 m) apart. Results showed that adjusting distance between traps according to orchard architecture gave good apple maggot suppression compared with a control with no traps. Suppression using traps was less effective in blocks with larger, less vigorously pruned trees. Suppression was also less effective on apple-maggot-susceptible apple varieties.

Cherry Fruit Flies

Western cherry fruit fly (*Rhagoletis indifferens*), eastern cherry fruit fly (*R. cingulate*), and spotted wing drosophila (*Drosophila suzukii*) are the species of concern in North America. Cherry fruit flies lay eggs on all varieties of cultivated and wild cherries. Eggs laid under the skin of fruits hatch into tiny maggots inside cherries. Larvae are creamy white, without a visible head, and legless. Adults are clear-winged flies with a distinctive pattern of dark bands on the transparent wings. They are weak fliers and will travel no farther than necessary to find a host tree. But where trees are close together, they can spread rapidly. Flies overwinter as pupae in the soil. Cherry fruits are not susceptible to egg laying until they ripen to a pinkish blush color.

MINIMAL-IMPACT INTERVENTION

Create blooming habitat for predators and parasites, but manage the ground cover directly beneath

Figure 13.2. Grass and clover ground cover around the base of fruit trees can physically suppress fruit fly larvae from burrowing into the soil to complete development and emerge.

cherry trees. Selectively mow beneath the trees just as the first fruits begin to color, and let the area dry out, because fruit fly egg laying is greatly enhanced by high humidity and moisture.

Ground covers and mulches around the base of trees can physically suppress larvae from burrowing into the soil to complete development and emerge. The best vegetation covers include grasses and other clovers with extensive, dense root systems.

Yellow sticky traps with an external bait of ammonium carbonate are a moderately effective monitoring tool, but not reliable to catch enough flies to decrease populations or even to catch the first flies entering your orchard.

For small home orchards, grow early-maturing varieties such as Chelan to escape fruit fly damage.

HEAVIEST-IMPACT INTERVENTION

Research in the western US suggests that applications of spinosad 10 to 21 days apart, starting when developing cherries begin to lighten in color and continuing through harvest, provide good suppression of fruit flies.

Codling Moths

Adult codling moths (*Cydia pomonella*) are ½ to ¾ inch (13–19 mm) long and gray with a distinctive dark band at the tip of their wings. Larvae are cream-colored, ¼- to ½-inch-long (6–13 mm) caterpillars with dark heads. Codling moths overwinter as larvae underneath apple tree bark or occasionally in the soil at the base of apple and pear trees. Adults emerge in the spring when temperatures reach 50°F (10°C). Optimal temperature for codling moth development is 68 to 81°F (20–27°C). Cool temperatures (below 55°F [13°C]) and rainy weather limit codling moth flight and thus mating. High temperatures (above 92°F [33°C]) also limit codling moth flight. Generally, mating is most prevalent when sunset temperatures are above 62°F (17°C). Moths are usually active only a few hours before and after sunset. Hence this is the best time of day to target treatments aimed at adult moths.

After mating, codling moth adults lay eggs singly on or near the fruit. Each moth lays between 30 and 130 eggs. Eggs hatch, and tiny (⅛ inch [3 mm] long) white larvae feed on apple fruit surfaces producing shallow, brown, sawdusty spots. Within 4 to 48 hours larvae begin to tunnel into the fruit. Thus there is only a narrow window of time to treat the larval stage before they are protected within the flesh. Larvae grow inside a fruit, then leave it to crawl down the tree trunk to pupate on tree bark or in the soil. There are normally two to four generations of codling moths per year depending on climate (more in warm areas and fewer in cold regions). Orchardists are worried that climate change may increase the average annual number of codling moth generations.

Figure 13.3. Codling moth damage to apple.

HOSTS

Apple and pear; rarely, plum.

MINIMAL-IMPACT INTERVENTION

An integrated approach is necessary for this challenging pest, and timing is everything. If you are planting new trees or replacing old ones, choose firm-fruited varieties, which are somewhat less susceptible to damage.

Create undisturbed habitat for codling moth predators and parasites, including birds, spiders, and carabid beetles. Predation is low for the larvae feeding inside fruit, but eggs on fruit surfaces, newly hatched larvae, and mature larvae and overwintering pupae are at a greater risk of predation. Predation can remove up to 20 percent of codling moth eggs laid in unsprayed orchards. Major predators of the eggs are spiders, birds, ground beetles, native parasitic wasps, minute pirate bugs, predatory thrips, assassin bugs, and green lacewing larvae. Wolf spiders, running crab spiders, harvestmen spiders, earwigs, and carabid beetles were found to be effective codling moth egg and larva predators in Washington State orchards. *Trichogramma* wasp releases have been used as one tool in an integrated codling moth control program in that state.

Remove and use or destroy apples that drop to the ground or are left on the tree. Hand-thin to remove all infested fruit (see figure 13.3 for the appearance of infested fruit) during each generation, before worms leave the fruit.

Take advantage of the fact that both mating of adult moths and larval survival are negatively affected by rainfall, especially in the evening. In one experiment 4 hours of overhead irrigation each

Figure 13.4. This wolf spider (family Lycosidae) in the orchard living mulch is an agile predator of codling moth. These spiders don't spin webs, but instead hunt down their prey.

Figure 13.5. Codling moth caught in a blacklight trap, which ran only from sundown until 11 P.M. Even so, other nontarget moth species were caught.

evening during moth flight and larval hatch decreased fruit injury by 90 percent. Remember, however, that evening moisture also encourages diseases like fire blight and apple scab!

Cardboard tree bands placed around the base of trees and covered with a sticky substance like Tanglefoot have been reported to trap as much as 65 percent of overwintering codling moths. Unfortunately, fruit damage was still 43 to 57 percent if tree banding was the only control. Also, if bands are not removed and burned after each generation, the bands can actually increase codling moth populations by giving them a nice, safe place to pupate. Banding works best on smooth-barked apple varieties such as Red Delicious.

Traps can be effective. There are several types:

Molasses and palm sugar bait traps with a 10 percent solution placed in yellow-jacket domes attract adult moths. The traps are hung in the lower canopy of trees. More female than male moths are attracted to these traps.

Codling moth pheromone traps catch only male moths; they will not control female moths in most situations. Pheromone traps are important, however, in order to know when the first codling moth adults appear and thus predict the first larval hatch. They are available online and at nurseries and garden supply stores.

Kairomone traps, sold commercially as the DA lure for codling moth, catch both female and male moths. The traps are available alone and in combination with pheromones (combo lure).

Blacklight traps used only from sundown until 11 P.M. reduced codling moth populations by 39 percent without catching large numbers of nontarget night-flying moths in one study. But nontarget moths are at risk with this method, so use only during codling moth flights based on degree-day calculations (see below).

Timing interventions based on degree-day calculations are explained in the "Degree-Day Models" section on page 128. This helps to time treatments and sprays using codling moth pheromone traps. The trap tells you when each generation or flight begins. The degree-day calculation lets you know just when egg hatch will occur and when the next generation should begin to fly. You can calculate degree days with a maximum-minimum thermometer and a degree-day chart, or you can use online university degree-day calculation tools, which are available in most states and local extension service areas where apples are grown. Codling moth egg hatch occurs at 250 degree days, base 50°F (10°C), after moths have been caught in traps for 3 consecutive days (evenings) in a row. For commercial growers, if large numbers of moths are being caught in traps each evening, do not wait until 200 to 250 degree days to treat. Instead apply your first spray earlier, at the beginning of egg hatch (160 degree days). Degree-day calculations are pretty accurate, but climate events (wind, rain, and warmer winters) and changing climates affect their accuracy.

Mating disruption. For apple blocks of more than 4 acres (1.6 ha), special codling moth pheromone twist-tie dispensers can be purchased and applied to each tree (approximately 200 to 400 per acre [0.4 ha]) to confuse and prevent male moths from finding and mating with female moths. Mating disruption is generally not a good management tool for less than 4 acres of fruit production. Mating disruption has resulted in successful codling moth control in commercial apple and pear orchards in California, Oregon, Washington, and Virginia when codling moth density was low to moderate.

Codling moth granulosis virus sprayed repeatedly to coincide with egg hatch reduced injury by 60 to 80 percent in California tests. University of California studies report that, in home garden trees, granulosis virus sprayed weekly during egg hatch throughout the season was as effective as chemical sprays for controlling codling moth. In Colorado organic orchards, granulosis virus was effective when sprayed at first hatch and then bimonthly through second-generation larval hatch. Colorado orchardists also use heavier impact sprays (see below) alternated with granulosis bimonthly.

***Bacillus thuringiensis* (*Bt*)** is generally not effective for codling moth control because it targets only larvae, and the window when larvae are vulnerable is very small. Because it degrades rapidly and the window is small to catch codling moth larvae, *Bt* must be sprayed several times (every 3 to 5 days) to coincide with first- and second-generation larval hatch. It requires a lot of sprays and perfect timing. Carl and I were never able to make *Bt* work well for codling moth control, certainly not well enough to justify the expense.

MODERATE-IMPACT INTERVENTION

Kaolin clay is effective to suppress egg laying by first-generation codling moths, applied when apple blossom petals fall and fruit begins to form. For second and third generations, kaolin clay should be used in conjunction with other sprays; it is inadequate if used alone.

Narrow-range or lightweight oils (petroleum- or vegetable-based) as part of an integrated approach are somewhat effective to smother codling moth eggs. They must be reapplied frequently during the egg-laying period, when adult moths are flying and being caught in traps. Oils can burn leaves if applied within a few weeks of a sulfur spray or at higher rates when air temperature is above 90°F (32°C). Oils can also injure soft-bodied beneficial insect larvae when reapplied frequently.

HEAVIEST-IMPACT INTERVENTION

According to several studies, spinosad sprayed bimonthly when codling moth larvae first hatch has been most effective when used in combination with bimonthly alternating sprays of granulosis virus. The first spring generation usually requires three sprays applied at 10-day intervals beginning at or when the first stings are found or at first larval hatch, calculated by degree days. For later generations, two spinosad sprays have been shown to be effective, with the first spray applied at the beginning of each new egg hatch and the second spray applied 10 to 14 days later. No more than six spinosad sprays should be applied per season. It is a big commitment to spray for codling moth, and you can begin to understand why Carl and I decided to accept some codling moth damage rather than devote so much energy to a spray program.

In studies conducted in North Carolina apple orchards, two insect growth regulators, methoxyfenozide and tebufenozide, showed some activity against codling moth eggs.[3] Insect growth regulators disrupt the normal molting process and slowly kill codling moth larvae before they become mating adults. These substances may have lower environmental impact than sprays such as spinosad, but available products are synthetic and not allowable for use on certified organic farms.

Earwigs

Earwigs (*Forficula auricularia*) are about 3/4 inch (2 cm) long, dark reddish brown, and have two long pincers on the rear. They are able to fly, but rarely do. Most of their hunting and foraging is done at night. They usually spend days hiding in dark, moist cracks and crevices, which is why you might find them at the stem end of fruits. Earwigs are harmless to humans and generally do little damage to crops. In fact, earwigs are beneficial insects because they feed on pests such as codling moth, mites and aphids. Earwigs are frequently blamed for damage that is actually being caused by other pests. I find them regularly in the tops of peaches, but usually they are eating mites or aphids. They will make holes in or enlarge holes made by other pests in soft fruit. Orchard researchers in Washington State report that earwigs are strong predators of codling moth and have been observed preying on codling moth pupae.

HOSTS

Decomposing organic residue; but they can be found on fruit and inside fruit cracks.

MINIMAL-IMPACT INTERVENTION

If you must, or if earwigs are causing damage to ripe fruit, trap them by placing rolled-up newspapers at the base of trees or in lower branches (they crawl into small places to hide.) You can also trap earwigs by setting out shallow cans at the base of trees and filling each one 1/2 inch (13 mm) deep with a vegetable-oil-and-water mix.

Eriophyid Mites

These microscopic mites (Eriophyidae family) cause russeting and discoloration of fruits as well as yellow raised spots and rusty-colored blisters on leaves. Fruit may be misshapen with brown, scabby scars. As buds swell in the spring just before starting first bloom, mites begin to move to leaves and fruit buds as they open. Most blister development occurs in the spring. At fall leaf drop, mites migrate to buds where they usually overwinter as eggs. The damage they cause usually looks worse than it actually is, especially on older tress that can sustain a lot of blistered or rusty-spotted leaves. But very heavy infestations might reduce bloom, especially on young trees.

HOSTS

Apple, pear (pear russet and pear leaf blister mites), and sometimes stone fruits.

Figure 13.6. Eriophyid mite damage to pear leaves: raised spots and rusty-colored blisters.

MINIMAL-IMPACT INTERVENTION

Predators are usually very effective in suppressing eriophyid mites, including predaceous mites (*Phytoseiulus* species and western predatory mites); tiny, round, black beetles (*Stethorus* spp.) related to lady beetles; and a predaceous thrips species called six-spotted thrips.

MODERATE-IMPACT INTERVENTION

Treat with horticultural oil *or* with sulfur (but not both) as the first buds break and trees are coming out of dormancy in the spring. Do *not* mix sulfur and oil together or apply them within 2 weeks of each other. Dormant bud oil sprays are not particularly effective; wait until buds are swelling and ready to break. If eriophyid mite feeding has been heavy (more than 50 percent of leaves damaged), consider a second oil or insecticidal soap application when the first leaves begin to expand in the evening. Do not spray on tender, young leaves if temperatures are higher than 85°F (29°C). Early-spring oil sprays can injure early-emerging soft-bodied predators such as lady beetle and syrphid fly larvae.

European Apple Sawflies

Apple trees are the only host plant for the European apple sawfly (*Hoplocampa testudinea*). Sawfly larvae resemble caterpillars but have long prolegs just below the head. Sawflies overwinter as larvae in the soil. The wasplike adults emerge during late pink stage and early bloom to lay eggs on the calyx end of developing fruit. Larvae feed on the fruit skin, creating a distinctive ribbon-like scar, and then bore into the core of the fruit.

MINIMAL-IMPACT INTERVENTION

When apples are at pink stage, place sticky white rectangle traps at head height near the edge of the canopy on the south side of trees along the orchard periphery. Place traps near a large group of blossoms at a density of one trap per 3 to 5 acres (1–2 ha). These traps can be used for monitoring: Threshold levels for intervention are five flies per trap between pink stage and petal fall.

Parasitic nematodes applied to irrigated soil at petal fall have been used successfully for this pest.

HEAVIEST-IMPACT INTERVENTION

One spray of spinosad can be applied after petal fall if monitoring shows that sawflies have reached a treatment threshold.

Leafhoppers

White apple leafhopper (*Typhlocyba pomaria*) and rose leafhopper (*Edwardsiana rosae*) excrete honeydew, resulting in white stippling on leaves and sticky exudate on fruit and leaves. Adults are small, winged, pale green to whitish insects ⅛ inch (3 mm) long. Populations need to be very high, with more than 50 percent of leaves showing stippling, before treatment is necessary.

HOSTS

Apple, cherry, plum, peach, and sometimes pear.

MINIMAL-IMPACT INTERVENTION

Yellow sticky traps can attract and decrease leafhopper populations. But yellow sticky traps also catch beneficial insects.

MODERATE-IMPACT INTERVENTION

Insecticidal soap and neem alone or in combination sprayed on leaf undersides in the early morning or later evenings, when leafhopper adults do not move so fast, can provide suppression if applied every 5 to 7 days. Using neem repeatedly can increase its ecological impact. Best suppression occurs if you discover an infestation early and treat when the young wingless nymphs are present, before adults emerge.

Leafrollers

Several species can infest fruit trees, including fruittree leafroller (*Archips argyrospila*), oblique-banded leafroller (*Choristoneura rosaceana*), variegated leafroller (*Platynota flavedana*), and pandemis leafroller (*Pandemis pyrusana*). Larvae chew on buds and roll up and tie leaves together with silk webbing. Sawdusty frass is present inside the rolled leaves. Fruits and fruits may have scarring from feeding damage. Pale green to whitish-colored larvae wriggle vigorously when disturbed and may drop from trees on a silken thread. The adult moths can fly several miles to find fruit trees.

HOSTS

Apple, pear, plum, cherry.

MINIMAL-IMPACT INTERVENTION

Maintain blooming habitat for predators and parasites beneath trees and in row middles. Several native parasitic insects provide adequate suppression of oblique-banded leafrollers, including a species of tachinid fly and species of braconid, ichneumonid, and chalcid wasps. Spiders, pathogens, and other predators suppressed oblique-banded leafrollers in a University of Wisconsin study.

Monitor plants weekly for signs of infestation. *Bacillus thuringiensis* can be sprayed just after apple blossom petal fall when larvae are ⅛ to ½ inch (3–13 mm) long; sprays must reach inside the rolled leaves. *Bacillus thuringiensis* var. *kurstaki* (*Btk*) or *B.t.* var. *Berliner* (*BtB*) is less effective on larger larvae.

Oriental Fruit Moths

Oriental fruit moths (*Grapholita molesta*) thrive in both humid and dry peach growing areas throughout the world. They overwinter as worms on the tree or in the orchard ground cover. The adult moths attack new, succulent terminals of rapidly growing branches in spring and early summer. Fruit are usually attacked when they are starting to ripen. Oriental fruit moth worms have a light brown head and are white, gradually turning pinkish white as they mature. They can be confused with peach twig borer worms, but the twig borer worms have a dark brown head and distinctive bands of color on their body.

HOSTS

Mainly peach and nectarine, also cherry, apricot; rarely, apple.

MINIMAL-IMPACT INTERVENTION

Maintain blooming habitat for predators and parasites beneath trees and in row middles. *Macrocentrus ancylivorus* is a common parasite of Oriental fruit moth worms. In unsprayed orchards with good habitat, parasitism can reach levels of 80 to 90 percent by late summer and early fall and provide long-term suppression. Research in California reported that Oriental fruit moth was suppressed by planting a small (0.3 to 0.5 acre [0.1–0.2 ha]) plot of sunflowers near or in the orchard, and replanting two or three times, so that there is a continual bloom all summer. The sunflowers provided *Macrocentrus* with an overwintering host, which allowed the parasite populations to build more rapidly the next season.

Mating disruption with Oriental fruit moth sex pheromones applied to every tree in the orchard is most effective in orchards with low moth populations, but efficacy is reduced by small orchard size (less than 4 acres [1.6 ha]). Mating disruption needs to be deployed in your orchard as soon as the first adult moths are captured in pheromone traps in early spring.

There are specific adaptations of codling moth granulosis virus for Oriental fruit moth that are somewhat effective to reduce populations when applied in the spring.

Figure 13.7. Oriental fruit moth damage to peach fruit.

HEAVIEST-IMPACT INTERVENTION

Spinosad sprayed at late bloom in the evening is somewhat effective. Timing of sprays can be determined with degree-day calculations (as explained in the "Degree-Day Models" section on page 128). The developmental thresholds of Oriental fruit moth are a lower threshold of 45°F (7°C) and an upper threshold of 90°F (32°C)

Peach Twig Borers

Peach twig borer (*Anarsia lineatella*) worms overwinter on fruit trees and emerge in early spring, just before and during bloom, to attack newly emerged leaves, blossoms, and shoots. They attack fruits as they mature and ripen. It is hard to tell the difference between peach twig borer larvae and Oriental fruit moth worms. Peach twig borer worms have a dark brown head and distinctive alternating dark and light brown bands on their body, whereas Oriental fruit moth worms have a lighter brown head and white or pinkish white body.

HOSTS

Peach, apricot, nectarine, and plum.

MINIMAL-IMPACT INTERVENTION

Macrocentrus ancylivorus is an effective parasite of peach twig borer worms and is encouraged by blooming habitat for predators and parasites beneath trees and in row middles. Planting sunflowers as described in "Oriental Fruit Moth" on page 312 helps encourage *Macrocentrus* and suppress peach twig borer. Another tiny wasp, *Pentalitomastix pyralis*, is a parasitoid of peach twig borer eggs.

Mating disruption with peach twig borer sex pheromones placed in every tree in early spring is most effective when moth populations are low, but efficacy is reduced by small orchard size (less than 4 acres [1.6 ha]).

Bacillus thuringiensis (*Bt*) sprayed at late bloom can be somewhat effective, but the timing must be perfect.

HEAVIEST-IMPACT INTERVENTION

Dormant oil sprays can be helpful, as is spinosad sprayed in the evening at late bloom. Use degree-day models to help time sprays. Doing degree-day calculations (as explained in the "Degree-Day Models" section on page 128) is important for timing sprays. The lower developmental threshold of peach twig borer is 50°F (10°C). The degree-day look-up table for codling moth can be used for peach twig borer, too.

Peach Tree Borers

Peach tree borer (*Synanthedon exitiosa*) is a native North American pest whose larvae are small (up to 1 inch [2.5 cm] long) whitish to brown worms that feed beneath the bark of the lower trunk near the soil surface and on larger roots. Adult moths lay eggs on the bark of the lower trunk and in soil cracks near the tree base in mid- to late summer. When eggs hatch, young larvae tunnel into the sapwood through bark cracks and wounds, where they continue to feed and develop. They spend the winter belowground under the bark. In the spring larvae resume feeding. Most injury occurs in mid- to late spring as the larvae mature.

Figure 13.8. Peach tree borers feeding on trunks can be spotted by dark-colored gum and sawdusty insect excrement.

HOSTS

Peach, nectarine, cherry, plum, apricot.

MINIMAL-IMPACT INTERVENTION

Avoid mechanical injury to the trunks. Trunk injury greatly increases attractiveness to peach tree borers. Healthy, adequately watered trees are less likely to invite infestation. A USDA study in Georgia found that both spring and fall applications of beneficial nematodes provided good control of peach tree borer.

Mating disruption is an option for stone fruit orchards greater than 1 acre (0.4 ha) in size.

MODERATE-IMPACT INTERVENTION

Paint tree trunks and exposed roots with a paste of kaolin clay in early spring. Coat the base of tree, exposed roots and trunk to a height of 12 inches (30 cm).

HEAVIEST-IMPACT INTERVENTION

Spray with neem oil on the lower trunk and around the crown of the tree in late spring or when adult moths are anticipated in your area (ask local extension agents and nursery owners). Saturate both bark and soil thoroughly with neem oil.

Plum Curculio

Plum curculio (*Conotrachelus nenuphar*) is not a pest in the western US, but it can do serious damage in the Midwest and eastern US. These small, brown-gray snout weevils have definite fruit variety, site, and weather preferences. They do not like cold snaps in early spring, high winds, or low humidity. Holistic orchardist Michael Phillips, who spent many years observing plum curculio behavior in his New England orchard, noted that warm, early springs favor early emergence on stone fruits, but cold springs often result in a stronger focus on apple.

Plum curculio adults overwinter in deciduous leaf litter, usually to the south of the orchard. At pink bud in early spring, adults begin to migrate into orchards and feed on developing fruit. Females make small holes in fruit skins and lay eggs. Eggs hatch and the grubs immediately begin to feed inside fruits, moving toward the center. Infested apples and plums fall to the ground about the same time as natural June drop, when fruit trees shed extra tiny fruits they don't want to support to maturity. Monitor the outer rows of an orchard regularly for fresh feeding or egg-laying scars that look like oozing punctures or a small half-moon. The larva is a grayish white, legless, C-shaped grub with a small brown head and is ⅓ inch (8 mm) long when fully grown. The grub is found inside infested fruits in early to midsummer.

HOSTS

Apple, pear, quince, and stone fruits such as plum, cherry, peach, and apricot. Formosa and Santa Rosa plums are especially attractive to plum curculio.

MINIMAL-IMPACT INTERVENTION

Hedgerows that include wild plums, native crab apples, and hawthorn are very attractive to plum curculios and can act as trap crops. However, plum curculios may use these hedgerows to migrate into orchards (see "Moderate-Impact Intervention").

In one study researchers at the Southeastern Fruit and Tree Nut Research Station in Georgia found that soil applications of the nematode *Steinernema riobrave* can suppress plum curculio larvae by 78 to 100 percent.

MODERATE-IMPACT INTERVENTION

You can spray wild plum hedgerows to reduce migrating plum curculio, but if you want to avoid spraying, another slower way to suppress the insect is to apply the parasitic fungus *Beauveria bassiana*. Research in Michigan involves infecting grains of rice and then placing them into the soil beneath susceptible trees prior to larval drop from fruit. The *Beauveria bassiana* slowly suppresses pupation of larvae in the soil.

Spray with kaolin clay in early spring. Spray perimeter trees first and most heavily (trees should look white). Apples need clay coverage in place at petal fall. Clay coverage on peaches, apricots, cherries, and plums is usually needed even sooner, just after full bloom at first petal fall.

Panel and pyramid traps baited with attractive synthetic odor and placed on perimeter orchard trees near adjacent woods or plum hedgerows can be used to predict the onset and magnitude of plum curculio migration into orchards. The most attractive lure developed for northern populations of plum curculios is a synthetic combination of benzaldehyde, a volatile component of both apple flowers and plum odor, and grandisoic acid, the plum curculio aggregation pheromone.[4]

HEAVIEST-IMPACT INTERVENTION

Pyrethrin and spinosad have been used on plum curculio. With diligent monitoring and an early kaolin clay spray, you may be able to limit application of these high-impact materials to perimeter orchard rows only. Spraying selected perimeter-row baited trap trees with spinosad or pyrethrum may be a good suppression tool. The trap trees contain traps baited with a combination of benzaldehyde and grandisoic acid.

Scale

Scale is mainly a pest of fruit tree branches and twigs, but it can feed on fruit, leaving pinprick red spots.

San Jose scale (*Quadraspidiotus perniciosus*) are small, hard, gray to black, and cone-shaped, with a tiny white knob in the center surrounded by rings or grooves. Oystershell scale (*Lepidosaphes ulmi*) look like tiny oysters. Scale insects overwinter on the tree and in late spring, tiny crawlers emerge and move to feeding sites on the bark, where they become stationary for the rest of the season or until there is a second generation in late summer. Crawlers are very small, oval, whitish to orange insects with six legs.

HOSTS

Apple is the main host; also peach, plum, pear, and cherry. Stone fruits can host several species of scale insects.

MINIMAL-IMPACT INTERVENTION

Many organisms prey on scale insects, including lady beetles. Larvae of green lacewings are aggressive scale predators. Several parasites prey on scale, including *Encarsia perniciosi*, *E. berlesi*, and *Aphytis proclia*. Scale are usually only a problem if your farm or garden agroecosystem is out of balance as a result of spraying insecticides that kill beneficial insects.

MODERATE-IMPACT INTERVENTION

To best time a spray, use two-sided sticky tape wrapped on small limbs in infested areas to determine when crawlers are active. Or if you're not monitoring for crawlers, spray a 2 percent solution of insecticidal soap alone or neem, or horticultural oil about 10 days after full petal fall, as long as temperatures are less than 85°F (29°C) and the trees are not drought-stressed. Repeated neem applications have a heavier ecological impact.

Spider Mites

Spider mites are mainly a pest of apples, but may be seen on other kinds of fruit trees. Mites are more closely related to spiders than insects; they have two body regions and four pairs of legs. Their very small size makes them difficult to detect, identify, and monitor. The most common of the many species of tree fruit leaf mites are European red spider mites and two-spotted spider mites. They can complete a generation in as little as 10 days at high temperatures (above 80°F [27°C]). Mites prefer warm temperatures, dry and dusty conditions, and low humidity. They thrive with excess nitrogen fertilization. Mites are usually a problem only if your farm or garden agroecosystem is out of balance from spraying insecticides that kill beneficial insects.

MINIMAL-IMPACT INTERVENTION

Use of sprinkler irrigation can diminish mite populations. Ensuring adequate irrigation is also important. Drought and fluctuating wet/dry soil conditions can stress plants and cause spider mite populations to increase. Avoid excessive nitrogen fertilization, which can enhance mite populations.

Many organisms prey on spider mites, including minute pirate bugs, plant bugs, *Stethorus* beetles, predatory mites, predatory thrips, and dustywings (*Coniopterygidae*), which are related to lacewings but are smaller and have wings covered with a white, powdery wax. Maintaining and managing cover crop ground cover in orchard reduces pest spider mite populations. In a 5-year study at Michigan State University, researchers evaluated ground cover experiments in established commercial orchards and in a new tart cherry orchard at the Northwest Horticultural Research Station. They monitored orchards with compost, mulch, or cover crops, including various species of clover, crown vetch, indigo vetch, alfalfa, rye, annual ryegrass, hard fescue, and buffalo grass. They found significantly higher season-long populations of beneficial mites in orchards and plots with a red clover cover crop ground cover. On the other hand, season-long vegetation-free strips (using either herbicide or mulch) had significantly higher pest mite populations. My experience bears out this research. I have not seen spider mites on the foliage of my fruit trees in any of my orchards, except as a curiosity, for the past 20 years.

MODERATE-IMPACT INTERVENTION

Spray a 2 percent solution of insecticidal soap, but only if temperatures are less than 85°F (29°C) and if the plants are not in direct sun or drought-stressed. Mix sulfur with soap sprays (1 tablespoon sulfur per gallon [4 L]) for higher populations; spray in the evening, and repeat applications every 5 to 7 days in hot weather.

Stink Bugs

Stink bugs (several species) cause catfacing or dimpling of stone fruit. Damage on apples looks a bit like bitter pit, but bitter pit damage tends to be around the bottom of the fruit, while stink bug damage is typically around the top. Stink bug adults are shield-shaped, ½ to ⅔ inch (13–17 mm) long, brown or green in color, with a triangle on the body below the head. Nymphs look like smaller versions of the adults, with more rounded, brightly patterned black, red, white, and green bodies. Stink bugs overwinter as adults and become active in the spring. It is easy to remove fruit damaged early during thinning. However, stink bugs feed throughout the fruit growing season; adults may migrate to orchards in late summer during dry years when the surrounding vegetation starts to dry.

HOSTS

Apple, pear, cherry, peach, apricot, and plum.

MINIMAL-IMPACT INTERVENTION

Natural enemies of stink bugs include birds (wrens, bluebirds, cardinals), spiders, daddy longlegs, assassin bugs, predatory stink bugs, earwigs, green lacewing larvae, ground beetles, praying mantises, and parasitic wasps. *Telenomus podisi* and *T. utahensis* are two scelionid wasp egg parasitoids that have suppressed stink bug populations in Washington State orchards. Researchers have observed egg mortality as high as 50 to 60 percent.

Tarnished Plant Bugs

Adult tarnished plant bugs (*Lygus pratensis*) are oval, brown, ¼-inch-long (6 mm) bugs with a white triangle on the back; nymphs are smaller, yellow-green, and have black spots on the back. Their feeding leads to russeted and deformed fruit.

These plant bugs do not reproduce on fruit trees, but overwinter as adults in weeds, such as mullein and legumes, in orchard ground cover. Adults fly to fruit trees in the spring to feed on developing flower buds. They feed on apple trees between early pink bud development until just after petal fall. On peach trees adults feed on buds at the pink stage and are most abundant at petal fall. Adults leave trees soon after petal fall to feed on blooming weeds and ground cover plants.

Tarnished plant bugs prefer to lay eggs on plants about to flower and will stay in the ground cover if it is undisturbed and there is profuse and sequential bloom, especially during fruit tree bloom. Development occurs as long as temperatures exceed about 50°F (10°C), but slows or stops when temperatures exceed 94°F (34°C).

Feeding on fruits leads to indentations, bumps, or russeting of the flesh. Tarnished plant bugs remain in weeds and ground cover unless the plants start to dry out or are mowed. Weed hosts include wild carrots and other umbelliferous plants, redroot pigweed (and other amaranths), lambsquarters, mustards, shepherd's purse, goldenrod, and mullein. Alfalfa is a favorite of tarnished plant bug, and harvesting or mowing alfalfa often causes the bugs to move into tree fruit crops.

HOSTS

Apple, pear, peach, and apricot.

MINIMAL-IMPACT INTERVENTION

When planning new orchard plantings, avoid siting vulnerable fruit tree species adjacent to alfalfa hay fields or strawberry fields, which are preferred hosts for tarnished plant bugs. Tarnished plant bugs stay in the orchard ground cover while temperatures are cool and move into the tree during the heat of the day. Predators such as lady beetles, many kinds of predatory bugs, and crab spiders prey on nymphs. The braconid wasp *Peristenus pallipes* attacks nymphs while another parasitic wasp, *P. digoneutis*, reportedly can cause up to 50 percent mortality of adult bugs. These predators and parasites feed on tarnished plant bugs on fruit trees and on the orchard ground cover.

To encourage predators and parasites, maintain undisturbed orchard ground cover. In particular, to avoid stimulating adult bugs to fly up into trees, do not mow orchard ground cover from bloom through petal fall.

White sticky traps may be used to control low-level infestations in small gardens.

HEAVIEST-IMPACT INTERVENTION

If populations are high and damaging, you can try spraying PyGanic, which is a commercial form of concentrated pyrethrin allowed for control of tarnished plant bugs on certified organic farms. PyGanic is hard on predators and parasites. Before choosing to spray, assess whether the economic loss due to bug damage will really outweigh the impact of spraying on overall orchard health. If you choose to spray, avoid doing so during bloom periods.

Thrips

Western flower thrips (*Frankliniella occidentalis*) are most damaging to peaches and nectarines, where they cause surface scarring, russeting and catfacing on fruit. Nectarine is more susceptible than peach to thrips damage. On apples, fruit damage is marginal. It appears as russeting and small white patches on fruit surfaces.

Western flower thrips overwinter in leaf litter and/or flowers. They emerge early, when peaches are at pre-bloom popcorn stage. Then thrips immediately seek out flowers, where they feed on pollen and nectar; they lay eggs in leaves or flowers. Thrips eat pollen and nectar of close to 250 species from 62 plant families. With that in mind, your job is to encourage thrips to feed on flowers other than the ones on your fruit trees!

HOSTS

Peach, nectarine, and apple.

MINIMAL-IMPACT INTERVENTION

An integrated approach is necessary for this insect.

- Keep fruit trees well irrigated, and avoid excessive applications of nitrogen fertilizers, which may promote higher populations of thrips.
- Create blooming, undisturbed habitat for thrips in the orchard ground cover or on orchard perimeters to lure them away from fruit blossoms in the spring. Bloom timing is everything. The bloom period of this habitat must begin before peach bloom and last several weeks after petal fall. Verbena and other flowers in colors that are low-ultraviolet-reflective (white, blue-violet, blue, and yellow) are especially attractive to western flower thrips. In the spring western flower thrips also showed a preference for flowers located at ground level.[5]

- Minute pirate bugs, predatory thrips, green lacewings, mites, and certain parasitic wasps feed on adult thrips.
- Create blooming habitat for predators and parasites. Avoid creating dust, since thrips like dust but the predators do not. Straw, living mulches, and interplanting crops with buckwheat encourage predators and parasites and may also repel certain pests. The timing of mowing living mulches is very important with thrips pests. Thrips tend to live in flowering living mulches or buckwheat strips and move to crops when these are mowed. We learned to avoid mowing the orchard at bloom to discourage thrips from moving from blooming ground cover into peach and nectarine blooms. This selective mowing greatly reduced thrips populations in the trees along with thrips damage. Two species of predatory mites, *Amblyseius cucmeris* and *A. swirskii*, are very effective at controlling thrips larvae in greenhouses, but not adult thrips.
- White sticky traps are best for thrips monitoring and collection because yellow traps attract a large number of beneficial insects that prey on thrips.
- Plant-derived essential oils, such as rosemary oil, have shown efficacy to repel thrips.[6] I plan to experiment with adding rosemary oil to my Mineral Mix Bloom Spray for peaches and nectarines.

Figure 13.09. Western flower thrips damage to nectarine fruit.

MODERATE-IMPACT INTERVENTION

Insecticidal soap can be used for thrips, especially nymph stages, but it is not recommended for use on yellow-skinned nectarine varieties.

HEAVIEST-IMPACT INTERVENTION

Both neem oil and pyrethrin suppress thrips if sprayed at petal fall. Spinosad performs well against thrips, but can be toxic to bees. If you decide to use spinosad, do not spray until petal fall.

CHAPTER 14

Fruit Tree Diseases and Interventions

Many tree fruit diseases are found on fruit trees worldwide, including fire blight, brown rot, leaf curl, and apple scab. It turns out that when we spread fruit trees throughout the world, we also spread the disease organisms that have co-evolved with them.

In this chapter I describe the most common tree fruit diseases in North America as well as interventions to help prevent occurrence and spread of disease. As with other descriptions of interventions in this book, I first list those that have minimal ecological impact, then those with moderate impact, and lastly the heaviest-impact materials. I also include a discussion of nutrient deficiencies and physiological disorders at the end of the chapter.

The best thing to suppress fruit tree diseases is to give your trees plenty of space and allow sun and air to readily penetrate through the branches. Most diseases are encouraged by dark, damp conditions. I present more specific biological information, including the optimal temperatures and moisture conditions each disease requires to cause infection and spread rapidly. I also present details on new research that shows the disease suppressive effects of spraying or amending soil with certain beneficial microbes. This is to remind us how important it is to create habitat for native beneficial soil and leaf microorganisms, as discussed in chapter 6.

Apple Scab

Apple scab is very easy to identify. The first symptom of apple scab (*Venturia inaequalis*) on leaves is small olive-green circles with velvety margins that enlarge and darken to brown or black. Spots on fruit are dark and may be cracked. Yellowing, browning, and death of leaves may result. Early drop may occur. Scab fungi overwinter on fallen leaves and also in apple buds and twigs. Overwintering spores become wet and are forcibly discharged and disseminated by the wind. Spores land on newly emerging leaf tissue and germinate in a film of moisture. After germination the relative humidity must be 95 percent or greater for infection to continue. Secondary infections occur throughout the summer if weather is wet and warm. Optimal temperatures for infection are 60 to 70°F (16–21°C). Apple scab spread is both temperature- and moisture-dependent. For example, it takes 12 hours at 53°F (12°C) before infection occurs, while only 9 hours of leaf wetness are required for infection at 65°F (18°C).

HOSTS

Apple and crab apple. Pear is host to another species of *Venturia*.

MINIMAL-IMPACT INTERVENTION

Practice good sanitation, pruning, and thinning. Use less and slower-release nitrogen fertilizer. If you have had heavy apple scab infection in a small garden area, you might rake up and destroy infected leaves in the fall.

There are many research studies on apple scab intervention.

- Studies have shown that orchards with a mix of scab-susceptible and scab-resistant varieties develop less scab when compared with monocultures.[1]
- Use of the microbe *Cladosporium cladosporioides* H39 reduced spore production and severity of scab on leaves and fruit. [2]
- In some studies compost tea sprays were effective as a scab preventive, while in others compost tea had no effect.
- There is some evidence that rock dusts sprayed on foliage can suppress apple scab. Carl's Mineral Mix Spray contains rock dusts, and we had not seen any apple scab in our orchard, but we never set up a trial comparing our mineral mix on apples with an unsprayed control.
- Potassium bicarbonate was reported to provide good protection against apple scab when applied repeatedly as a 1 percent solution during a multiple-year study in Belgium. This is really more of a moderate-impact intervention however.

HEAVIEST-IMPACT INTERVENTION

Sulfur can be sprayed before apple scab infection periods, when the weather is wet and warm (60 to 70°F [16–21°C]), to protect leaves from infection. Sulfur must be reapplied when precipitation (or irrigation) reaches greater than 1 inch (2.5 cm) since the previous application.

For heavy scab infection, apply lime sulfur as buds start to break in the spring. Use oil-free lime sulfur to avoid bud injury. Lime sulfur can be sprayed on foliage, but it can cause burning. It can also injure blooms, so use carefully to avoid bloom. Use it only when apple scab temperature and moisture conditions favor extended infection periods.

Armillaria Root Disease

Armillaria root rot (*Armillaria* spp.), also called oak root rot, causes weak, slow growth, yellow to brown foliage, branch dieback, and eventual death of the tree. White mycelial fans are present in the cambium and bark of roots and stems. Honey-colored mushrooms (the fruiting bodies) may be found at the base of trees during wet periods in the fall. Fungal spores are windblown to other areas. The fungus also spreads from diseased trees or stumps by way of infected root systems coming together and by fungal mycelium growing through the soil from infected to susceptible tree roots. The fungus continues to spread along the root until it reaches the root collar where it spreads to other primary roots. *Armillaria* spores survive on dead roots and stumps for 20 to 30 years. Contact with old roots can infect new trees.

HOSTS

Most fruit trees; some small fruits such as raspberry, strawberry, and grape; many conifers and hardwoods.

MINIMAL-IMPACT INTERVENTION

Plant on raised beds or berms if you have heavy, wet soils. Protect plants against injuries to the lower trunk and upper portion of the root system. If using drip irrigation or micro-sprinklers, move emitters and sprinklers away from tree trunks after trees are well established to avoid formation of wet areas right next to the trunk.

Remove and destroy severely infected trees, being careful to remove as much root material as possible from soil. If practical, do not replant where infected trees have been removed. If you must replant in the same spot, choose resistant fruit species such as persimmon, fig, pear on OHF rootstocks, and mulberry. The plum rootstock Marianna 2624 is reported to have *Armillaria* tolerance, but it suckers profusely and has a shallow root system.

Drying the *Armillaria* fungus kills it. There are some reports that *Armillaria* growth can be slowed or stopped by removing the soil from around the base of an infected tree and allowing the area to dry completely.

Bacterial Blight or Canker

Bacterial blight disease (*Pseudomonas syringae*) is also known as bacterial canker, gummosis, blossom blast, dieback, spur blight, and twig blight. Infected flowers often open, but then collapse, turn brown, and exude amber-colored gum. Infected leaves are spotted and yellow. Leaf spots often have yellow halos. The centers of leaf spots fall out, leaving a shot-hole appearance. Infected twigs turn black. Infection is usually limited to new growth; stems 1 year or older seldom have lesions. Amber-colored gum occurs on bark surfaces around sunken cankers. Underneath cankers the inner bark is reddish brown and streaked, as shown in figure 14.1.

Figure 14.1. Bacterial blight disease damage to a sweet cherry tree trunk.

The bacterium is spread by splashing rain and wind. Bacterial canker requires high moisture, splashing rain, and low temperatures for infection. Affected limbs usually fail to leaf out or produce new growth in the spring, or new growth dies soon after temperatures start to get hot.

Bacterial canker is more prevalent when trees are stressed by cold/freeze injury, low soil pH, phosphorus and micronutrient deficiency, high ring nematode populations, drought, and both too-low and excessive nitrogen. Vigorous trees are less susceptible, but young trees (2 to 8 years old) are most susceptible.

HOSTS

Peach, apricot, plum, and sweet and sour cherries; apple is an occasional host.

MINIMAL-IMPACT INTERVENTION

When planting new trees, choose canker-tolerant varieties or rootstocks. Less-susceptible or resistant rootstocks are becoming available for cherry, peach, and plum.

In one California study, Lovell peach rootstock provided the greatest protection from bacterial canker as measured by disease incidence and tree mortality. In another, Lovell and Viking rootstocks were more canker-tolerant than Nemaguard. MP-29 rootstock, a clonal plum × peach interspecific hybrid, shows promise as a bacterial-resistant rootstock for warmer climates, such as the southern US.

For cherry rootstock, Mazzard seedlings (particularly the F12-1) appear to be resistant to bacterial canker in Oregon studies. In California studies, Mahaleb is the most tolerant. The cherry cultivars

Corum, Regina, Rainier, Sam, and Sue appear to have sufficient tolerance to canker to be grown commercially without serious tree loss.

Avoid excessive nitrogen fertilization, but make sure phosphorus and micronutrient levels are sufficient. Avoid overhead sprinklers and keep irrigation water off tree trunks and branches. Timing of pruning is vital to suppress bacterial blight. Prune only in the spring, as late as possible before full bloom. *Do not* prune during cool, wet periods in the spring and fall. Prune infected branches at least 12 inches (30 cm) below cankers, and dispose of branches by burning or removing and letting them dry out. Disinfect pruning tools after each cut by dipping in a 10 percent bleach solution or a denatured alcohol dilution for at least 30 seconds.

Grow diverse ground covers, which will increase soil microbial diversity, minimizing in turn the proliferation of soilborne pathogens and pathogens that come in contact with the soil after pruning or leaf drop.

Research shows some of the suppressive effects that amending soil with microbes can have against bacterial blight:

- *Trichoderma harzianum* T-22 as a soil amendment was observed to reduce canker in two of three orchard experiments.[3]
- In one study, trees inoculated with arbuscular mycorrhizal fungi had fewer cankers than un-inoculated trees.[4]
- Application of plant-growth-promoting rhizobacterial products containing a bacterial mix reduced canker compared to the un-amended control.
- Organic products containing *Bacillus amyloliquefaciens* strain D747 and *B. subtilis* strain QST 713 are reported to suppress this bacterium.

MODERATE-IMPACT INTERVENTION

Apply Mineral Mix Bloom Spray containing micronutrients, potassium, and sulfur as described in chapter 7.

HEAVIEST-IMPACT INTERVENTION

Copper-containing sprays are sometimes advocated for bacterial canker management, but copper-based products have not worked well under conditions favorable for disease development. Also, in many areas, copper-resistant strains of the bacteria are present (see "Copper and Copper Hydroxide" on page 149 for the ecological impacts of copper).

Bacterial Leaf Spot

Bacterial leaf spot (*Xanthomonas campestris* pv. *pruni*) affects foliage, new shoots, and fruit. Water-soaked spots on leaves are concentrated near the tips of the foliage, along the mid-vein, and at leaf edges. Angular spots enlarge, darken, and become purplish black. Spots drop out, leaving shot holes, which can be confused with fungal shot-hole disease (compare figures 14.2 and 14.7). Fruit spots are sunken and crack as fruit expands.

Figure 14.2. Bacterial leaf spot disease can appear as a water-soaked area and black spot at the leaf tip along the mid-vein.

Bacterial spot is favored by windy, sandy sites, and infection occurs during warm, humid, wet weather. Optimal temperatures for disease development are 80 to 86°F (27–30°C) with water present in the form of dew, rain, or sprinkler irrigation. Transmission occurs via insects, water, and tools. The bacteria overwinter in dark, gummy branch tips and diseased twigs.

HOSTS

Nectarine, peach, plum, and apricot. Nectarines are generally more susceptible than peaches. Sweet and tart cherries are occasional hosts.

MINIMAL-IMPACT INTERVENTION

Avoid wetting leaves or water in the morning to encourage rapid drying. Low nutrient content in soil and in plants aggravates this disease. Removal of gummy branch tips and diseased twigs during spring pruning may help to reduce inoculum levels. Ground cover has been shown to suppress this disease compared with bare soil beneath trees, especially in windy locations. In fact, in windy locations one strategy is to "hide" more-susceptible varieties on the inside of orchard blocks and to plant varieties that are most resistant at exposed orchard edges.

SUPPRESSIVE SOIL MICROORGANISMS

An organic preventive biofungicide containing *Bacillus amyloliquefaciens* strain D747 has been shown in some studies to suppress fungal and bacterial plant diseases, including *Xanthomonas* species that cause bacterial leaf spot disease.

HEAVIEST-IMPACT INTERVENTION

Spray copper at first green tip only if you have had severe problems with bacterial leaf spot in the past and the weather is wet. Copper is ineffective once the disease is well established and can itself cause spotting on fruit tree leaves.

Brown Rot

Disease symptoms appear in the spring at first bloom if weather is wet. Flowers wilt, turn brown, and are often covered with fuzzy, brownish gray spores. Fruit infections start as soft brown spots which rapidly expand and produce a fuzzy mass. Whole fruits rot, then dry and shrink into a wrinkled "mummy." *Monilinia fructicola*, the fungus that causes brown rot blossom and twig blight, requires rain, humid weather, and temperatures ranging from 60 to 70°F (16–21°C) for infection and disease development. Development of brown rot (also called blossom and twig blight) is very slow at cool temperatures. For example, at 50°F (10°C) blossoms must remain wet for 18 hours before infection occurs. But only 5 hours of wetness are needed for infection at 77°F (25°C). Hot, dry weather stops new disease infection and slows disease development.

HOSTS

Apricot (and aprium), sweet and sour cherries, nectarine, peach, plum (and pluot/plumcot).

MINIMAL-IMPACT INTERVENTION

Practice good orchard sanitation, pruning, and thinning, and use less and slower-release nitrogen fertilizer. Brown rot has been suppressed in some California studies by maintaining moderately low levels of leaf nitrogen, adding slow-release nitrogen fertilizers (such as high-carbon compost), and removing mummified fruits from trees.[5] Higher ascorbic acid (vitamin C) content in fruit decreased peach susceptibility to brown rot in one study.[6]

Microorganisms *Aureobasidium pullulans* strains DSM 14940 and 14941, *Ulocladium oudemansii* strain U3, *Bacillus amyloliquefaciens* strain D747, and *B. subtilis* strain QST 713 are reported to suppress brown rot. Harglow, Tilton, Puget Gold, Harojoy, Harostar, Harcot, Harlayne, Harogem, Tomcot, and Hargrand have some brown rot resistance.

MODERATE-IMPACT INTERVENTION

Applying a Mineral Mix Bloom Spray that contains micronutrients, potassium, and sulfur helps to suppress the disease organism and boost plant health, thus decreasing the chance of infection.

HEAVIEST-IMPACT INTERVENTION

Spray sulfur spray at 20 to 40 percent bloom and again at 80 to 100 percent bloom on very susceptible varieties (such as O'Henry, Red Top, and Royal Glo peaches) to prevent infection, or for any variety if heavy rainfall and other conditions are occurring that result in high susceptibility to infection.

Figure 14.3. Brown rot disease damage to peach.

Cytospora Canker Disease

Leucocytospora cincta is a fungal disease that infects trees by entering areas of the bark that are weak, injured, or stressed due to drought or cold. It first appears as yellow-orange and black regions that later ooze a gummy secretion that darkens as the disease progresses. Cytospora canker is a warm-season disease with optimal growth occurring at 60 to 80°F (16–27°C) and peak development at 90°F (32°C). Canker infection is highest when temperatures are high and tree growth activity is low (late summer/early fall). Water stress, potassium deficiency, high pH-induced iron deficiency, insufficient thinning, shallow or heavy-textured (clay) soils, and high levels of ring nematodes in the soil attacking plant roots increase tree susceptibility to canker development.

HOSTS

Peach, apricot, plum, and sweet and sour cherries; apple, poplar, willow, birch, and aspen.

MINIMAL-IMPACT INTERVENTION

Avoid excessive nitrogen fertilization and potassium deficiency, both of which makes trees more susceptible to the disease. Water during dry weather to prevent drought stress, but avoid overhead sprinkler use. Use permanent ground covers to avoid the root injury that occurs during weed cultivation.

Prune only in early spring after trees begin growth and before full bloom. Avoid leaving pruning stubs. Prune out and remove all cankers on small branches, cutting at least 4 inches (10 cm) below the margin of the canker. If there are long, elongate cankers on the trunk or scaffold branches, you can remove these cankers. To do this, cut away all diseased tissue, both bark and wood, with a knife and then disinfect the blade by swabbing it in a solution of 7 parts denatured alcohol and 3 parts water. Paint over the excision area with a mixture of 4 cups (1 L) white latex paint and 4 ounces (113 g) powdered sulfur. Sometimes I add finished compost to the paint/sulfur mix, but it is harder to apply this mix.

MODERATE-IMPACT INTERVENTION

Control peach tree borer to avoid trunk damage where the fungus can enter the tree.

Fire Blight

Fire blight (*Erwinia amylovora*) bacteria overwinter in cankers on branches and trunks. When temperatures reach about 65°F (18°C), bacteria are active and

Figure 14.4. Fire blight symptoms on apple leaves and shoots. Photo by Sebastian Stabinger.

become visible as drops of clear to amber-colored ooze. Bacteria are spread to blossoms by rain and wind. Blossom-to-blossom transmission is carried out by bees and other insects as they pollinate flowers. Insects may also transmit bacteria to growing shoots. Conditions that promote fire blight are temperatures of 65°F or above, relative humidity 60 percent or more, rain, wind, and hail. At 75°F (24°C) blossom blight and shoot blight infections become visible in just 4 to 5 days as bacterial ooze. This ooze creates additional sources of bacteria for new infections. In dry, hot weather, when temperatures are above 85°F (29°C), new fire blight infection is halted.

HOSTS

Apple and pear; ornamentals in the Rosaceae family.

MINIMAL-IMPACT INTERVENTION

Avoid excessive nitrogen fertilization, which makes trees more susceptible to the disease. Water during dry weather to prevent drought stress, but avoid use of overhead sprinklers.

Select resistant apple varieties (see table 12.4). Be most cautious with very susceptible apple varieties such as Jonathan and Rome Beauty or with Bartlett and Bosc pears.

Blackened, wilted new shoots or spurs should be pruned off as soon as you see them. Remove shoots 8 to 12 inches (20–30 cm) below the last signs of browning. Remember to disinfect pruning tools between cuts with a 10 percent bleach solution or denatured alcohol, because contaminated tools can spread the bacteria. Prune out all cankers in limbs at least 8 inches below external evidence of the canker on apples, and at least 12 inches below on pears. If temperatures are below 40°F (4°C), it is not as important to disinfect tools between each cut. Prune in late winter while trees are dormant to decrease the risk of infection, but avoid excessive winter pruning, which may stimulate too much vegetative growth the following season.

MICROBIAL TREATMENTS

Organic products containing the yeast *Aureobasidium pullulans* sprayed at early bloom have been shown to be effective in suppressing fire blight. In fact, in six field trials conducted in Germany, an organic product containing *A. pullulans* sprayed two to four times during bloom decreased fire blight incidence by 82 percent.[7] (See "The Power of Bloom Sprays" on page 280 for more on native *A. pullulans*.) Woodleaf Farm and other researchers have found that *A. pullulans* is killed or suppressed by sulfur and lime sulfur sprays. There is some evidence that some formulations of sprays containing *A. pullulans* sometimes cause fruit to russet.

Another organic product that may have some activity against fire blight when sprayed at late bloom contains *Bacillus subtilis*. It works differently than products containing *Aureobasidium pullulans*, and some sources recommend spraying both products—*A. pullulans* at early bloom, followed by *B. subtilis* at late bloom. We tried these products in our Woodleaf California orchard, but had better results with our Mineral Mix Bloom Spray on apples and pears. *Pseudomonas fluorescens* is a biological control also sprayed at bloom that has shown some promise as a blossom blight preventive.

HEAVIEST-IMPACT INTERVENTION

Lime sulfur or copper can be sprayed when trees are dormant, just before buds break, or just as you see the first green leaf tips. Dormant sprays are good for fire-blight-infected cankers if you have a lot in your orchard. If infection periods occur after petal fall, you can apply a soluble copper spray. (Copper can cause fruit and foliage injury and has serious ecological impacts. See the details about spraying copper in "Copper and Copper Hydroxide" on page 149.) In experiments at Oregon State University, lime sulfur used to thin apples and sprayed at bloom, followed by organic products containing the yeast *Aureobasidium pullulans*, resulted in good suppression of fire blight.

Peach Leaf Curl

Leaf curl (*Taphrina deformans*) fungus first appears in spring as reddish areas on new leaves. These areas become thickened, puckered, curled, and deformed. Later in the summer these infected leaves drop. Leaf curl fungus overwinters on surfaces and buds of peach and nectarine trees and is favored by cool, wet weather during spring. Optimal temperatures for this disease are 68 to 79°F (20–26°C); the minimum is 48°F (9°C), and the maximum is 87°F (31°C). However, infection can occur under cool conditions, 46 to 53°F (8–12°C), when wetting periods more than 10 hours long occur with rainfall greater than ½ inch (13 mm), or when trees are wet from rain, dew, or irrigation for

Figure 14.5. Peach leaf curl symptoms on peach leaves and shoots with deformed, curled, red leaves.

2 or more days. In fact, cool spring weather can prolong disease development by favoring the pathogen while also slowing tree growth. Peach leaf curl disease stops when young leaves mature or the when weather turns dry and warmer than 87°F (31°C). Diseased leaves drop by midsummer.

MINIMAL-IMPACT INTERVENTION

Practice good sanitation, and maintain and periodically mow ground cover to stimulate microbial activity and rapidly decompose diseased falling leaves. Prune as needed to maintain open trees with good airflow.

MODERATE-IMPACT INTERVENTION

Applying a Mineral Mix Spray as described in chapter 7 at early bloom can suppress but not eliminate leaf curl infection, especially when fall and spring weather is rainy and humid.

HEAVIEST-IMPACT INTERVENTION

Spray with lime sulfur, Bordeaux mixture, or copper when trees are dormant in the early or late winter. See chapter 7 for ecological warnings about copper and Bordeaux mixture.

Phytophthora Crown and Root Rot

When species of *Phytophthora* fungi infect fruit trees, leaves wilt but remain attached to the tree, cankers form on the trunk at soil level, and a dark discoloration develops in the bark, which may also be slimy with gum or dark sap. If you peel back the outer bark of the trunk, the normally green cambium is orange-brown and there is a sharp line of demarcation is between the healthy and diseased (orange-brown) tissue. Winter injury can cause similar symptoms to phytophthora crown rot, but with winter injury the orange-brown color will not be present.

Phytophthera fungal species overwinter as spores in soil or diseased plant material. Species that cause root and crown rots enter host plants near the root collar via wounds or through the succulent parts of small roots. Fungal spores move in water and are attracted to the root exudates from stressed plants. These spores can survive for years in moist soil without host plant roots. However, if the soil is completely dried out, these spores are less likely to survive for more than a few months. *Phytophthora* can be spread in splashing rain or irrigation water. Flooded and saturated soil favor the spread.

HOSTS

Apple, cherry, peach, and apricot.

MINIMAL-IMPACT INTERVENTION

The most important factor in reducing phytophthora rot is good water management. Avoid prolonged saturation of the soil or standing water. Farmers can plant into berms, and in smaller gardens you can plant in raised beds to provide for good drainage. Check each crop's water requirements and group them according to their specific irrigation needs. Soils with high organic matter and low compaction are less likely to stay at overly wet levels for long periods. Many soil bacteria and fungi in an active soil organic matter system are antagonistic to root rot fungi. *Trichoderma*, *Streptomyces lydicus*, *Bacillus amyloliquefaciens*, and *B. subtillus* have been reported to reduce soilborne disease infections.

There is no treatment for *Phytophthora* once infection is present. No apple varieties are resistant to all strains of the pathogen. Choose resistant rootstocks. For apples, apricots, plums, and peaches, see table 12.1. For cherries, Mahaleb is the most susceptible to *Phytophthora* infection, but Mazzard, Morello, and Colt are more resistant and are recommended on heavier soils.

Powdery Mildew

Symptoms of powdery mildew (*Podosphaera leucotricha*) include distorted, curled leaves at the ends of new shoots with white, powdery areas on both upper and lower surfaces of leaves. Leaves turn brown and drop by midsummer. Most healthy trees can tolerate light infections, and the disease does not become serious.

Infection occurs in late spring, during warm days and cool nights, with high humidity, and night-time dew. However, no moisture is required for spore germination to occur. Optimal temperatures are 65 to 80°F (18–27°C). Temperatures below 50°F (10°C) slow disease development. Spores are spread by wind, dew, rain, and irrigation. The disease is inhibited at temperatures above 90°F (32°C) and in bright, sunny conditions. If disease pressure is high, the fungus can cause fruit russet.

HOSTS

Apple, cherry, nectarine, peach.

MINIMAL-IMPACT INTERVENTION

Plant fruit trees in the full sun and avoid shade. Avoid excess fertilizer. Remove infected shoots to reduce the inoculum in early spring.

Figure 14.6. Powdery mildew on pear leaves.

SUPPRESSIVE SOIL MICROORGANISMS

Organic products that contain *Bacillus subtilis* strain QST 713, *B.s.* strain IAB/BS0, *B. amyloliquefaciens* strain F727, and *Streptomyces lydicus* WYEC 108 can be applied as a foliar spray just before and at bloom and may suppress the microorganisms that cause powdery mildew.

MODERATE-IMPACT INTERVENTION

Potassium-bicarbonate-based fungicides have been used on this disease. Potassium bicarbonate has been reported to kill powdery mildew on contact by pulling water from spores and their growing strands.

JMS Stylet-Oil (a highly refined white mineral oil), insecticidal soap, and neem have been reported by growers and researchers to suppress powdery mildew.

HEAVIEST-IMPACT INTERVENTION

Sulfur is reported as moderately effective, but be careful because sulfur can burn tree fruit foliage.

Rust Diseases

Cedar apple rust (*Gymnosporangium juniperi-virginianae*) infection appears as round, orange leaf spots, defoliation, and fruit and twig deformities. This disease is usually more of a cosmetic problem on fruit crops grown in dry climates and does not generally seriously affect fruit quality or yield. The disease requires two hosts to cause infection, and you will see it on junipers before it infects your fruit crops. Monitor nearby junipers for infection; infections first appear as swellings or galls on twigs. These galls form distinctive gelatinous orange fruiting bodies that then infect fruit crops in the spring during rainy, wet periods. Optimal temperatures for rust infections are 64 to 70°F (18–21°C). Temperatures above 85°F (29°C) and dry weather discourage rust disease.

HOSTS

Because rust disease depends on two species to spread and develop, it occurs anywhere that an evergreen alternate or second host (juniper, cedar, and the like) grows near its tree crop hosts (apple, crab apple). If there are no alternate host junipers or cedars nearby, there is less chance of your apples being a host for and becoming infected by rust disease.

MINIMAL-IMPACT INTERVENTION

Irrigate early enough in the day that plant surfaces have time to dry before the cool of evening. Avoid irrigation water landing on leaves, branches, and trunk. Prune plants for good air circulation.

Organic products containing the beneficial bacteria *Bacillus amyloliquefaciens* strain D747 have been used on rust disease with variable results. Researchers report that it is a preventive rather than a cure.

HEAVIEST-IMPACT INTERVENTION

Sulfur sprays can be effective if sprayed preventively in the spring when rust infection periods occur (wet with temperatures 60 to 70°F [16–21°C]), but are rarely necessary unless you live in a wet, warm climate.

Shot-Hole Disease

Wilsonomyces carpophilus is the fungus that causes shot-hole disease. Infection results in leaf and fruit spots that are light brown with dark purple margins. On fruit, spots are usually clustered on the upper sides of fruit and become scabby and rough. Leaf spots fall out, eventually giving leaves the characteristic shot-hole appearance from which the disease takes its common name. It can sometimes be confused with bacterial leaf spot (compare figures 14.7 and 14.2).

The fungus overwinters on buds and on twigs. Fruit infection is favored by wet spring weather and splashing rain. This fungus can germinate and infect at temperatures as low as 36°F (2°C) if there is continuous rain for at least 24 hours. In the spring, only 6 hours of rain at 77°F (25°C) is needed for infection to occur.

Figure 14.7. Shot-hole disease caused by the fungus *Wilsonomyces carpophilus* results in dark, round leaf spots that later drop out, giving a shot-hole appearance.

HOSTS

Mostly apricot, but also plum, cherry, nectarine, and peach.

MINIMAL-IMPACT INTERVENTION

Irrigate early in the day so that plant surfaces dry as quickly as possible. Do not let irrigation water land on leaves, branches, or trunks. Prune plants for good air circulation and remove varnished-appearing (infected) buds and twigs with round, reddish spots.

HEAVIEST-IMPACT INTERVENTION

If this disease has been a serious problem in the past and weather conditions are favorable for infection, apply Bordeaux mixture or copper before buds break in the spring. See chapter 7 for ecological warnings about copper and Bordeaux mix. Sulfur-based products have been reported to be ineffective for this disease.

Physiological Disorders and Nutrient Deficiencies

Tree fruits can respond to environmental stress (like too much heat or cold) and to nutrient deficiencies by developing disease-like symptoms. Refer to "Deciding When a Nutrient Intervention Is Necessary" on page 82 for descriptions of nutrient deficiency signs in detail and in a larger ecosystem context. Fruit crops are less susceptible to nitrogen deficiencies than are vegetable crops, but they are usually more susceptible to different micronutrient problems.

BITTERPIT OR CALCIUM DEFICIENCY

Add gypsum or calcitic lime if your soil pH is below 6.2. Apply Mineral Mix Bloom Sprays as described in chapter 7.

BORON DEFICIENCY

Boron is especially important to promote fruit set in fruit trees. Reduced fruit set is a common symptom of boron deficiency in pears. In peaches, boron deficiency can cause internal and external browning of fruit. Be careful when applying boron, though, because a little goes a long way. While boron is essential for root growth and fruit development, it can become toxic if overapplied. Always test your soil first and apply only the recommended amount. Add borax to the soil in liquid form: 1 tablespoon in 1 gallon of water per 100 square feet (4 L/9.3 m^2). Or add 1 tablespoon Solubor in 1 gallon water. Or add kelp (10 pounds per 100 square feet [49 kg/100 m^2]) or wood ashes (wood ashes will also increase soil pH and soil potassium as well, sometimes excessively).

Spraying a foliar seaweed solution as soon as transplants are set and repeating several times until you see fruit formation may help to prevent deficiency problems. Or you can apply Mineral Mix Bloom Sprays as described in chapter 7.

IRON DEFICIENCY

Soil pH above 7.5 will lead iron to form compounds that are unusable by most fruit trees. Adding sulfur can lower soil pH. For an emergency intervention for a serious deficiency, apply a root drench or spray leaves with iron chelate. Iron that has been treated with lignin derived from woody plants is usually allowed by organic certification agencies. Iron that has been chelated with synthetic materials, such as EDTA, is not allowed. Apply 1 tablespoon chelated iron in 1 gallon (4 L) water as a soil drench or as a spray on new foliage, in the evening when humidity levels increase. As an alternative, apply Mineral Mix Bloom Sprays as described in chapter 7.

Figure 14.8. Iron deficiency symptoms on apple leaves.

NITROGEN DEFICIENCY

Nitrogen is a common deficiency of young, establishing fruit trees, but is less common in older fruit trees with a developed soil organic matter system. Plants grow slowly and new leaves are small and pale. Older leaves turn yellow and die. Conversely, too much nitrogen causes rapid growth of leaves with little root growth and low flower and fruit production.

Apply liquid nitrogen sources as a root drench as soon as you see the first deficiency symptoms. Mix ¼ pound (113 g) of alfalfa meal plus ¼ pound soybean meal in 20 gallons (76 L) of water and let sit at 65 to 70°F (18–21°C) for 12 to 36 hours before application. See "Helen's Magic Mix Liquid Fertilizer Intervention" on page 87.

Figure 14.9. The clear, light-colored drops of liquid on this peach tree are due to winter injury. If the cause of this gummosis were a disease, the gum would be a darker, amber color.

PHOSPHORUS DEFICIENCY

Cool spring soils make phosphorus unavailable to plant roots. Warm soil with plastic mulch. As soon as you see deficiency symptoms, add phosphorus in liquid form as a root drench, such as 2 cups (475 grams) rock phosphate mixed in 10 gallons (38 L) water and left to sit at 65 to 70°F (18–21°C) for 12 to 36 hours before application. See also "Helen's Magic Mix Liquid Fertilizer Intervention" on page 87.

POTASSIUM DEFICIENCY

When plants are deficient in potassium, leaf margins curl upward and leaf edges look scorched. Reddish patches develop, particularly on leaves near fruits. Potassium deficiency is usually associated with poorly drained, sandy soils.

Fruits are strong sinks for potassium as they grow and ripen. The ratio of potassium to calcium is very important; take care not to add too much potassium unless your soil is deficient and/or your trees are showing deficiency symptoms. However, according to Dr. Gerry Neilsen, soil scientist with Agriculture and Agri-Food Canada in British Columbia, where soils are deficient in potassium, applications do not increase apple fruit susceptibility to bitter pit (a condition caused by insufficient calcium in the fruit). Preventively apply Mineral Mix Bloom Sprays as described in chapter 7.

WINTER INJURY

Plums and other stone fruits often respond to cold and wind stress by developing tiny holes in their leaves. This symptom could be mistaken for bacterial leaf spot or fungal leaf spot. However, leaf spots from environmental stress are not purple-black-colored and do not have distinctive lighter-colored centers.

Peach trees often respond to winter stress with a clear bark gummosis as shown in figure 14.9. The lower section of the trunk of an injured tree is slightly sunken at the soil line, with a dark brown to black color.

One precaution to lessen the risk of winter injury is to avoid pruning or fertilizing with nitrogen late in the summer, which might stimulate growth that fails to harden for winter properly.

WET ROOTS

Saturated soils can reduce the soil oxygen level to the point where roots die. Roots killed by "wet roots" or drowning have an overall brown color, and there may be a decaying or fermenting smell. If soils are heavy, before you plant or replant, create raised beds or berms. Mix composted wood bark into the raised beds or berms to aerate soils.

ZINC DEFICIENCY

Zinc deficiency is more often a problem in sandy or high-pH soils, especially during cool, moist springs. The first symptoms of zinc deficiency are interveinal mottling and yellowing. Symptoms are usually seen first in the springtime on the previous year's growth. Often trees grow out of these symptoms as soon as soil and air temperatures increase.

Soil applications of zinc may not be the best option, because zinc is easily tied up in the soil and does not get taken up by fruit tree roots. Zinc is an important component of the Mineral Mix Bloom Spray I use (see the recipe in chapter 7). Washington State University recommends foliar zinc solutions sprayed at the following timing: late dormant for stone fruits, silver-tip for apples and pears, and post-harvest for all tree fruits except for apricot. Zinc deficiency is important to overcome for best tree health, but also because consuming fruit produced from zinc-deficient crops can lead to human zinc deficiency, which reduces immune function. Creating a microbially active soil organic matter system may help zinc uptake by fruit tree roots. Several studies with other crops, such as wheat, rice, and corn, showed that inoculation with mycorrhizal fungi improved the availability of micronutrients, particularly zinc in soils, due to root zone acidification and transport of nutrients through the mycorrhizal biomass.

Acknowledgments

Thanks to all my Missoula, Montana, Master Gardener students from 1992 through 2009, who asked questions I had to discover answers to and who helped with my on-farm research.

Thanks to all the organizations who helped to fund Carl's and my on-farm research: the Organic Farming Research Foundation, The Western Sustainable Agriculture Research and Education Grants Program, The High Stakes Foundation, The FruitGuys Community Fund, and Rodale Institute Your 2 Cents program.

Appendix

Forms for Monitoring and Planning Interventions

PROBLEM IDENTIFICATION FORM

Plant/Crop injured:

Date:

Description of damage:

Amount of damage (percent of total plants infested):

Amount of damage to a single plant (expressed as a percentage):

Progression of problem (change from previous week):

Signs of insect feeding/egg laying (type of damage to leaves and/or whole plant):

Signs of environmental stress on plants:

Weather conditions:

Appendix

PRE-INTERVENTION INFORMATION FORM

Pest Information

Pest life cycle: ____________________

Stage when pest is most vulnerable: ____________________

Development threshold temperatures for pest: ____________________

Is the pest at an economic threshold level? ____________________

Degree of crop susceptibility to pest: ____________________

Suppression Options

Natural, on-farm biological controls: ____________________

Purchased biological controls: ____________________

Spray Intervention Options

Material	Ecological Impact Level	Application Method/Timing	Cautions
Microbial insecticides			
Oils/soaps			
Minerals			
Botanical insecticides			

Notes

Introduction: My Farming Revolution

1. Guido Lingua et al., "Arbuscular Mycorrhizal Fungi and Plant Growth-Promoting Pseudomonads Increases Anthocyanin Concentration in Strawberry Fruits (*Fragaria × ananassa* var. Selva) in Conditions of Reduced Fertilization," *International Journal of Molecular Sciences* 14, no. 8 (2013): 16207–25, https://doi.org/10.3390/ijms140816207.
2. Daniel Garcia-Seco et al., "Application of *Pseudomonas fluorescens* to Blackberry under Field Conditions Improves Fruit Quality by Modifying Flavonoid Metabolism," *PLoS ONE* 10, no. 11 (2015): e0142639, https://doi.org/10.1371/journal.pone.0142639; Mansour Ghorbanpour et al., "Phytochemical Variations and Enhanced Efficiency of Antioxidant and Antimicrobial Ingredients in *Salvia officinalis* as Inoculated with Different Rhizobacteria," *Chemistry & Biodiversity* 13, no. 3 (2016): 319–30, https://doi.org/10.1002/cbdv.201500082.
3. Beatriz Ramos-Solano et al., "Bacterial Bioeffectors Delay Postharvest Fungal Growth and Modify Total Phenolics, Flavonoids and Anthocyanins in Blackberries," *Lebensmittel-Wissenschaft + Technologie* 61, no. 2 (2015): 437–43, https://doi.org/10.1016/j.lwt.2014.11.051.

Chapter 2: The Soil Revolution

1. Elisa Bona et al., "Arbuscular Mycorrhizal Fungi and Plant Growth-Promoting Pseudomonads Improve Yield, Quality and Nutritional Value of Tomato: A Field Study," *Mycorrhiza* 27, no. 1 (2017): 1–11, https://doi.org/10.1007/s00572-016-0727-y.
2. Matthew D. Whiteside et al., "Organic Nitrogen Uptake by Arbuscular Mycorrhizal Fungi in a Boreal Forest," *Soil Biology and Biochemistry* 55 (2012): 7–13, https://doi.org/10.1016/j.soilbio.2012.06.001.
3. Torgny Näsholm, Knut Kielland, and Ulrika Ganeteg, "Uptake of Organic Nitrogen by Plants," *New Phytologist* 182 (2009): 31–48, https://doi.org/10.1111/j.1469-8137.2008.02751.x.
4. Sasha B. Kramer et al., "Reduced Nitrate Leaching and Enhanced Denitrifier Activity and Efficiency in Organically Fertilized Soils," *Proceedings of the National Academy of Sciences of the United States of America* 103, no. 12 (2006): 4522–27, https://doi.org/10.1073/pnas.0600359103.
5. Matthias C. Rillig, "Arbuscular Mycorrhizae, Glomalin, and Soil Aggregation," *Canadian Journal of Soil Science* 84 (2004): 355–63, https://doi.org/10.4141/S04-003; Stavros D. Veresoglou, Baodong Chen, and Matthias C. Rillig, "Arbuscular Mycorrhiza and Soil Nitrogen Cycling," *Soil Biology and Biochemistry* 46 (2012): 53–62, https://doi.org/10.1016/j.soilbio.2011.11.018.
6. J. M. Mirás-Avalos et al., "The Influence of Tillage on the Structure of Rhizosphere and Root-Associated Arbuscular Mycorrhizal Fungal Communities," *Pedobiologia* 54 (2011): 235–41, https://doi.org/10.1016/j.pedobi.2011.03.005.
7. P. D. Millner and S. F. Wright, "Tools for Support of Ecological Research on Arbuscular Mycorrhizal Fungi," *Symbiosis* 33 (2002): 101–23. Available online as a pdf: https://dalspace.library.dal.ca/bitstream/handle/10222/77909/VOLUME%2033-NUMBER%202-2002-PAGE%20101.pdf?sequence=1.
8. Kristine A. Nichols and Sara F. Wright, "Contributions of Fungi to Soil Organic Matter in Agroecosystems," in *Soil Organic Matter in Sustainable Agriculture*, ed. Fred Magdoff and Ray R. Weil (Boca Raton: CRC Press, 2004), 179–98; S. F. Wright and K. Nichols, "Glomalin: Hiding Place for a Third of the World's Stored Soil Carbon," *Agricultural Research* 50 (2002): 4–7, https://doi.org/10.24925/turjaf.v9i1.191-196.3803.
9. Miranda Hart, David L. Ehret, Angelika Krumbein, Connie Leung, Susan Murch, Christina Turi, and Philipp Franken, "Inoculation with Arbuscular Mycorrhizal Fungi Improves the Nutritional Value of Tomatoes," *Mycorrhiza* 25, no. 5 (2015): 359–76, https://doi.org/10.1007/s00572-014-0617-0.

10. A. Chabbi and C. Rumpel, "Organic Matter Dynamics in Agro-Ecosystems: The Knowledge Gaps," *European Journal of Soil Science* 60, no. 2 (2009):153–57, https://doi.org/10.1111/j.1365-2389.2008.01116.x.
11. Laurie E. Drinkwater, Peggy Wagoner, and Marianne Sarrantonio, "Legume-Based Cropping Systems Have Reduced Carbon and Nitrogen Losses," *Nature* 396, no. 6708 (1998): 262–65, https://doi.org/10.1038/24376.
12. Ashley B. Jernigan et al., "Legacy Effects of Contrasting Organic Grain Cropping Systems on Soil Health Indicators, Soil Invertebrates, Weeds, and Crop Yield," *Agricultural Systems* 177 (2020): 102719, https://doi.org/10.1016/j.agsy.2019.102719.
13. C. E. Jones, "Liquid Carbon Pathway," *Australian Farm Journal* (2008): 15–17, https://amazingcarbon.com/PDF/JONES-LiquidCarbonPathway(July08).pdf.

Chapter 3: Soil System Management Principles for Farmers and Gardeners

1. S. Schlüter, T. Eickhorst, and C. W. Mueller, "Correlative Imaging Reveals Holistic View of Soil Microenvironments," *Environmental Science & Technology* 53, no. 2 (2019): 829–37, https://doi.org/10.1021/acs.est.8b05245.
2. Eric B. Brennan and R. F. Smith, "Cover Crop Frequency and Compost Effects on a Legume–Rye Cover Crop during Eight Years of Organic Vegetables," *Agronomy Journal* 109, no. 5 (2017): 2199–213, https://doi.org/10.2134/agronj2016.06.0354; Eric B. Brennan and V. Acosta-Martinez, "Cover Cropping Frequency Is the Main Driver of Soil Microbial Changes during Six Years of Organic Vegetable Production," *Soil Biology and Biochemistry* 109 (2017): 188–204, https://doi.org/10.1016/j.soilbio.2017.01.014.
3. J. A. Delgado et al., "Potential Use of Cover Crops for Soil and Water Conservation, Nutrient Management, and Climate Change Adaptation across the Tropics," in vol. 165, *Advances in Agronomy*, ed. D. L. Sparks (Cambridge, MA: Academic Press, 2021), 175–247, https://doi.org/10.1016/bs.agron.2020.09.003.
4. Michelle M. Wander et al., "Organic and Conventional Management Effects on Biologically Active Soil Organic Matter Pools," *Soil Science Society of America Journal* 58, no. 4 (1994): 1130–39, https://doi.org/10.2136/sssaj1994.03615995005800040018x.
5. S. Temple et al., "Evaluating Soil Quality in Organic, Low-Input, and Conventional Farming Systems," *Sustainable Agriculture Farming Systems Project* 2, no. 1 (1996): 1–4, http://safs.ucdavis.edu/newsletter/quality.pdf.
6. Timothy M. Bowles et al., "Soil Enzyme Activities, Microbial Communities, and Carbon and Nitrogen Availability in Organic Agroecosystems across an Intensively-Managed Agricultural Landscape," *Soil Biology and Biochemistry* 68 (2014): 252–62, https://doi.org/10.1016/j.soilbio.2013.10.004.
7. Franciska T. De Vries and Richard D. Bardgett, "Plant–Microbial Linkages and Ecosystem Nitrogen Retention: Lessons for Sustainable Agriculture," *Frontiers in Ecology and the Environment* 10, no. 8 (2012): 425–32, https://doi.org/10.1890/110162; Franciska T. De Vries et al., "Abiotic Drivers and Plant Traits Explain Landscape-Scale Patterns in Soil Microbial Communities," *Ecology Letters* 15, no. 11 (2012): 1230–39, https://doi.org/10.1111/j.1461-0248.2012.01844.x.
8. Personal communication.
9. Francisco J. Calderón et al., "Microbial Responses to Simulated Tillage in Cultivated and Uncultivated Soils," *Soil Biology and Biochemistry* 32, nos. 11–12 (2000): 1547–59, https://doi.org/10.1016/S0038-0717(00)00067-5.
10. S. J. Corak, W. W. Frye, and M. S. Smith, "Legume Mulch and Nitrogen Fertilizer Effects on Soil Water and Corn Production," *Soil Science Society of America Journal* 55, no. 5 (1991): 1395–400, https://doi.org/10.2136/sssaj1991.03615995005500050032x.
11. G. H. H. Hammouda and W. A. Adams, "The Decomposition, Humification and Fate of Nitrogen during the Composting of Some Plant Residues," in *Compost: Production, Quality and Use*, ed. M. De Bertoldi et al. (New York: Elsevier Science Publishing, 1987), https://eurekamag.com/research/029/314/029314326.php.
12. Denise M. Finney, Charles M. White, and Jason P. Kaye, "Biomass Production and Carbon/Nitrogen Ratio Influence Ecosystem Services from Cover Crop Mixtures," *Agronomy Journal* 108, no. 1 (2016): 39–52, https://doi.org/10.2134/agronj15.0182.
13. C. E. Jones, "Liquid Carbon Pathway," *Australian Farm Journal* (2008): 15–17, https://amazingcarbon.com/PDF/JONES-LiquidCarbonPathway(July08).pdf.

Chapter 4: Minimizing Tillage and Growing Your Own Fertilizer

1. P. T. Tomlinson, G. L. Buchschacher, and R. M. Teclaw, "Sowing Methods and Mulch Affect 1+0 Northern Red Oak Seedling Quality," *New Forests* 13 (1997): 193–208, https://doi.org/10.1023/A:1006534324616.
2. H. Yibirin, J. W. Johnson, and D. J. Eckert, "No-Till Corn Production as Affected by Mulch, Potassium Placement, and Soil Exchangeable Potassium," *Agronomy Journal* 85, no. 3 (1993): 639–44, https://doi.org/10.2134/agronj1993.00021962008500030022x.
3. Guanglong Tian, Biauw Tjwan Kang, and Lijbert Brussaard, "Effect of Mulch Quality on Earthworm Activity and Nutrient Supply in the Humid Tropics," *Soil Biology and Biochemistry* 29, nos. 3–4 (1997): 369–73, https://doi.org/10.1016/S0038-0717(96)00099-5.
4. Jayne M. Zajicek and James L. Heilman, "Transpiration by Crape Myrtle Cultivars Surrounded by Mulch, Soil, and Turfgrass Surfaces," *HortScience* 26, no. 9 (1991): 1207–10, https://doi.org/10.21273/HORTSCI.26.9.1207.

Chapter 5: Strengthening the "Immune System" of Your Farm or Garden

1. Miguel A. Altieri and Clara I. Nicholls, "Soil Fertility Management and Insect Pests: Harmonizing Soil and Plant Health in Agroecosystems," *Soil and Tillage Research* 72, no. 2 (2003): 203–11, https://doi.org/10.1016/S0167-1987(03)00089-8.
2. Lora A. Morandin, Rachael F. Long, and Claire Kremen, "Hedgerows Enhance Beneficial Insects on Adjacent Tomato Fields in an Intensive Agricultural Landscape," *Agriculture, Ecosystems & Environment* 189 (2014): 164–70, https://doi.org/10.1016/j.agee.2014.03.030; D. J. Skirvin et al., "The Effect of Within-Crop Habitat Manipulations on the Conservation Biological Control of Aphids in Field-Grown Lettuce," *Bulletin of Entomological Research* 101, no. 6 (2011): 623–31, https://doi.org/10.1017/S0007485310000659; Rachael Freeman Long et al., "Beneficial Insects Move from Flowering Plants to Nearby Crops," *California Agriculture* 52, no. 5: 23–26, https://doi.org/10.3733/ca.v052n05p23; M. B. Thomas, S. D. Wratten, and N. W. Sotherton, "Creation of 'Island' Habitats in Farmland to Manipulate Populations of Beneficial Arthropods: Predator Densities and Emigration," *Journal of Applied Ecology* 28, no. 3 (1991): 906–17, https://doi.org/10.2307/2404216; M. B. Thomas et al., "Habitat Factors Influencing the Distribution of Polyphagous Predatory Insects between Field Boundaries," *Annals of Applied Biology* 120, no. 2 (1992): 197–202, https://doi.org/10.1111/j.1744-7348.1992.tb03417.x; M. B. Thomas, S. D. Wratten, and N. W. Sotherton, "Creation of 'Island' Habitats in Farmland to Manipulate Populations of Beneficial Arthropods: Predator Densities and Species Composition," *Journal of Applied Ecology* 29, no. 2 (1992): 524–31, https://doi.org/10.2307/2404521; Susan Thomas et al., "Botanical Diversity of Beetle Banks: Effects of Age and Comparison with Conventional Arable Field Margins in Southern UK," *Agriculture, Ecosystems & Environment* 93, nos. 1–3 (2002): 403–12, https://doi.org/10.1016/S0167-8809(01)00342-5.
3. Matthias Tschumi et al., "High Effectiveness of Tailored Flower Strips in Reducing Pests and Crop Plant Damage," *Proceedings of the Royal Society B: Biological Sciences* 282, no. 1814 (2015): 20151369, https://doi.org/10.1098/rspb.2015.1369.
4. Keith Sunderland and Ferenc Samu, "Effects of Agricultural Diversification on the Abundance, Distribution, and Pest Control Potential of Spiders: A Review," *Entomologia Experimentalis et Applicata* 95, no. 1 (2000): 1–13, https://doi.org/10.1046/j.1570-7458.2000.00635.x.
5. Christine Haaland, Russell E. Naisbit, and Louis-Félix Bersier, "Sown Wildflower Strips for Insect Conservation: A Review," *Insect Conservation and Diversity* 4, no. 1 (2011): 60–80, https://doi.org/10.1111/j.1752-4598.2010.00098.x.
6. Céline E. Géneau et al., "Selective Flowers to Enhance Biological Control of Cabbage Pests by Parasitoids," *Basic and Applied Ecology* 13, no. 1 (2012): 85–93, https://doi.org/10.1016/j.baae.2011.10.005; Timothy D. Nafziger Jr. and Henry Y. Fadamiro, "Suitability of Some Farmscaping Plants as Nectar Sources for the Parasitoid Wasp, *Microplitis croceipes* (Hymenoptera: Braconidae): Effects on Longevity and Body Nutrients," *Biological Control* 56, no. 3 (2011): 225–29, https://doi.org/10.1016/j.biocontrol.2010.11.005.
7. David G. James et al., "Beneficial Insects Attracted to Native Flowering Buckwheats (*Eriogonum* Michx) in Central Washington," *Environmental Entomology* 43, no. 4 (2014): 942–48, https://doi.org/10.1603/EN13342; David G. James et al., "Beneficial Insects Associated with Stinging Nettle, *Urtica dioica* Linnaeus, in Central

Washington State," *Pan-Pacific Entomologist* 91, no. 1 (2015): 82–90, https://doi.org/10.3956/2014-91.1.082.

8. Meg Ballard, Judith Hough-Goldstein, and Douglas Tallamy, "Arthropod Communities on Native and Nonnative Early Successional Plants," *Environmental Entomology* 42, no. 5 (2013): 851–59, https://doi.org/10.1603/EN12315.
9. Miguel A. Altieri, Clara I. Nicholls, and Marlene Fritz, *Manage Insects on Your Farm: A Guide to Ecological Strategies*, vol. 7, *SARE Handbook Series* (College Park, MD: Sustainable Agriculture Research and Education, 2005), https://www.sare.org/wp-content/uploads/Manage-Insects-on-Your-Farm.pdf.
10. W. O. C. Symondson, K. D. Sunderland, and M. H. Greenstone, "Can Generalist Predators Be Effective Biocontrol Agents?," *Annual Review of Entomology* 47 (2002): 561–94, https://doi.org/10.1146/annurev.ento.47.091201.145240.

Chapter 6: Working with Pest and Beneficial Insects

1. T. K. Wilkinson and D. A. Landis, "Habitat Diversification in Biological Control: The Role of Plant Resources," in *Plant-Provided Food for Carnivorous Insects*, ed. F. L. Wäckers, P. C. J. van Rijn, and J. Bruin (Cambridge, U.K.: Cambridge University Press, 2005), 305–25, https://doi.org/10.1017/CBO9780511542220.011.
2. Rebecca A. Schmidt-Jeffris et al., "Nontarget Impacts of Herbicides on Spiders in Orchards," *Journal of Economic Entomology* 115, no. 1 (2022): 65–73, https://doi.org/10.1093/jee/toab228.
3. Stano Pekár, "Spiders (Araneae) in the Pesticide World: An Ecotoxicological Review," *Pest Management Science* 68, no. 11 (2012): 1438–46, https://doi.org/10.1002/ps.3397.
4. Keith Sunderland, "Mechanisms Underlying the Effects of Spiders on Pest Populations," *Journal of Arachnology* 27, no. 1 (1999): 308–16, http://www.jstor.org/stable/3706002.
5. S. Bogya and P. J. M. Mols, "The Role of Spiders as Predators of Insect Pests with Particular Reference to Orchards: A Review," *Acta Phytopathologica et Entomologica Hungarica* 31 (1996): 83–159, https://library.wur.nl/WebQuery/wurpubs/38269; A. L. Knight, J. E. Turner, and B. Brachula, "Predation on Eggs of Codling Moth (Lepidoptera: Tortricidae) in Mating Disrupted and Conventional Orchards in Washington," *Journal of the Entomological Society of British Columbia* 94 (1997): 67–74, https://biostor.org/reference/204821; David R. Horton et al., "Diversity and Phenology of Predatory Arthropods Overwintering in Cardboard Bands Placed in Pear and Apple Orchards of Central Washington State," *Annals of the Entomological Society of America* 95, no. 4 (2002): 469–80, https://doi.org/10.1603/0013-8746(2002)095[0469:DAPOPA]2.0.CO;2; Marc Isaia et al., "Spiders as Biological Controller in Apple Orchards Infested by Cydia spp.," in *XXIV European Congress of Arachnology* (Bern: Natural History Museum, 2010), 79–88; C. Boreau de Roincé et al., "Early-Season Predation on Aphids by Winter-Active Spiders in Apple Orchards Revealed by Diagnostic PCR," *Bulletin of Entomological Research* 103, no. 2 (2013): 148–54, https://doi.org/10.1017/S0007485312000636; Stano Pekár et al., "Biological Control in Winter: Novel Evidence for the Importance of Generalist Predators," *Journal of Applied Ecology* 52, no. 1 (2015): 270–79, https://doi.org/10.1111/1365-2664.12363; Thomas R. Unruh et al., "Gut Content Analysis of Arthropod Predators of Codling Moth in Washington Apple Orchards," *Biological Control* 102 (2016): 85–92, https://doi.org/10.1016/j.biocontrol.2016.05.014; Hazem Dib, Myriam Siegwart, and Yvan Capowiez, "Spiders (Arachnida: Araneae) in Organic Apple (Rosaceae) Orchards in Southeastern France," *Canadian Entomologist* 152, no. 2 (2020): 224–36, https://doi.org/10.4039/tce.2020.5; László Mezőfi et al., "Beyond Polyphagy and Opportunism: Natural Prey of Hunting Spiders in the Canopy of Apple Trees," *PeerJ* 8 (2020): e9334, https://doi.org/10.7717/peerj.9334.
6. Eugene R. Miliczky, Carrol O. Calkins, and David R. Horton, "Spider Abundance and Diversity in Apple Orchards under Three Insect Pest Management Programs in Washington State, U.S.A.," *Agricultural and Forest Entomology* 2, no. 3 (2000): 203–15, https://doi.org/10.1046/j.1461-9563.2000.00067.x.
7. Schmidt-Jeffris et al., "Nontarget Impacts of Herbicides."
8. Susan E. Riechert, "The Hows and Whys of Successful Pest Suppression by Spiders: Insights from Case Studies," *Journal of Arachnology* (1999): 387–96, https://www.jstor.org/stable/3706011; Miliczky, Calkins, and Horton, "Spider Abundance and Diversity in Apple Orchards."
9. Juraj Halaj, Alan B. Cady, and George W. Uetz, "Modular Habitat Refugia Enhance Generalist

Predators and Lower Plant Damage in Soybeans," *Environmental Entomology* 29, no. 2 (2000): 383–93, https://doi.org/10.1093/ee/29.2.383.

Chapter 7: Working with Beneficial and Disease-Causing Microorganisms

1. Janine Bartelt-Ryser et al., "Soil Feedbacks of Plant Diversity on Soil Microbial Communities and Subsequent Plant Growth," *Perspectives in Plant Ecology, Evolution and Systematics* 7, no. 1 (2005): 27–49, https://doi.org/10.1016/j.ppees.2004.11.002.
2. L. E. Datnoff, W. H. Elmer, D. M. Huber, eds., *Mineral Nutrition and Plant Disease* (St. Paul: APS Press, 2007).
3. D. M. Huber, "The Role of Mineral Nutrition in Defense," in *How Plants Defend Themselves*, vol. 5, *Plant Disease: An Advanced Treatise*, ed. J. G. Horsfall and E. B. Cowling (New York: Academic Press, 1980), 381–406, https://doi.org/10.1016/B978-0-12-356405-4.50028-9; D. M. Huber and D. C. Arny, "Interactions of Potassium with Plant Disease," in *Potassium in Agriculture*, ed. Robert D. Munson (Madison: ASA, CSSA, and SSSA Books, 1985): 467–88, https://doi.org/10.2134/1985.potassium.c20.
4. Huber, "The Role of Mineral Nutrition in Defense."
5. Mark Mazzola, "Manipulation of Rhizosphere Bacterial Communities to Induce Suppressive Soils," *Journal of Nematology* 39, no. 3 (2007): 213–20, https://pubmed.ncbi.nlm.nih.gov/19259490/.
6. Bryony E. A. Dignam et al., "Effect of Land Use and Soil Organic Matter Quality on the Structure and Function of Microbial Communities in Pastoral Soils: Implications for Disease Suppression," *PLoS ONE* 13, no. 5 (2018): e0196581, https://doi.org/10.1371/journal.pone.0196581.
7. Hongwei Liu et al., "Evidence for the Plant Recruitment of Beneficial Microbes to Suppress Soil-Borne Pathogens," *New Phytologist* 229, no. 5 (2021): 2873–85, https://doi.org/10.1111/nph.17057.
8. Anas Raklami et al., "Use of Rhizobacteria and Mycorrhizae Consortium in the Open Field as a Strategy for Improving Crop Nutrition, Productivity and Soil Fertility," *Frontiers in Microbiology* 10 (2019): 1106, https://doi.org/10.3389/fmicb.2019.01106.
9. Nathan Vannier, Matthew Agler, and Stéphane Hacquard, "Microbiota-Mediated Disease Resistance in Plants." *PLoS Pathogens* 15, no. 6 (2019): e1007740, https://doi.org/10.1371/journal.ppat.1007740.
10. B. Reinhold-Hurek et al., "Roots Shaping Their Microbiome: Global Hotspots for Microbial Activity," *Annual Review of Phytopathology* 53 (2015): 403–24, https://doi.org/10.1146/annurev-phyto-082712-102342; J. Sasse, E. Martinoia, and T. Northen, "Feed Your Friends: Do Plant Exudates Shape the Root Microbiome?," *Trends in Plant Science* 23, no. 1 (2018): 25–41, https://doi.org/10.1016/j.tplants.2017.09.003.
11. T. N. Temple et al., "Floral Colonization Dynamics and Specificity of *Aureobasidium pullulans* Strains Used to Suppress Fire Blight of Pome Fruit," *Plant Disease* 104, no. 1 (2020): 121–28, https://doi.org/10.1094/PDIS-09-18-1512-RE.
12. Maurice Lourd and D. Bouhot, "Research and Characterization of Soils Resistant to *Pythium* spp. in the Brazilian Amazon," *EPPO Bulletin* 17 (1987): 569–75, https://doi.org/10.1111/j.1365-2338.1987.tb00076.x.
13. A. Palojärvi et al., "Tillage System and Crop Sequence Affect Soil Disease Suppressiveness and Carbon Status in Boreal Climate," *Frontiers in Microbiology* 11 (2020): 534786, https://doi.org/10.3389/fmicb.2020.534786.
14. W. Zhang, W. A. Dick, and H. A. J. Hoitink, "Compost-Induced Systemic Acquired Resistance in Cucumber to Pythium Root Rot and Anthracnose," *Phytopathology* 86, no. 10 (1996): 1066–70.
15. D. E. Spring et al., "Suppression of the Apple Collar Rot Pathogen in Composted Hardwood Bark," *Phytopathology* 70, no. 12 (1980): 1209–12, https://doi.org/10.1094/phyto-70-1209; Harry J. Hoitink, Y. Inbar, and M. J. Boehm, "Status of Compost-Amended Potting Mixes Naturally Suppressive to Soil-Borne Diseases of Floricultural Crops," *Plant Disease* 75, no. 9 (1991): 869–73.
16. Fekede Workneh et al., "Variables Associated with Corky Root and Phytophthora Root Rot of Tomatoes in Organic and Conventional Farms," *Phytopathology* 83 (1993): 581–89, https://doi.org/10.1094/PHYTO-83-581.
17. M. J. Boehm, L. V. Madden, and H. A. J. Hoitink, "Effect of Organic Matter Decomposition Level on Bacterial Species Diversity and Composition in Relationship to Pythium Damping-Off Severity," *Applied and Environmental Microbiology* 59, no. 12 (1993): 4171–79, https://doi.org/10.1128/aem.59.12.4171-4179.1993.
18. K. N. Asirifi, W. C. Morgan, and D. G. Parbey, "Suppression of *Sclerotinia* Soft Rot of Lettuce with Organic Amendments," *Australian Journal of*

Experimental Agriculture 34, no. 1 (1994): 131–36, https://doi.org/10.1071/EA9940131.

19. J. B. Ristaino, G. Parra, and C. L. Campbell, "Suppression of Phytophthora Blight in Bell Pepper by a No-Till Wheat Cover Crop," *Phytopathology* 87, no. 3 (1997): 242–49, https://doi.org/10.1094/PHYTO.1997.87.3.242.
20. X. Qian et al., "Effects of Living Mulches on the Soil Nutrient Contents, Enzyme Activities, and Bacterial Community Diversities of Apple Orchard Soils," *European Journal of Soil Biology* 70 (2015): 23–30, https://doi.org/10.1016/j.ejsobi.2015.06.005.
21. J. Yang et al., "Long-Term Cover Cropping Seasonally Affects Soil Microbial Carbon Metabolism in an Apple Orchard," *Bioengineered* 10, no. 1 (2019): 207–17, https://doi.org/10.1080/21655979.2019.1622991.
22. J. R. Davis et al., "Ecological Relationships of Verticillium Wilt Suppression of Potato by Green Manures," *American Journal of Potato Research* 87 (2010): 315–26, https://doi.org/10.1007/s12230-010-9135-6.
23. N. Ntahimpera et al., "Effects of a Cover Crop on Splash Dispersal of *Colletotrichum acutatum* Conidia," *Phytopathology* 88, no. 6 (1998): 536–43, https://doi.org/10.1094/PHYTO.1998.88.6.536.
24. C. A. Wyenandt et al., "Fall- and Spring-Sown Cover Crop Mulches Affect Yield, Fruit Cleanliness, and Fusarium Fruit Rot Development in Pumpkin," *HortTechnology* 21, no. 3 (2011): 343–54, https://doi.org/10.21273/HORTTECH.21.3.343.
25. Wyenandt et al., "Fall- and Spring-Sown Cover Crop Mulches."
26. Kathryne L. Everts, "Reduced Fungicide Applications and Host Resistance for Managing Three Diseases in Pumpkin Grown on a No-Till Cover Crop," *Plant Disease* 86, no. 10 (2002): 1134–41, https://doi.org/10.1094/PDIS.2002.86.10.1134.
27. A. P. Keinath et al., "Cover Crops of Hybrid Common Vetch Reduce Fusarium Wilt of Seedless Watermelon in the Eastern United States," *Plant Health Progress* 11, no. 1 (2010), https://doi.org/10.1094/PHP-2010-0914-01-RS.
28. M. G. Lloyd, N. McRoberts, and T. R. Gordon, "Cryptic Infection and Systemic Colonization of Leguminous Crops by *Verticillium dahliae*, the Cause of Verticillium Wilt," *Plant Disease* 103, no. 12 (2019): 3166–71, https://doi.org/10.1094/PDIS-04-19-0850-RE.
29. G. Qi et al., "Cover Crops Restore Declining Soil Properties and Suppress Bacterial Wilt by Regulating Rhizosphere Bacterial Communities and Improving Soil Nutrient Contents," *Microbiological Research* 238 (2020): 126505, https://doi.org/10.1016/j.micres.2020.126505.
30. Robert P. Larkin and Timothy S. Griffin, "Control of Soilborne Potato Diseases Using Brassica Green Manures," *Crop Protection* 26, no. 7 (2007): 1067–77, https://doi.org/10.1016/j.cropro.2006.10.004.
31. A. Gamliel and J. J. Stapleton, "Characterization of Antifungal Volatile Compounds Evolved from Solarized Soil Amended with Cabbage Residues," *Phytopathology* 83, no. 9 (1993): 899–905.
32. Robert P. Larkin, C. Honeycutt, and M. Olanya, "Management of Verticillium Wilt of Potato with Disease-Suppressive Green Manures and as Affected by Previous Cropping History," *Plant Disease* 95, no. 5 (2011): 568–76, https://doi.org/10.1094/PDIS-09-10-0670.
33. J. C. Hansen et al., "Soil Microbial Biomass and Fungi Reduced with Canola Introduced into Long-Term Monoculture Wheat Rotations," *Frontiers in Microbiology* 10 (2019): 1488, https://doi.org/10.3389/fmicb.2019.01488.
34. L. J. Thomson and A. A. Hoffmann, "Field Validation of Laboratory-Derived IOBC Toxicity Ratings for Natural Enemies in Commercial Vineyards," *Biological Control* 39, no. 3 (2006): 507–15, https://doi.org/10.1016/j.biocontrol.2006.06.009.
35. Eric Vukicevich et al., "Cover Crops to Increase Soil Microbial Diversity and Mitigate Decline in Perennial Agriculture: A Review," *Agronomy for Sustainable Development* 36, no. 3 (2016): 1–14, https://doi.org/10.1007/s13593-016-0385-7.

Chapter 8: Working with Plant Competition

1. A. H. Bunting, "Some Reflections on the Ecology of Weeds," in *Biology of Weeds*, ed. John L. Harper (Oxford: Blackwell Scientific Publications, 1960), 11–26.
2. Katherine L. Gross and Patricia A. Werner, "Colonizing Abilities of 'Biennial' Plant Species in Relation to Ground Cover: Implications for Their Distributions in a Successional Sere," *Ecology* 63, no. 4 (1982): 921–31, https://doi.org/10.2307/1937232.
3. R. S. Fawcett and F. W. Slife, "Effects of Field Applications of Nitrate on Weed Seed Germination and Dormancy," *Weed Science* 26, no. 6 (1978): 594–96, http://www.jstor.org/stable/4042936.

4. Fawcett and Slife, "Effects of Field Applications of Nitrate."
5. H. A. Roberts and Patricia M. Feast, "Emergence and Longevity of Seeds of Annual Weeds in Cultivated and Undisturbed Soil," *Journal of Applied Ecology* 10, no. 1 (1973): 133–43, https://doi.org/10.2307/2404721.
6. P. D. Ominski and M. H. Entz, "Eliminating Soil Disturbance Reduces Post-Alfalfa Summer Annual Weed Populations," *Canadian Journal of Plant Science* 81, no. 4 (2001): 881–84, https://doi.org/10.4141/P01-042.
7. T. V. Toai and D. L. Linscott, "Phytotoxic Effect of Decaying Quackgrass (*Agropyron repens*) Residues," *Weed Science* 27, no. 6 (1979): 595–98, http://www.jstor.org/stable/4043075.
8. Eric R. Gallandt, "How Can We Target the Weed Seedbank?," *Weed Science* 54, no. 3 (2006): 588–96, https://doi.org/10.1614/WS-05-063R.1.
9. Matt Liebman and Adam S. Davis, "Integration of Soil, Crop and Weed Management in Low-External-Input Farming Systems," *Weed Research* 40, no. 1 (2000): 27–47, https://doi.org/10.1046/j.1365-3180.2000.00164.x.
10. David Weisberger, Virginia Nichols, and Matt Liebman, "Does Diversifying Crop Rotations Suppress Weeds? A Meta-Analysis," *PLoS ONE* 14, no. 7 (2019): e0219847, https://doi.org/10.1371/journal.pone.0219847.
11. Gallandt, "How Can We Target the Weed Seedbank?"
12. Eric R. Gallandt et al., "Effects of Pest and Soil Management Systems on Weed Dynamics in Potato," *Weed Science* 46, no. 2 (1998): 238–48, http://www.jstor.org/stable/4045943.
13. Michelle J. du Croix Sissons et al., "Depth of Seedling Recruitment of Five Weed Species Measured *In Situ* in Conventional- and Zero-Tillage Fields," *Weed Science* 48, no. 3 (2000): 327–32, https://doi.org/10.1614/0043-1745(2000)048[0327:DOSROF]2.0.CO;2.
14. Matt Liebman and Adam S. Davis, "Managing Weeds in Organic Farming Systems: An Ecological Approach," in *Organic Farming: The Ecological System*, vol. 54, *Agronomy Monographs*, ed. Charles Francis (Madison: American Society of Agronomy, 2009): 173–96, https://doi.org/10.2134/agronmonogr54.c8.
15. Elizabeth Dyck, Matt Liebman, and M. S. Erich, "Crop-Weed Interference as Influenced by a Leguminous or Synthetic Fertilizer Nitrogen Source: II. Rotation Experiments with Crimson Clover, Field Corn, and Lambsquarters," *Agriculture, Ecosystems & Environment* 56, no. 2 (1995): 109–20, https://doi.org/10.1016/0167-8809(95)00644-3.
16. Yvonne E. Lawley, John R. Teasdale, and Ray R. Weil, "The Mechanism for Weed Suppression by a Forage Radish Cover Crop," *Agronomy Journal* 104, no. 2 (2012): 205–14, https://doi.org/10.2134/agronj2011.0128.
17. Marianne Sarrantonio and Eric R. Gallandt, "The Role of Cover Crops in North American Cropping Systems," *Journal of Crop Production* 8 (2003): 53–74, https://doi.org/10.1300/J144v08n01_04.
18. Nancy G. Creamer, Mark A. Bennett, and Benjamin R. Stinner, "Evaluation of Cover Crop Mixtures for Use in Vegetable Production Systems," *HortScience* 32, no. 5 (1997): 866–70, https://doi.org/10.21273/HORTSCI.32.5.866.
19. T. Ohno et al., "Phytotoxic Effects of Red Clover Amended Soils on Wild Mustard Seedling Growth," *Agriculture, Ecosystems & Environment* 78, no. 2 (2000): 187–92, https://doi.org/10.1016/S0167-8809(99)00120-6.
20. C. L. Mohler et al., "Reduction in Weed Seedling Emergence by Pathogens Following the Incorporation of Green Crop Residue," *Weed Research* 52, no. 5 (2012): 467–77, https://doi.org/10.1111/j.1365-3180.2012.00940.x.
21. John R. Teasdale, "Contribution of Cover Crops to Weed Management in Sustainable Agricultural Systems," *Journal of Production Agriculture* 9, no. 4 (1996): 475–79, https://doi.org/10.2134/jpa1996.0475.
22. Joseph M. Di Tomaso, "Approaches for Improving Crop Competitiveness through the Manipulation of Fertilization Strategies," *Weed Science* 43, no. 3 (1995): 491–97, https://doi.org/10.1017/S0043174500081522.
23. Liebman and Davis, "Integration of Soil, Crop and Weed Management."
24. R. E. Blackshaw, L. J. Molnar, and H. H. Janzen, "Nitrogen Fertilizer Timing and Application Method Affect Weed Growth and Competition with Spring Wheat," *Weed Science* 52, no. 4 (2004): 614–22, https://doi.org/10.1614/WS-03-104R.

Chapter 9: Vegetable Crop Growing and Troubleshooting Guide

1. Harry J. Hoitink and Peter C. Fahy, "Basis for the Control of Soilborne Plant Pathogens with Composts," *Annual Review of Phytopathology* 24 (1986): 93–114, https://doi.org/10.1146/annurev.py.24.090186.000521; A. M. Litterick et al., "The Role of Uncomposted

Materials, Composts, Manures, and Compost Extracts in Reducing Pest and Disease Incidence and Severity in Sustainable Temperate Agricultural and Horticultural Crop Production—A Review," *Critical Reviews in Plant Sciences* 23, no. 6 (2004): 453–79, https://doi.org/10.1080/07352680490886815.

2. Harry J. Hoitink and Marcella E. Grebus, "Status of Biological Control of Plant Diseases with Composts," *Compost Science & Utilization* 2, no. 2 (1994): 6–12, https://doi.org/10.1080/1065657X.1994.10771134.
3. Stefan J. Green et al., "Succession of Bacterial Communities during Early Plant Development: Transition from Seed to Root and Effect of Compost Amendment," *Applied and Environmental Microbiology* 72, no. 6 (2006): 3975–83, https://doi.org/10.1128/AEM.02771-05.
4. Ehud Inbar et al., "Competing Factors of Compost Concentration and Proximity to Root Affect the Distribution of Streptomycetes," *Microbial Ecology* 50, no. 1 (2005): 73–81, http://www.jstor.org/stable/25153227; M.-S. Benitez et al., "Multiple Statistical Approaches of Community Fingerprint Data Reveal Bacterial Populations Associated with General Disease Suppression Arising from the Application of Different Organic Field Management Strategies," *Soil Biology and Biochemistry* 39, no. 9 (2007): 2289–301, https://doi.org/10.1016/j.soilbio.2007.03.028.
5. Allison L. H. Jack et al., "Choice of Organic Amendments in Tomato Transplants Has Lasting Effects on Bacterial Rhizosphere Communities and Crop Performance in the Field," *Applied Soil Ecology* 48, no. 1 (2011): 94–101, https://doi.org/10.1016/j.apsoil.2011.01.003.
6. A. M. Hammermeister et al., "Nutrient Supply from Organic Amendments Applied to Unvegetated Soil, Lettuce and Orchardgrass," *Canadian Journal of Soil Science* 86, no. 1 (2006): 21–33, https://doi.org/10.4141/S05-021.

Chapter 10: Vegetable Crop Insect Pests and Interventions

1. Alexandria Bryant, "Influence of Living and Non-Living Habitat Complexity on Arthropods in Strip-Tilled Cabbage Fields" (master's thesis, Michigan State University, 2013), https://d.lib.msu.edu/etd/1086.
2. Michael J. Costello, "Broccoli Growth, Yield and Level of Aphid Infestation in Leguminous Living Mulches," *Biological Agriculture & Horticulture* 10, no. 3 (1994): 207–22, https://doi.org/10.1080/01448765.1994.9754669; Michael J. Costello and Miguel A. Altieri, "Abundance, Growth Rate and Parasitism of *Brevicoryne brassicae* and *Myzus persicae* (Homoptera: Aphididae) on Broccoli Grown in Living Mulches," *Agriculture, Ecosystems & Environment* 52, nos. 2–3 (1995): 187–96, https://doi.org/10.1016/0167-8809(94)00535-M; K.-H. Wang, C. R. R. Hooks, and S. P. Marahatta, "Can Using a Strip-Tilled Cover Cropping System Followed by Surface Mulch Practice Enhance Organisms Higher Up in the Soil Food Web Hierarchy?," *Applied Soil Ecology* 49 (2011): 107–17, https://doi.org/10.1016/j.apsoil.2011.06.008; Cerruti R. R. Hooks, H. R. Valenzuela, and J. Defrank, "Incidence of Pests and Arthropod Natural Enemies in Zucchini Grown with Living Mulches," *Agriculture, Ecosystems & Environment* 69, no. 3 (1998): 217–31, https://doi.org/10.1016/s0167-8809(98)00110-8.
3. Robert T. Allen, "The Occurrence and Importance of Ground Beetles in Agricultural and Surrounding Habitats," in *Carabid Beetles*, ed. Terry L. Erwin et al. (Dordrecht: Springer, 1979), 485–505, https://doi.org/10.1007/978-94-009-9628-1_27.
4. Cerruti R. R. Hooks and Marshall W. Johnson, "Using Undersown Clovers as Living Mulches: Effects on Yields, Lepidopterous Pest Infestations, and Spider Densities in a Hawaiian Broccoli Agroecosystem," *International Journal of Pest Management* 50, no. 2 (2004): 115–20, https://doi.org/10.1080/09670870410001663462; Cerruti R. R. Hooks and Marshall W. Johnson, "Population Densities of Herbivorous Lepidopterans in Diverse Cruciferous Cropping Habitats: Effects of Mixed Cropping and Using a Living Mulch," *BioControl* 51, no. 4 (2006): 485–506, https://doi.org/10.1007/s10526-006-9009-5.
5. Bryant, "Influence of Living and Non-Living Habitat Complexity on Arthropods."
6. Catherine E. Bach, "Effects of Plant Density and Diversity on the Population Dynamics of a Specialist Herbivore, the Striped Cucumber Beetle, Acalymma Vittata (Fab)," *Ecology* 61, 6 (1980): 1515–30, https://doi.org/10.2307/1939058.
7. Gary R. Cline et al., "Organic Management of Cucumber Beetles in Watermelon and Muskmelon Production," *HortTechnology* 18, no. 3 (2008): 436–44, https://doi.org/10.21273/HORTTECH.18.3.436.

8. John M. Luna and L. Xue, "Aggregation Behavior of Western Spotted Cucumber Beetle (Coleoptera: Chrysomelidae) in Vegetable Cropping Systems," *Environmental Entomology* 38, no. 3 (2009): 809–14, https://doi.org/10.1603/022.038.0334.
9. L. S. Adler and R. V. Hazzard, "Comparison of Perimeter Trap Crop Varieties: Effects on Herbivory, Pollination, and Yield in Butternut Squash," *Environmental Entomology* 38, no. 1 (2009): 207–15, https://doi.org/10.1603/022.038.0126.
10. Hanna M. Kahl, Alan W. Leslie, and Cerruti R. R. Hooks, "Effects of Red Clover Living Mulch on Arthropod Herbivores and Natural Enemies, and Cucumber Yield," *Annals of the Entomological Society of America* 112, no. 4 (2019): 356–64, https://doi.org/10.1093/aesa/say036.
11. Hanna M. Kahl, Alan W. Leslie, and Cerruti R. R. Hooks, "Consumptive and Non-Consumptive Effects of Wolf Spiders on Cucumber Beetles and Cucumber Plant Damage," *Annals of Applied Biology* 178, no. 1 (2021): 109–20, https://doi.org/10.1111/aab.12643.
12. Juraj Halaj and David H. Wise, "Impact of a Detrital Subsidy on Trophic Cascades in a Terrestrial Grazing Food Web," *Ecology* 83, no. 11 (2002): 3141–51, https://doi.org/10.2307/3071849.
13. Miguel A. Altieri and Stephen R. Gliessman, "Effects of Plant Diversity on the Density and Herbivory of the Flea Beetle, *Phyllotreta cruciferae* Goeze, in California Collard (*Brassica oleracea*) Cropping Systems," *Crop Protection* 2, no. 4 (1983): 497–501, https://doi.org/10.1016/0261-2194(83)90071-6.
14. Jane Breen Pierce et al., "Predation of Sentinel Eggs in Cotton and Sorghum in New Mexico," in *Proceedings 71st Beltwide Cotton Conference, National Cotton Council of America, Dallas, TX* (2017): 536–41.
15. Susan E. Riechert and Leslie Bishop, "Prey Control by an Assemblage of Generalist Predators: Spiders in Garden Test Systems," *Ecology* 71, no. 4 (1990): 1441–50, https://doi.org/10.2307/1938281.
16. Jan Petersen et al., "Weed Suppression by Release of Isothiocyanates from Turnip-Rape Mulch," *Agronomy Journal* 93, no. 1 (2001): 37–43, https://doi.org/10.2134/agronj2001.93137x.
17. B. J. Cartwright, C. Palumbo, and W. S. Fargo. "Influence of Crop Mulches and Row Covers on the Population Dynamics of the Squash Bug (Heteroptera: Coreidae) on Summer Squash," *Journal of Economic Entomology* 83, no. 5 (1990): 1988–993, https://doi.org/10.1093/jee/83.5.1988.
18. Jennifer Tillman, Ajay Nair, Mark Gleason, and Jean Batzer, "Rowcovers and Strip Tillage Provide an Alternative to Plasticulture Systems in Summer Squash Production," *HortScience* 50, no. 12 (2015): 1777–83, https://doi.org/10.21273/HORTSCI.50.12.1777.
19. C. Welty and J. Jasinski, *Squash Vine Borer Control on Zucchini in Ohio* (Columbus: Ohio State University, 2008), 1–3, https://cpb-us-west-2-juc1ugur1qwqqqo4.stackpathdns.com/u.osu.edu/dist/1/8311/files/2014/12/OhioZucchini2008_FinalReport-1owok1p.pdf.
20. Lorenzo Furlan, "An IPM Approach Targeted against Wireworms: What Has Been Done and What Has to Be Done," *IOBC/WPRS Bulletin* 28, no. 2 (2005): 91–100, https://www.researchgate.net/publication/311424588_An_IPM_approach_targeted_against_wireworms_what_has_been_done_and_what_has_to_be_done.

Chapter 11: Vegetable Crop Diseases and Interventions

1. Anupama Babu et al., "Improvement of Growth, Fruit Weight and Early Blight Disease Protection of Tomato Plants by Rhizosphere Bacteria Is Correlated with Their Beneficial Traits and Induced Biosynthesis of Antioxidant Peroxidase and Polyphenol Oxidase," *Plant Science* 231 (2015): 62–73, https://doi.org/10.1016/j.plantsci.2014.11.006.
2. Most. Ferdousi Begum, M. A. Rahman, and M. Firoz Alam, "Biological Control of Alternaria Fruit Rot of Chili by *Trichoderma* Species under Field Conditions," *Mycobiology* 38, no. 2 (2010): 113–17, https://www.tandfonline.com/doi/abs/10.4489/MYCO.2010.38.2.113.
3. L. Liu, J. W. Kloepper, and S. Tuzun, "Induction of Systemic Resistance in Cucumber against Bacterial Angular Leaf Spot by Plant Growth-Promoting Rhizobacteria," *Phytopathology* 85, no. 8 (1995): 843–47, https://www.apsnet.org/publications/phytopathology/backissues/Documents/1995Articles/Phyto85n08_843.PDF; G. Wei, J. W. Kloepper, and S. Tuzun, "Induced Systemic Resistance to Cucumber Diseases and Increased Plant Growth by Plant Growth-Promoting Rhizobacteria under Field Conditions," *Phytopathology* 86, no. 2 (1996): 221–24, https://www.apsnet.org/publications/phytopathology/backissues/Documents/1996Articles/Phyto86n02_221.PDF.

4. Andrés M. P. Nuñez, Gabriel A. A. Rodríguez, Fernando P. Monteiro, Amanda F. Faria, Julio C. P. Silva, Ana C. A. Monteiro, Carolina V. Carvalho, et al., "Bio-Based Products Control Black Rot (*Xanthomonas campestris* pv. *campestris*) and Increase the Nutraceutical and Antioxidant Components in Kale," *Scientific Reports* 8, no. 1 (2018): 1–11, https://doi.org/10.1038/s41598-018-28086-6.
5. Ednar G. Wulff et al., "Biological Control of Black Rot (*Xanthomonas campestris* pv. *campestris*) of Brassicas with an Antagonistic Strain of *Bacillus subtilis* in Zimbabwe," *European Journal of Plant Pathology* 108, no. 4 (2002): 317–25, https://doi.org/10.1023/A:1015671031906.
6. R. L. Bargabus et al., "Screening for the Identification of Potential Biological Control Agents That Induce Systemic Acquired Resistance in Sugar Beet," *Biological Control* 30, no. 2 (2004): 342–50, https://doi.org/10.1016/j.biocontrol.2003.11.005.
7. Stefania Galletti et al., "*Trichoderma* as a Potential Biocontrol Agent for Cercospora Leaf Spot of Sugar Beet," *BioControl* 53, no. 6 (2008): 917–30, https://doi.org/10.1007/s10526-007-9113-1.
8. L. Bavaresco and R. Eibach, "Investigations on the Influence of N Fertilizer on Resistance to Powdery Mildew (*Oidium tuckeri*) Downy Mildew (*Plasmopara viticola*) and on Phytoalexin Synthesis in Different Grapevine Varieties," *Vitis* 26, no. 4 (1987), 192–200; Afiniki B. Zarafi, A. M. Emechebe, A. D. Akpa, and O. Alabi, "Effect of Fertilizer Levels on Grain Yield, Incidence and Severity of Downy Mildew in Pearl Millet," *Archives of Phytopathology and Plant Protection* 38, no. 1 (2005): 11–17, https://doi.org/10.1080/00222930400007480.
9. E. Gatch, "Organic Seed Treatments and Coatings," in *Organic Seed Resource Guide* (eOrganic: 2016): 1–6, https://www.eorganic.org/node/749.
10. Waseem Raza et al., "Success Evaluation of the Biological Control of *Fusarium* Wilts of Cucumber, Banana, and Tomato Since 2000 and Future Research Strategies," *Critical Reviews in Biotechnology* 37, no. 2 (2017): 202–12, https://doi.org/10.3109/07388551.2015.1130683; Tina Netzker et al., "Bacterial Volatile Compounds: Functions in Communication, Cooperation, and Competition," *Annual Review of Microbiology* 74 (2020): 409–30, https://doi.org/10.1146/annurev-micro-011320-015542.
11. Hongwei Wang et al., "Enhanced Biocontrol of Cucumber Fusarium Wilt by Combined Application of New Antagonistic Bacteria *Bacillus amyloliquefaciens* B2 and Phenolic Acid-Degrading Fungus *Pleurotus ostreatus* P5," *Frontiers in Microbiology* (2021): 2371, https://doi.org/10.3389/fmicb.2021.700142.
12. Rui Li et al., "Chemical, Organic and Bio-Fertilizer Management Practices Effect on Soil Physicochemical Property and Antagonistic Bacteria Abundance of a Cotton Field: Implications for Soil Biological Quality," *Soil and Tillage Research* 167 (2017): 30–38, https://doi.org/10.1016/j.still.2016.11.001.
13. Silas D. Silva et al., "Evaluation of *Pochonia chlamydosporia* and *Purpureocillium lilacinum* for Suppression of *Meloidogyne enterolobii* on Tomato and Banana," *Journal of Nematology* 49, no. 1 (2017): 77, https://doi.org/10.21307/jofnem-2017-047.
14. Shuwu Zhang, Yantai Gan, and Bingliang Xu, "Biocontrol Potential of a Native Species of *Trichoderma longibrachiatum* against *Meloidogyne incognita*," *Applied Soil Ecology* 94 (2015): 21–29, https://doi.org/10.1016/j.apsoil.2015.04.010.
15. Elizabeth Herrera-Parra, Jairo Cristóbal-Alejo, and José A. Ramos-Zapata, "*Trichoderma* Strains as Growth Promoters in *Capsicum annuum* and as Biocontrol Agents in *Meloidogyne incognita*," *Chilean Journal of Agricultural Research* 77, no. 4 (2017): 318–24, https://doi.org/10.4067/S0718-58392017000400318.
16. Pia Euteneuer et al., "Earthworms Affect Decomposition of Soil-Borne Plant Pathogen *Sclerotinia sclerotiorum* in a Cover Crop Field Experiment," *Applied Soil Ecology* 138 (2019): 88–93, https://doi.org/10.1016/j.apsoil.2019.02.020.
17. Mohammad Reza Ojaghian et al., "Brassica Green Manure Rotation Crops Reduce Potato Stem Rot Caused by *Sclerotinia sclerotium*," *Australasian Plant Pathology* 41, no. 4 (2012): 347–49, https://doi.org/10.1007/s13313-012-0142-6.
18. M. C. Saure, "Why Calcium Deficiency Is Not the Cause of Blossom-End Rot in Tomato and Pepper Fruit—A Reappraisal," *Scientia Horticulturae* 174 (2014): 151–54, https://doi.org/10.1016/j.scienta.2014.05.020.
19. Michaela Schmitz-Eiberger, Roland Haefs, and Georg Noga, "Calcium Deficiency—Influence on the Antioxidative Defense System in Tomato Plants,"

Journal of Plant Physiology 159, no. 7 (2002): 733–42, https://doi.org/10.1078/0176-1617-0621.

20. United States Department of Agriculture National Organic Certification Program, National List of Allowed and Prohibited Substances, https://www.ams.usda.gov/rules-regulations/national-list-allowed-and-prohibited-substances.

Chapter 12: Fruit Tree Growing and Troubleshooting Guide

1. C. C. Mundt, "Use of Multiline Cultivars and Cultivar Mixtures for Disease Management," *Annual Review of Phytopathology* 40 (2002): 381–410, https://doi.org/10.1146/annurev.phyto.40.011402.113723.
2. Nian-Feng Wan, Xiang-Yun Ji, and Jie-Xian Jiang, "Testing the Enemies Hypothesis in Peach Orchards in Two Different Geographic Areas in Eastern China: The Role of Ground Cover Vegetation," *PLoS ONE* 9, no. 6 (2014): e99850, https://doi.org/10.1371/journal.pone.0099850.
3. Miguel Altieri, R. C. Wilson, and L. L. Schmidt, "The Effects of Living Mulches and Weed Cover on the Dynamics of Foliage- and Soil-Arthropod Communities in Three Crop Systems," *Crop Protection* 4, no. 2 (1985): 201–13, https://doi.org/10.1016/0261-2194(85)90018-3; Lessando M. Gontijo, Elizabeth H. Beers, and William E. Snyder, "Flowers Promote Aphid Suppression in Apple Orchards," *Biological Control* 66, no. 1 (2013): 8–15, https://doi.org/10.1016/j.biocontrol.2013.03.007.
4. David R. Horton et al., "Effects of Mowing Frequency on Densities of Natural Enemies in Three Pacific Northwest Pear Orchards," *Entomologia Experimentalis et Applicata* 106, no. 2 (2003): 135–45, https://doi.org/10.1046/j.1570-7458.2003.00018.x.
5. G. Tonon et al., "Fate of 15N Derived from Soil Decomposition of Abscised Leaves and Pruning Wood from Apple (*Malus domestica*) Trees," *Soil Science and Plant Nutrition* 53, no. 1 (2007): 78–85, https://doi.org/10.1111/j.1747-0765.2007.00112.x; Claudia Neto et al., "Nitrogen Distribution, Remobilization and Re-Cycling in Young Orchard of Non-Bearing 'Rocha' Pear Trees," *Scientia Horticulturae* 118, no. 4 (2008): 299–307, https://doi.org/10.1016/j.scienta.2008.06.023; M. Ventura et al., "Nutrient Release during Decomposition of Leaf Litter in a Peach (*Prunus persica L.*) Orchard," *Nutrient Cycling in Agroecosystems* 87, no. 1 (2010): 115–25, https://doi.org/10.1007/s10705-009-9317-0.
6. Corina Carranca, "Nitrogen Use Efficiency by Annual and Perennial Crops," in *Farming for Food and Water Security*, ed. Eric Lichtfouse (Dordrecht: Springer, 2012), 57–82, https://doi.org/10.1007/978-94-007-4500-1_3.
7. Matthew D. Whiteside et al., "Organic Nitrogen Uptake by Arbuscular Mycorrhizal Fungi in a Boreal Forest," *Soil Biology and Biochemistry* 55 (2012): 7–13, https://doi.org/10.1016/j.soilbio.2012.06.001.
8. Sacha K. Heath and Rachael F. Long, "Multiscale Habitat Mediates Pest Reduction by Birds in an Intensive Agricultural Region," *Ecosphere* 10, no. 10 (2019): e02884.
9. Goonewardene et al., "Resistance to European Red Mite, *Panonychus ulmi* (Koch), in Apple," *Journal of the American Society for Horticultural Science* 101, no. 5 (1976): 532–37; H. F. Goonewardene et al., "A 'No Choice' Study for Evaluating Resistance of Apple Fruits to Four Insect Pests," *HortScience* (1979).
10. Neelendra K. Joshi, Edwin G. Rajotte, Clayton T. Myers, Greg Krawczyk, and Larry A. Hull, "Development of a Susceptibility Index of Apple Cultivars for Codling Moth, *Cydia pomonella* (L.) (Lepidoptera: Tortricidae) Oviposition," *Frontiers in Plant Science* 6 (2015): 992, https://doi.org/10.3389/fpls.2015.00992.
11. Henry W. Hogmire and Stephen S. Miller, "Relative Susceptibility of New Apple Cultivars to Arthropod Pests," *HortScience* 40, no. 7 (2005): 2071–75, https://doi.org/10.21273/HORTSCI.40.7.2071.
12. Hogmire and Miller, "Relative Susceptibility of New Apple Cultivars."
13. Hogmire and Miller, "Relative Susceptibility of New Apple Cultivars."
14. Ahmet Esitken et al., "Effects of Floral and Foliar Application of Plant Growth Promoting Rhizobacteria (PGPR) on Yield, Gowth and Nutrition of Sweet Cherry," *Scientia Horticulturae* 110, no. 4 (2006): 324–27, https://doi.org/10.1016/j.scienta.2006.07.023.
15. Corina Carranca, G. Brunetto, and M. Tagliavini, "Nitrogen Nutrition of Fruit Trees to Reconcile Productivity and Environmental Concerns," *Plants* 7, no. 1 (2018): 4, https://doi.org/10.3390/plants7010004.
16. Mark Wilson and Steven E. Lindow, "Coexistence among Epiphytic Bacterial Populations Mediated through Nutritional Resource Partitioning," *Applied*

and Environmental Microbiology 60, no. 12 (1994): 4468–77, https://doi.org/10.1128/aem.60.12.4468-4477.1994; M. Wilson and S. E. Lindow, "Interactions between the Biological Control Agent *Pseudomonas fluorescens* A506 and *Erwinia amylovora* in Pear Blossoms," *Phytopathology* 83, no. 1 (1993): 117–23, https://www.apsnet.org/publications/phytopathology/backissues/Documents/1993Articles/Phyto83n01_117.PDF.

Chapter 13: Fruit Tree Insect Pests and Interventions

1. Henry W. Hogmire and Stephen S. Miller, "Relative Susceptibility of New Apple Cultivars to Arthropod Pests," *HortScience* 40, no. 7 (2005): 2071–75, https://doi.org/10.21273/HORTSCI.40.7.2071.
2. Muhammad Usman et al., "Potential of Entomopathogenic Nematodes against the Pupal Stage of the Apple Maggot *Rhagoletis pomonella* (Walsh) (Diptera: Tephritidae)," *Journal of Nematology* 52 (2020): 1–9, https://doi.org/10.21307/jofnem-2020-079.
3. D. M. Borchert et al., "Toxicity and Residual Activity of Methoxyfenozide and Tebufenozide to Codling Moth (Lepidoptera: Tortricidae) and Oriental Fruit Moth (Lepidoptera: Tortricidae)," *Journal of Economic Entomology* 97, no. 4 (2004): 1342–52, https://doi.org/10.1093/jee/97.4.1342.
4. Jaime C. Piñero and Ronald J. Prokopy,. "Field Evaluation of Plant Odor and Pheromonal Combinations for Attracting Plum Curculios," *Journal of Chemical Ecology* 29, no. 12 (2003): 2735–48, https://doi.org/10.1023/B:JOEC.0000008017.16911.aa.
5. Isobel A. Pearsall, "Flower Preference Behaviour of Western Flower Thrips in the Similkameen Valley, British Columbia, Canada," *Entomologia Experimentalis et Applicata* 95, no. 3 (2000): 303–13, https://doi.org/10.1046/j.1570-7458.2000.00669.x.
6. J. Bennison et al., "Towards the Development of a Push-Pull Strategy for Improving Biological Control of Western Flower Thrips on Chrysanthemum," in *Proceedings of the 7th International Symposium on Thysanoptera: Thrips and Tospoviruses*, ed. Rita Marullo and Laurence Mounds (Reggio Calabria, Italy: CSIRO, 2002), 199–206, https://www.ento.csiro.au/thysanoptera/Symposium/Section7/30-Bennison-et-al.pdf.

Chapter 14: Fruit Tree Diseases and Interventions

1. Frederique Didelot, Laurent Brun, and Luciana Parisi, "Effects of Cultivar Mixtures on Scab Control in Apple Orchards," *Plant Pathology* 56, no. 6 (2007): 1014–22, https://doi.org/10.1111/j.1365-3059.2007.01695.x.
2. J. Köhl et al., "Toward an Integrated Use of Biological Control by *Cladosporium cladosporioides* H39 in Apple Scab (*Venturia inaequalis*) Management," *Plant Disease* 99, no. 4 (2015): 535–43, https://doi.org/10.1094/PDIS-08-14-0836-RE.
3. Lucas Alexander Shuttleworth. "Alternative Disease Management Strategies for Organic Apple Production in the United Kingdom," *CABI Agriculture and Bioscience* 2, no. 1 (2021): 1–15, https://doi.org/10.1186/s43170-021-00054-7.
4. Despina Berdeni, T. E. Anne Cotton, Tim J. Daniell, Martin I. Bidartondo, Duncan D. Cameron, and Karl L. Evans, "The Effects of Arbuscular Mycorrhizal Fungal Colonisation on Nutrient Status, Growth, Productivity, and Canker Resistance of Apple (*Malus pumila*)," *Frontiers in Microbiology* 9 (2018): 1461, https://doi.org/10.3389/fmicb.2018.01461.
5. Kent Daane et al., "Excess Nitrogen Raises Nectarine Susceptibility to Disease and Insects," *California Agriculture* 4, no. 4 (July 1995): 13–18, https://doi.org/10.3733/ca.v049n04p13.
6. V. Obi et al., "Is the Tolerance of Commercial Peach Cultivars to Brown Rot Caused by *Monilinia laxa* Modulated by Its Antioxidant Content?," *Plants* 9, no. 5 (2020): 589, https://doi.org/10.3390/plants9050589.
7. S. Kunz, A. Schmitt, and P. Haug, "Development of Strategies for Fire Blight Control in Organic Fruit Growing," *Acta Horticulturae* 5 (2011): 896, https://doi.org/10.17660/ActaHortic.2011.896.62.

Index

Note: Page numbers in *italics* refer to photographs and illustrations. Page numbers followed by *t* refer to tables.

Index

Index

Index

Index

Index

Index

Index

Index

Index

Q

R

Index

Index

T

U

V

W

Y

Z

About the Author

Cindy Haugen

Helen Atthowe has worked for 35 years to connect farming, food systems, land stewardship, and conservation. She bought a new farm in Montana in 2023, but still consults for and monitors on-going research at Woodleaf Farm in eastern Oregon. She serves as a consultant with farmers across the United States and internationally. Helen and her late husband, Carl Rosato, co-owned and operated a certified organic orchard in California where they pioneered methods for raising apples, peaches, and other tree fruits without the use of any type of pesticides. Her on-farm research includes ecological weed and insect management, organic minimal soil disturbance systems for vegetable and orchard crops, and managing living mulches for soil and habitat building. She is a contributing writer to *The Organic Gardener's Handbook of Natural Pest and Disease Control* and other books. She has served as a board member for the Organic Farming Research Foundation and adviser for the Wild Farm Alliance. Atthowe has a master's degree in horticulture from Rutgers University and has worked in education and research at the University of Arkansas, Rutgers University, and Oregon State University, and served as a horticulture extension agent in Montana, where she annually taught an organic Master Gardener Course.